T0237607

Undergraduate Lecture Notes in Physics

Undergraduate Lecture Notes in Physics (ULNP) publishes authoritative texts covering topics throughout pure and applied physics. Each title in the series is suitable as a basis for undergraduate instruction, typically containing practice problems, worked examples, chapter summaries, and suggestions for further reading.

ULNP titles must provide at least one of the following:

- An exceptionally clear and concise treatment of a standard undergraduate subject.
- A solid undergraduate-level introduction to a graduate, advanced, or non-standard subject.
- A novel perspective or an unusual approach to teaching a subject.

ULNP especially encourages new, original, and idiosyncratic approaches to physics teaching at the undergraduate level.

The purpose of ULNP is to provide intriguing, absorbing books that will continue to be the reader's preferred reference throughout their academic career.

More information about this series at http://www.springer.com/series/8917

Masud Chaichian · Hugo Perez Rojas ·
Anca Tureanu

Basic Concepts in Physics

From the Cosmos to Quarks

Second Edition

 Springer

Masud Chaichian
Department of Physics
University of Helsinki
Helsinki, Finland

Hugo Perez Rojas
Department of Theoretical Physics
ICIMAF
La Habana, Cuba

Anca Tureanu
Department of Physics
University of Helsinki
Helsinki, Finland

ISSN 2192-4791   ISSN 2192-4805   (electronic)
Undergraduate Lecture Notes in Physics
ISBN 978-3-662-62315-2   ISBN 978-3-662-62313-8   (eBook)
https://doi.org/10.1007/978-3-662-62313-8

This Springer imprint is published by the registered company Springer-Verlag GmbH, DE part of
Springer Nature.
The registered company address is: Heidelberger Platz 3, 14197 Berlin, Germany

Preface to the Second Edition

The praise of the first edition of the book by many readers encouraged us to prepare the present second edition. We express our deep gratitude to all those readers for their remarks and suggestions – in this edition we have tried to take into account all of them as much as possible, and as well to come up with their wishes to include some problems to be solved, together with their solutions or at least sufficient hints to solve them.

As its previous edition, this book is intended for undergraduate students, physics teachers, students in high schools, researchers and general readers interested to know what physics is about together with its latest developments and discoveries.

Thinking about the book to be useful also as a textbook, totally or in part, we have added several new topics with the latest findings in those fields. For instance, the recent discovery of gravitational waves, as one of the most important achievements of modern physical sciences, is presented in Chap. 10. At the end of Chaps. 1–11 some problems are included with their solutions or hints how to solve them given at the end of the book. Those problems are useful for a complementary understanding of the theories and their implication. However, for non-specialized readers it is recommended to bypass, at least in their first-time reading, the problems as well as the mathematical details.

The added new topics also provide connections among the subjects treated in different chapters. For instance, the wobble of some stars interacting with their planets, as explained by the two body Kepler problem, helps to detect invisible companions, by using Doppler spectroscopy of the star light. The Clapeyron–Clausius equation helps to understand the development of life at dark, deep and hot oceanic vents at high pressures, as well as why the hot Earth nucleus is solid. The creation of the magnetosphere is explained as due to the deviation of the solar wind by the Earth magnetic field. A reference to the former experiments is made in order to resolve the loophole appeared there and to support, thanks to more recent experiments, the occurrence of quantum entanglement, and to show the validity of the violation of Bell inequalities as a genuine quantum phenomenon. Gravitational lensing, as well as the correction of time for GPS satellites, as the technical applications of special and general relativity, are explained. Some earlier figures have been improved and new ones were added.

Our special thanks go to François Englert, Igal Galili, and Markku Oksanen for their valuable comments and advice.

Helsinki, Finland   Masud Chaichian
La Habana, Cuba   Hugo Perez Rojas
Helsinki, Finland   Anca Tureanu
May 2021

Preface to the First Edition

This book is the outcome of many lectures, seminars, and colloquia the authors have given on different occasions to different audiences in several countries over a long period of time and the experience and feedback obtained from them. With a wide range of readers in mind, some topics have been presented in twofold form, both descriptively and more formally.

This book is intended not only for first to second year undergraduate students, as a complement to specialized textbooks but also for physics teachers and students in high schools. At the same time, it is addressed to researchers and scientists in other fields, including engineers and general readers interested in acquiring an overview of modern physics. A minimal mathematical background, up to elementary calculus, matrix algebra and vector analysis, is required. However, mathematical technicalities have not been stressed, and long calculations have been avoided. The basic and most important ideas have been presented with a view to introducing the physical concepts in a pedagogical way. Since some specific topics of modern physics, particularly those related to quantum theory, are an important ingredient of student courses nowadays, the first five chapters on classical physics are presented keeping in mind their connection to modern physics whenever possible.

In most chapters, historical facts are included. Several themes are discussed which are sometimes omitted in basic courses on physics. For instance, the relation between entropy and information, exchange energy and ferromagnetism, superconductivity and the relation between phase transitions and spontaneous symmetry breaking, chirality, the fundamental C, P, and T invariances, paradoxes of quantum theory, the problem of measurement in quantum mechanics, quantum statistics and specific heat in solids, quantum Hall effect, graphene, general relativity and cosmology, CP violation, Casimir and Aharonov–Bohm effects, causality, unitarity, spontaneous symmetry breaking and the Standard Model, inflation, baryogenesis, and nucleosynthesis, ending with a chapter on the relationship between physics and life, including biological chiral symmetry breaking.

To non-specialized readers it is recommended to bypass, at least on a first reading, the mathematical content of sections and subsections 1.8, 1.9, 2.5, 3.11, 4.5, 6.7, 6.8.1, 7.3, 7.4.1, 8.2, 10.3, and 10.5.

During the preparation of this book the authors have benefited greatly from discussions with many of their colleagues and students, to whom we are indebted. It is a pleasure to express our gratitude in particular to Cristian Armendariz-Picon, Alexander D. Dolgov, François Englert, Josef Kluson, Vladimir M. Mostepanenko, Viatcheslav Mukhanov, Markku Oksanen, Roberto Sussmann, and Ruibin Zhang for their stimulating suggestions and comments, while our special thanks go to Tiberiu Harko, Peter Prešnajder and Daniel Radu, to whom we are most grateful for their valuable advice in improving an initial version of the manuscript.

Helsinki, Finland
La Habana, Cuba
Helsinki, Finland
March 2013

Masud Chaichian
Hugo Perez Rojas
Anca Tureanu

Contents

Chapter 1
Gravitation and Newton's Laws

Our Sun is a star of intermediate size with a set of major planets describing closed orbits around it. These are Mercury, Venus, Earth, Mars, Jupiter, Saturn, Uranus, and Neptune. Pluto, considered the Solar System's ninth planet until 2006, was reclassified by the International Astronomical Union as a dwarf planet, due to its very small mass, together with other trans-Neptunian objects (Haumea, Makemake, Eris, Sedna, and others) recently discovered in that zone, called the Kuiper belt. Except for Mercury and Venus, all planets and even certain dwarf planets have satellites. Some of them, like the Moon and a few of the Jovian satellites, are relatively large. Between Mars and Jupiter, there are a lot of small planets or asteroids moving in a wide zone, the largest one being Ceres, classified as a dwarf planet. Other distinguished members of the Solar System are the comets, such as the well-known comet bearing the name of Halley. It seems that most comets originate in the Kuiper belt.

The Sun is located approximately 30,000 light-years (1 light-year $= 9.4 \times 10^{12}$ km) from the Galactic Centre, around which it makes a complete turn at a speed of nearly 250 km/s in approximately 250 million years. The number of stars in our galaxy is estimated to be of the order of 10^{11}, classified by age, size, and state of evolution: young, old, red giants, white dwarfs, etc. (Fig. 1.1).

In fact, our galaxy, the Milky Way, is one member of a large family estimated to contain of the order of 10^{13} galaxies. These are scattered across what we call the visible Universe, which seems to be in expansion after some initial event. The galaxies are moving away from each other like dots painted on an inflating rubber balloon.

At the present time, our knowledge of the Universe and the laws governing it is increasing daily. Today we possess a vast knowledge of our planetary system, stellar evolution, and the composition and dynamics of our own galaxy, not to mention millions of other galaxies. Even the existence of several extra-solar planetary systems has been deduced from the discovery of planets orbiting around 51 Pegasi, 47 Ursae Majoris, and several other stars. But barely five centuries ago, we only knew about the existence of the Sun, the Moon, five planets (Mercury, Venus, Mars, Jupiter, and Saturn), some comets, and the visible stars. For thousands of years, people had gazed

© The Author(s), under exclusive license to Springer-Verlag GmbH, DE, part of Springer Nature 2021
M. Chaichian et al., *Basic Concepts in Physics*, Undergraduate Lecture Notes in Physics, https://doi.org/10.1007/978-3-662-62313-8_1

Fig. 1.1 The Andromeda galaxy, at a distance of two million light-years from our own galaxy. They are similar in size.

intrigued at those celestial objects, watching as they moved across the background of fixed stars, without knowing what they were, nor why they were moving like that.

The discovery of the mechanism underlying the planetary motion, the starting point for our knowledge of the fundamental laws of physics, required a prolonged effort, lasting several centuries. Sometimes scientific knowledge took steps forward, but subsequently went back to erroneous concepts. However, fighting against the established dogma and sometimes going against their own prior beliefs, passionate scholars finally discovered the scientific truth. In this way, the mechanism guiding planetary motions was revealed, and the first basic chapter of physics began to be written: the science of mechanics.

1.1 From Pythagoras to the Middle Ages

Pythagoras of Samos (c. 580–c. 500 BCE) was the founder of a mystic school, where philosophy, science, and religion were blended together. For the Pythagorean school, numbers had a magical meaning. The Cosmos for Pythagoras was formed by the spherical Earth at the centre, with the Sun, the Moon and the planets fixed to concentric spheres which rotated around it. Each of these celestial bodies produced a specific musical sound in the air, but only the master, Pythagoras himself, had the gift of hearing the music of the spheres.

Philolaus (c. 470–c. 385 BCE), a disciple of Pythagoras, attributed to the Earth one motion, not around its axis, but around some external point in space, where there was a central fire. Between the Earth and the central fire, Philolaus assumed the existence of an invisible planet, Antichthon, a sort of "counter-Earth". Antichthon revolved in such a way that it could not be seen, because it was always away from the Greek hemisphere. The central fire could not be seen from the Greek world either, and with its shadow Antichthon protected other distant lands from being burned. Antichthon, the Earth, the Sun, the Moon, and the other known planets Mercury, Venus, Mars, Jupiter, and Saturn revolved in concentric orbits around the central fire. The fixed stars were located on a fixed sphere behind all the above celestial bodies.

Heraclides of Pontus (c. 390–c. 310 BCE) took the next step in the Pythagorean conception of the Cosmos. He admitted the rotation of the Earth around its axis, and that the Sun and the Moon revolved around the Earth in concentric orbits. Mercury and Venus revolved around the Sun, and beyond the Sun, Mars, Jupiter, and Saturn also revolved around the Earth (Fig. 1.2).

Around the year when Heraclides died, Aristarchus (c. 310–c. 230 BCE) was born in Samos. From him, only a brief treatise has reached us: *On the Sizes and Distances from the Sun and the Moon*. In another book, Aristarchus claimed that the centre of the Universe was the Sun and not the Earth. Although this treatise has been lost, the ideas expressed in it are known through Archimedes and Plutarch. In one of his books Archimedes states: "He [Aristarchus] assumed the stars and the Sun as fixed,

Fig. 1.2 The system of Heraclides.

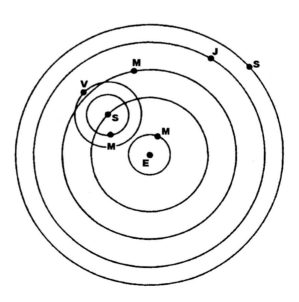

but that the Earth moves around the Sun in a circle, the Sun lying in the middle of the orbit." Plutarch also quotes Aristarchus as claiming that: "The sky is quiet and the Earth revolves in an oblique orbit, and also revolves around its axis."

Aristarchus was recognized by posterity as a very talented man, and one of the most prominent astronomers of his day, but in spite of this, his heliocentric system was ignored for seventeen centuries, supplanted by a complicated and absurd system first conceived by Apollonius of Perga in the third century BCE, later developed by Hipparchus of Rhodes in the next century, and finally completed by Ptolemy of Alexandria (c. 70–c. 147 CE).

The Earth's sphericity was accepted as a fact from the time of Pythagoras, and its dimensions were estimated with great accuracy by another Greek scholar Eratosthenes of Cyrene, in the third century BCE. He read in a papyrus scroll that, in the city of Swenet (known nowadays as Aswan), almost on the Tropic of Cancer, in the south of Egypt, on the day corresponding to our 21 June (summer solstice), a rod nailed vertically on the ground did not cast any shadow at noon. He decided to see whether the same phenomenon would occur in Alexandria on that day, but soon discovered that this was not the case: at noon, the rod did cast some shadow. If the Earth had been flat, neither rods would have cast a shadow on that day, assuming the Sun rays to be parallel. But if in Alexandria the rod cast some shadow, and in Swenet not, the Earth could not be flat, but had to be curved.

It is believed that Eratosthenes paid some money to a man to measure the distance between Swenet and Alexandria by walking between the two cities. The result was equivalent to approximately 800 km. On the other hand, if we imagine the rods to extend down to the Earth's centre, the shadow indicated that the angle α between them was about 7° (Fig. 1.3). Then, establishing the proportionality

$$\frac{360}{7} = \frac{x}{800} \, ,$$

the result is approximately $x = 40,000$ km, which would be the length of the circumference of the Earth if it were a perfect sphere. The value obtained by Eratosthenes was a little less (0.5% smaller).

It is astonishing that, using very rudimentary instruments, angles measured from the shadows cast by rods nailed on the ground, and lengths measured by the steps of a man walking a long distance (but having otherwise an exceptional interest in observation and experimentation), Eratosthenes was able to obtain such an accurate result for the size of the Earth, and so long ago, in fact, twenty-two centuries ago. He was the first person known to have measured the size of the Earth. We know at present that, due to the flattening of the Earth near the poles, the length of a meridian is shorter than the length of the equator. Later, Hipparchus measured the distance from the Moon to the Earth as 30.25 Earth diameters, making an error of only 0.3%.

But let us return to Ptolemy's system (Fig. 1.4). The reasons why it prevailed over Aristarchus' heliocentric system, are very complex. Some blame can probably be laid on Plato and Aristotle, but mainly the latter. Aristotle deeply influenced philosophical and ecclesiastic thinking up to modern times. Neither Plato nor Aristotle had a

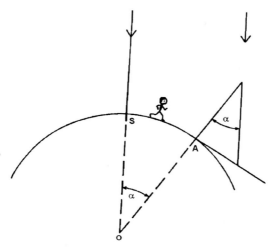

Fig. 1.3 Eratosthenes concluded that the shape of the Earth was a sphere. He used the fact that, when two rods were nailed vertically on the ground, one in the ancient Swenet and the other in Alexandria, at the noon of the day corresponding to our 21 June, the second cast a shadow while the first did not.

Fig. 1.4 The system of the world according to Ptolemy. The Earth was the centre of the Universe and the planets were fixed to spheres, each one rotating around some axis, which was supported on another sphere which in turn rotated around some axis, and so on.

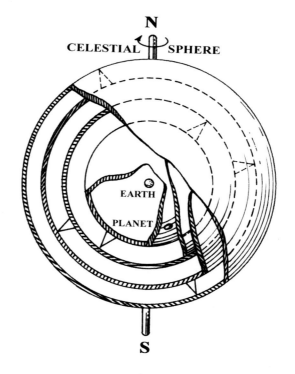

profound knowledge of astronomy, but they adopted the geocentric system because it was in better agreement with their philosophical beliefs, and their preference for a pro-slavery society. Their cosmology was subordinated to their political and philosophical ideas: they separated mind from matter and the Earth from the sky. And these ideas remained, and were adopted by ecclesiastic philosophy, until the work begun by Copernicus, Kepler, and Galileo and completed by Newton imposed a new way of thinking, where the angels who moved the spheres were no longer strictly necessary.

The system proposed by Ptolemy (Fig. 1.4) needed more than 39 wheels or spheres to explain the complicated motion of the planets and the Sun. When the king Alphonse X of Castile, nicknamed the Wise (1221–1284 CE), who had a deep interest in astronomy, learned about the Ptolemaic system, he exclaimed: "If only the Almighty had consulted me before starting the Creation, I would have recommended something simpler."

In spite of this, the tables devised by Ptolemy for calculating the motion of the planets were very precise and were used, together with the fixed stars catalog of Hipparchus, as a guide for navigation by Christopher Columbus and Vasco da Gama. This teaches us an important lesson: an incorrect theory may be useful within the framework of its compatibility with the results of observation and experimentation.

In the Middle Ages, most knowledge accumulated by the Ancient Greeks had been forgotten, with very few exceptions, and even the idea of the Earth's sphericity was effaced from people's minds.

1.2 Copernicus, Kepler, and Galileo

In the fifteenth century, a Polish astronomer, Nicolaus Copernicus (1473–1543) brought Ptolemy's system to crisis by proposing a heliocentric system. Copernicus assumed the Sun (more exactly, a point near the Sun) to be the centre of the Earth's orbit and the centre of the planetary system. He considered that the Earth (around which revolved the Moon), as well as the rest of the planets, rotated around that point near the Sun describing circular orbits (Fig. 1.5). Actually, he rediscovered the system that Aristarchus had proposed in ancient times. Copernicus delayed the publication of his book containing the details of his system until the last few days of his life, apparently so as not to contradict the official science of the ecclesiastics. His system allowed a description of the planetary motion that was at least as good as the one which was based on Ptolemaic spheres. But his work irritated many of his contemporaries. The Catholic Church outlawed his book in 1616, and also Martin Luther rejected it, as being in contradiction with the Bible.

The next step was taken by Johannes Kepler, born in 1571 in Weil, Germany. Kepler soon proved to be gifted with a singular talent for mathematics and astronomy, and became an enthusiastic defender of the Copernican system. One day in the year of 1595, he got a sudden insight. From the Ancient Greeks, it was known

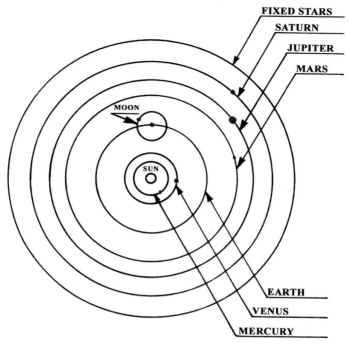

Fig. 1.5 The system of the world according to Copernicus. The Sun was at the centre of the planetary system, and around a point very near to it revolved the Earth and the rest of the planets, all describing circular orbits.

that there are five regular polyhedra: tetrahedron, cube, octahedron, dodecahedron, and icosahedron—the so-called "Platonic solids" of antiquity. Each of these can be inscribed in a sphere. Similarly, there were five spaces among the known planets. Kepler guessed that the numbers might be related in some way. That idea became fixed in his mind and he started to work to prove it.

He conceived of an outer sphere associated with Saturn, and circumscribed in a cube. Between the cube and the tetrahedron came the sphere of Jupiter. Between the tetrahedron and the dodecahedron was the sphere of Mars. Between the dodecahedron and the icosahedron was the sphere of Earth. Between the icosahedron and the octahedron, the sphere of Venus. And finally, within the octahedron came the sphere of Mercury (Fig. 1.6). He soon started to compare his model with observational data. As it was known at that time that the distances from the planets to the Sun were not fixed, he imagined the planetary spheres as having a certain thickness, so that the inner wall corresponded to the minimum distance and the outer wall to the maximum distance.

Kepler was convinced a priori that the planetary orbits must fit his model. So when he started to do the calculations and realized that something was wrong, he

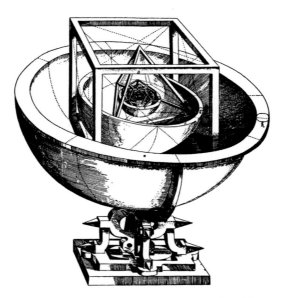

Fig. 1.6 Kepler's system of spheres and inscribed regular Platonic solids.

attributed the discrepancies to the poor reliability of the Copernican data. Therefore he turned to the only man who had more precise data about planetary positions: the Danish astronomer Tycho Brahe (1546–1601), living at that time in Prague, who had devoted 35 years to performing exact measurements of the positions of the planets and stars.

Tycho Brahe conceived of a system which, although geocentric, differed from that of Ptolemy and borrowed some elements from the Copernican system. He assumed that the other planets revolved around the Sun, but that the Sun and the Moon revolved around the Earth (Fig. 1.7).

In an attempt to demonstrate the validity of his model, he made very accurate observations of the positions of the planets with respect to the background of fixed stars. Brahe was a first-rate experimenter and observer. For more than 20 years he gathered the data of his observations, which were finally used by Kepler to deduce the laws of planetary motion.

Kepler believed in circular orbits, and to test his model, he used Brahe's observations of the positions of Mars. He found agreement with the circle up to a point, but the next observation did not fit that curve. So Kepler hesitated. The difference was 8 min of arc. What was wrong? Could it be his model? Could it be the observations made by Brahe? In the end, he accepted the outstanding quality of Brahe's measurements, and after many attempts, finally concluded that the orbit was elliptical. At this juncture, he was able to formulate three basic laws of planetary motion:

1. All planets describe ellipses around the Sun, which is placed at one of the foci;

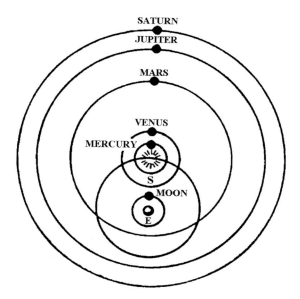

Fig. 1.7 Tycho Brahe's system. The Earth is the centre of the Universe, but the other planets rotate around the Sun, while this in turn moves around the Earth.

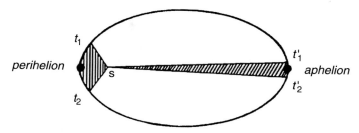

Fig. 1.8 The radius vector or imaginary line joining a planet with the Sun, sweeps out equal areas in equal intervals of time; when the planet is near the Sun, at perihelion, it moves faster than when it is at the other extreme of the orbit, at aphelion.

2. The radius vector or imaginary line which joins a planet to the Sun sweeps out equal areas in equal intervals of time. Consequently, when the planet is nearest to the Sun (at the point called *perihelion*), it moves faster than when it is at the other extreme of the orbit, called *aphelion* (Fig. 1.8);
3. The squares of the periods of revolution of planets around the Sun are proportional to the cubes of the semi-major axis of the elliptical orbit.

Galileo Galilei (1564–1642) was a contemporary of Kepler and also a friend. At the age of 26, he became professor of mathematics at Pisa, where he stayed until 1592. His disagreement with Aristotle's ideas, and especially the claim that a heavy body falls faster than a light one, caused him some personal persecution, and he moved to the University of Padua as professor of mathematics. Meanwhile, his fame as a teacher spread all over Europe. In 1608, Hans Lippershey, a Dutch optician, invented a rudimentary telescope, as a result of a chance observation by an apprentice. Galileo learnt about this invention in 1609, and by 1610, he had already built a telescope. The first version had a magnifying factor of 3, but he improved it in time to a factor of 30. This enabled him to make many fundamental discoveries. He observed that the number of fixed stars was much greater than what could be seen by the naked eye, and he also found that the planets appeared as luminous disks.

In the case of Venus, Galileo discovered phases like those of the Moon. And he found that four satellites revolved around Jupiter. Galileo's observations with the telescope provided definite support for the Copernican system. He became famous also for his experiments with falling bodies and his investigations into the motion of a pendulum.

Galileo's work provoked a negative reaction, because it had brought Ptolemy's system into crisis. This left only two alternatives for explaining the phases of Venus: either Brahe's geocentric system or the Copernican system. The latter definitely went against the ecclesiastical dogma. The Church had created scholasticism, a mixture of religion and Aristotelian philosophy, which claimed to support the faith with elements of rational thinking.

But the Church also had an instrument of repression in the form of the Holy Inquisition, set up to punish any crime against the faith. When Galileo was 36, in 1600, the Dominican friar and outstanding scholar Giordano Bruno (1548–1600) was burned at the stake. He had committed the unforgivable crimes of declaring that he accepted the Copernican ideas of planetary motion, and holding opinions contrary to the Catholic faith (Figs. 1.9, 1.10).

Fig. 1.9 Nicolaus Copernicus. His model was presented in his book *De Revolutionibus Orbium Coelestium (On the Revolutions of Celestial Spheres)*, published thanks to the efforts of his collaborator Rheticus. This book was considered by the Church as heresy, and its publication was forbidden because it went against Ptolemy's system and its theological implications.

Fig. 1.10 Johannes Kepler was named "legislator of the firmament" for his laws of planetary motion, deduced as a result of long and patient work, using the extremely precise data gathered by Tycho Brahe.

When Galileo made his first astronomical discoveries, Bruno's fate was still fresh in his mind. Now he was becoming more and more convinced of the truth of the Copernican system, even though it was in conflict with official science, based on Ptolemy's system. The reaction of the Florentine astronomer Francesco Sizzi, when he learned about the discovery of Jupiter's satellites, was therefore no surprise: *The satellites are not visible to the naked eye, and for that reason they cannot influence the Earth. They are therefore useless, so they do not exist.*

On the one hand, Galileo's discoveries put him in a position of high prestige among many contemporaries, but on the other, he was attracting an increasing number of opponents. The support given by his discoveries to the Copernican theory and his attacks on Aristotelian philosophy aroused the anger of his enemies. In 1616, possibly under threat of imprisonment and torture, he was ordered by the Church "to relinquish altogether the said opinion that the Sun is the centre of the world and immovable [. . .] not henceforth to hold, teach or defend it in any way." Galileo acquiesced before the decrees and was allowed to return to Pisa. The Church was afraid to weaken its position by accepting facts opposed to the established Christian–Aristotelian–Ptolemaic doctrine.

In 1623, one of his friends, Cardinal Maffeo Barberini, became Pope Urban VIII, and Galileo received assurances of *pontifical good will*. Considering that the decree of 1616 would no longer be enforced, he wrote his book *Dialogues on the Ptolemaic and Copernican Systems*. But he faced an ever increasing number of enemies, and even the Pope became convinced that Galileo had tricked him. Galileo was called for trial under suspicion of heresy before the Inquisition at the age of 67. He was forced to retract under oath his beliefs about the Copernican system (Fig. 1.11).

Fig. 1.11 Tycho Brahe. Although his system of planetary motion was wrong, his very precise observations of the planetary positions enabled Kepler to formulate his laws.

Later, a legend was concocted that Galileo, after abjuring, pronounced in low voice the words *And yet it moves*, referring to the Earth's motion around the Sun. That is, in spite of any court and any dogma, it was not possible to deny this physical fact, the objective reality of Earth's motion. However, it is interesting that Galileo never accepted the elliptical orbits discovered by Kepler; he believed only in circular orbits.

Among the most important achievements of Galileo, one must mention his laws of falling bodies, which can be resumed in two statements:

1. All bodies fall in vacuum with the same acceleration. That is, if we let one sheet of paper, one ball of lead, and a piece of wood fall simultaneously in vacuum, they will fall with the same acceleration;
2. All bodies fall in vacuum with uniformly accelerated motion. This means that their acceleration is constant, that is, their velocity increases in proportion to the time elapsed from the moment the bodies started to fall.

The work initiated by Copernicus, Kepler, and Galileo was completed by Isaac Newton. He was born in 1642, the year in which Galileo died, and lived until 1727 (Fig. 1.12).

1.3   Newton and Modern Science

One day, Edmund Halley visited his friend Newton after a discussion with Robert Hooke and Christopher Wren, in which Hooke had claimed that he was able to

Fig. 1.12 Galileo Galilei. He discovered, among other things, four satellites of Jupiter and the phases of Venus, using a telescope of his own improved design. He enunciated the basic laws of falling bodies. His works stirred the antagonistic attitude of the ecclesiastical authorities, and he was forced to stand trial and to abjure his beliefs about the Copernican system.

explain planetary motions on the basis of an attractive force, inversely proportional to the square of the distance. When asked his opinion about it, Newton replied that he had already demonstrated that the trajectory of a body under such a central force was an ellipse.

Newton subsequently sent his calculations to Halley, and after looking through the manuscript, Halley convinced Newton to write in detail about the problem, since it could provide an explanation for the complicated motion of the whole planetary system. And this is how Newton started to write his *Philosophiae Naturalis Principia Mathematica*, a monograph which produced a revolution in modern science.

In the first book Newton stated his laws of motion, which owed much to Galileo, and laid their mechanical foundations. He deduced Kepler's laws by assuming a force inversely proportional to the square of the distance, and demonstrated that according to this law the mass of a homogeneous sphere can be considered as concentrated at its centre.

The second book is devoted to motion in a viscous medium, and it is the first known study of the motion of real fluids. In this book Newton dealt with wave motion and even with wave diffraction.

In the third book Newton studied the motion of the satellites around their planets, and of the planets around the Sun, due to the force of gravity. He estimated the density of the Earth as between 5 and 6 times that of water (the presently accepted value is 5.5), and with this value he calculated the masses of the Sun and the planets. He went on to give a quantitative explanation for the flattened shape of the Earth. Newton demonstrated that, for that shape of the Earth, the gravitational force exerted by the Sun would not behave as if all its mass were concentrated at its centre, but that its axis would describe a conical motion due to the action of the Sun: this phenomenon is known as the precession of the equinoxes.

Fig. 1.13 Isaac Newton. His scientific work marks the beginning of physics as a modern science. His formulation of the laws of mechanics and universal gravitation laid the basis for explaining planetary motion and obtaining the Kepler laws. His work in optics, as well as in mathematics, was also remarkable, and he invented the differential and integral calculus independently of his contemporary Gottfried Leibniz.

Although Newton used the differential and integral calculus (which he invented himself, independently of Gottfried Leibniz) to get his results, he justified them in his book by using the methods of classical Greek geometry. One of the most practical consequences of his work was to supply a calculational procedure for determining the orbit of the Moon and the planets with much greater accuracy than ever before, using a minimum number of observations. Only three observations were enough to predict the future position of a planet over a long period of time. A confirmation of this was given by his friend Edmund Halley, who predicted the return of the comet which bears his name. Some other very important confirmations appeared in the nineteenth and twentieth centuries due to Le Verrier and Lowell, who predicted the existence of the then undiscovered planets Neptune and Pluto, deducing their existence from the perturbations they produced on other planetary motions.

The theory of gravitation conceived by Newton, together with all his other contributions to modern astronomy, marked the end of the Aristotelian world adopted by the scholastics and challenged by Copernicus. Instead of a Universe composed of perfect spheres moved by angels, Newton proposed a mechanism of planetary motion which was the consequence of a simple physical law, without need for the continuous application of direct holy action (Fig. 1.13).

1.4 Newton's Laws

Newton established the following three laws as the basis of mechanics:

1. Every body continues in its state of rest or in uniform motion in a straight line unless it is compelled to change that state by forces acting on it;
2. The rate of change of momentum is proportional to the applied force, and it is in the direction in which the force acts;
3. To every action there is always opposed an equal reaction.

In the second law, momentum is defined as the product of the mass and the velocity of the body.

1.4.1 Newton's First Law

Newton's first law is known as the law or *principle of inertia*. It can only be verified approximately, since to do it exactly, a completely free body would be required (without external forces), and this would be impossible to achieve. But in any case it has a great value, since it establishes a limiting law, that is, a property which, although never exactly satisfied, is nevertheless satisfied more and more accurately, as the conditions of experimentation or observation approach the required ideal conditions.

As an example, an iron ball rolling along the street would move forward a little way, but would soon come to a stop. However, the same ball rolling on a polished surface like glass, would travel a greater distance, and in the first part of its trajectory, it would move uniformly along a straight line. Furthermore, the length of its trajectory would be longer if the friction between the ball and the surface (and between the ball and air) could be reduced. The only applied force is friction (acting in the opposite direction to the motion of the ball). The weight of the body acts perpendicular to the surface, and it is balanced by the reaction force of the surface.

1.4.2 Newton's Second Law

Newton's second law, known also as the *fundamental principle of dynamics*, states the proportionality between the acceleration a and the force F acting on a given body:

$$\mathbf{F} = m\mathbf{a}. \tag{1.1}$$

The constant of proportionality m is called mass. The mass can be interpreted as a measure of the inertia of the body. The larger the mass, the larger the force required to produce a given acceleration on a given body. The smaller the mass of a body, the larger the acceleration it would get when a given force is applied, and obviously, the more quickly it would reach high speeds. In modern physics this is observed with elementary particles: much less energy (and force) is required to accelerate electrons than to accelerate protons or heavy nuclei. On the other hand, photons (light quanta) move at the highest possible velocity (the speed of light, which is about 300,000 km/s), since they behave as massless particles (see Chap. 5).

But let us return to the second law. Its extraordinary value is due essentially to the fact that, if the interaction law is known for two bodies, from the mathematical expression for the mutual forces exerted it is possible to obtain their trajectories.

For instance, in the case of the Sun and a planet, as mentioned above, Newton established that a mutual force of attraction is exerted between them, a manifestation of universal gravitation. That force is directed along the line joining their centres, and it is proportional to the product of their masses and inversely proportional to the square of the distance between them. That is,

$$\mathbf{F} = -\frac{GMm}{r^2}\mathbf{r}_0, \tag{1.2}$$

where M and m are the masses of the Sun and planet, respectively, r is the distance between their centres, G is a constant whose value depends on the system of units used, and $\mathbf{r}_0$ is a unit vector along $\mathbf{r}$. $\mathbf{F}$ is a central force, that is, its direction always passes through a point which is the so-called centre of forces (in this case, it is a point inside the Sun).

Then, taking into account the fact that acceleration is a measure of the instantaneous rate of change of velocity with respect to time (the time derivative of velocity) and that in turn velocity is the rate of change of the position of the planet (time derivative of position), we have a mathematical problem that is easily solved (at least in principle) using differential calculus. Since acceleration is the second derivative with respect to time of the position vector of the planet with respect to the Sun, we can write:

$$m\frac{d^2\mathbf{r}}{dt^2} = -\frac{GMm\mathbf{r}_0}{r^2}. \tag{1.3}$$

This differential equation can be solved using the fact that the solar mass M is much greater than that of the planet m. The solution tells us that the orbit described by the planet is a conic section in which the Sun is placed at one of the foci. The type of orbit depends on the total energy of the body.

If the energy is negative, we have elliptical orbits (in the case of a minimum energy value, the ellipse degenerates into a circular orbit). If the energy is zero, the orbits are parabolic. Here we consider the total energy as the sum of the potential and kinetic energies, so that the zero corresponds to the case in which these are equal in absolute value; as we shall see later, in this case the potential energy is negative. Finally, for positive energies we have hyperbolic orbits.

The known planets describe elliptic orbits, but some comets coming from outer space describe parabolic or hyperbolic orbits. In that case, they get close to the Sun, move around it, and later disappear for ever. For most known comets, like Halley's, the orbit is elliptical but highly eccentric (i.e., very flattened).

As pointed out earlier, the application of Newtonian mechanics to the study of planetary motion gave astronomers an exceptionally important tool for the calculation of planetary orbits. But from the methodological point of view, Newtonian mechanics was of transcendental importance in modern science, since for the first time in physics a theory was established from which it was possible to predict consequences compatible with the results of observation. In that sense, Newton closed a circle which was initiated by Brahe, and which was continued by Kepler when he derived the laws of planetary motion from the data of Brahe's observations. Newton showed that such laws could be obtained by starting from very general physical principles: the equations of mechanics and the gravitational force between bodies.

For observers at rest or in uniform motion along a straight line, the laws of mechanics are the same. But the validity of Newton's laws depends on the acceleration of the observer: they do not hold equally for observers who are accelerated in different ways. For that reason it became necessary to introduce the concept of frame of reference, in particular, the concept of inertial frame, in which Newton's laws are valid. An inertial frame is something more than a system of reference; it includes the time, i.e., some clock. A simple geometrical change of coordinates does not change the frame of reference. We shall return to inertial frames in Sect. 1.7.

Vectors. We have already spoken about vectors indicating the position of the planets, and when discussing forces, velocities, and accelerations. Implicitly we have referred to the vectorial nature of these quantities. In order to characterize vectors, it is not sufficient to use simple numbers or scalars indicating their magnitude or absolute value. For vectors, besides the magnitude or modulus, we need to indicate their direction. Vectors are represented by arrows whose length and direction represent the magnitude and direction of the vector, respectively.

For instance, when referring to the velocity of a body, it is not enough to say how many meters per second it moves. We must also specify in which direction it is moving. A body that falls has a velocity which increases proportionally to the time elapsed, and its direction is vertical, from up to down. We represent that velocity as a vertical vector of increasing magnitude, with its end pointing downward.

Sometimes vectors are used to indicate the position of a point that moves with respect to another one taken as fixed. This is the case of the radius vector, to which we referred when describing Kepler's laws. The origin of the radius vector is at the Sun and the end is at the planet that moves.

Two parallel vectors, $\mathbf{A}$ and $\mathbf{B}$, are simply summed, and the sum has the same direction as the added vectors. If they are parallel and of opposite directions, their sum is a vector of modulus equal to the difference of the moduli of the given vectors and its direction is that of the vector of larger modulus.

If two vectors $\mathbf{A}$ and $\mathbf{B}$ are not parallel, but have different orientations, their sum is geometrically a third vector obtained by displacing $\mathbf{B}$ parallel to itself so that its origin coincides with the end of $\mathbf{A}$, and then, by joining the origin of $\mathbf{A}$ with the end of $\mathbf{B}$ we get the sum $\mathbf{A} + \mathbf{B}$ of the two vectors.

Given a system of orthogonal coordinates $Oxyz$, the vector $\mathbf{A}$ can be written in terms of its three components along the coordinate axes, $\mathbf{A} = (A_x, A_y, A_z)$, obtained from the projection of the vector on them. The modulus of $\mathbf{A}$ is given by

$$A = \sqrt{A_x^2 + A_y^2 + A_z^2},$$

where $A_x = A \cos \alpha$, $A_y = A \cos \beta$, $A_z = A \cos \gamma$, with α, β, γ being the angles between A and the axes Ox, Oy, and Oz, respectively. Thus, a vector in three dimensions is defined by an ordered set of three numbers, which are its components. Let us define the unit vectors

$$\mathbf{i} = (1, 0, 0), \quad \mathbf{j} = (0, 1, 0), \quad \mathbf{k} = (0, 0, 1).$$

One can write

$$\mathbf{A} = A_x \mathbf{i} + A_y \mathbf{j} + A_z \mathbf{k}.$$

In the same way,

$$\mathbf{B} = B_x \mathbf{i} + B_y \mathbf{j} + B_z \mathbf{k},$$

and their sum is obviously

$$\mathbf{A} + \mathbf{B} = (A_x + B_x)\mathbf{i} + (A_y + B_y)\mathbf{j} + (A_z + B_z)\mathbf{k}.$$

An important vector is the *position vector*,

$$\mathbf{r} = x\mathbf{i} + y\mathbf{j} + z\mathbf{k},$$

of any arbitrarily chosen point with respect to the system of coordinates $Oxyz$.

Mechanical quantities such as displacements, velocities, accelerations, forces, etc., are to be summed in accordance with this procedure of vectorial or geometrical sum.

If two forces have opposite directions but equal moduli, their vector sum is a null vector, that is, a vector of modulus zero. However, that does not necessarily mean that the physical effect is canceled: if the forces are applied at different points, both of them will have a mechanical effect. Opposite forces are responsible for static equilibrium—for instance, for a body having the weight $\mathbf{G}$ lying on a table. The weight $\mathbf{G}$ is applied to the table and the reaction of the table $\mathbf{R} = -\mathbf{G}$ is applied to the body. Opposite forces of equal modulus also appear in dynamics, as in the case of the Sun and a planet: their mutual action is expressed by opposite forces, but the forces are applied at different points, on the Sun and on the planet: the vector sum of the forces is zero, nevertheless they produce the motion of the bodies.

Given two vectors $\mathbf{A}$ and $\mathbf{B}$, their *scalar product* is a number obtained by multiplying together the modulus of each vector by the cosine of the angle formed by their directions. Usually, the scalar product is represented by means of a dot between the two vectors:

$$\mathbf{A} \cdot \mathbf{B} = AB \cos \alpha. \tag{1.4}$$

The scalar product of two vectors can also be expressed as the product of the modulus of one of the vectors by the projection of the other on it. The scalar product is commutative, $\mathbf{A} \cdot \mathbf{B} = \mathbf{B} \cdot \mathbf{A}$. Moreover, $\mathbf{A} \cdot \mathbf{A} = A^2$, that is, the modulus squared of a vector is given by the scalar product of the vector with itself. If $\mathbf{A}$ and $\mathbf{B}$ are perpendicular, then $\mathbf{A} \cdot \mathbf{B} = 0$. If c is a number, it is obvious that $(c\mathbf{A}) \cdot \mathbf{B} = c(\mathbf{A} \cdot \mathbf{B})$.

The unit vectors satisfy the properties

$$\mathbf{i} \cdot \mathbf{i} = \mathbf{j} \cdot \mathbf{j} = \mathbf{k} \cdot \mathbf{k} = 1$$

and

$$\mathbf{i} \cdot \mathbf{j} = \mathbf{j} \cdot \mathbf{k} = \mathbf{k} \cdot \mathbf{i} = 0.$$

Then one can write the scalar product in the form

$$\mathbf{A} \cdot \mathbf{B} = A_x B_x + A_y B_y + A_z B_z. \tag{1.5}$$

The scalar product is particularly useful in expressing the work performed by a force on a particle that describes an arbitrary trajectory between two points, P_0 and P. At each point of the curve the force forms an angle with the tangent to the curve at the point. The total work performed by the force can be calculated in the following way: divide the curve into segments at the points 1, 2, 3, etc., and draw the corresponding chords $\Delta\mathbf{S}_1$, $\Delta\mathbf{S}_2$, $\Delta\mathbf{S}_3$ as vectors that join the points P_0, P_1, P_2, P_3, etc. Then take the value of the force at an arbitrary point inside each of these segments. Let $\mathbf{F}_1$, $\mathbf{F}_2$, $\mathbf{F}_3$, etc., be the values of the force at such points (Fig. 1.14). Then take the sum of the scalar products:

$$\mathbf{F}_1 \cdot \Delta\mathbf{S}_1 + \mathbf{F}_2 \cdot \Delta\mathbf{S}_2 + \ldots + \mathbf{F}_n \cdot \Delta\mathbf{S}_n. \tag{1.6}$$

When the number of the points of the division tends to infinity, such that the modulus of the largest of the vectors $\Delta\mathbf{S}_i$ tends to zero, the work done by the force is obtained as

$$W = \lim_{\Delta S_i \to 0} \sum_{i=1}^{\infty} \mathbf{F}_i \cdot \Delta\mathbf{S}_i. \tag{1.7}$$

Fig. 1.14 The scalar product is used, for example, for calculating the work performed by a force.

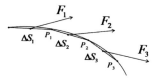

This is represented by the symbol

$$W = \int_{P_0 P} \mathbf{F} \cdot d\mathbf{S} , \qquad (1.8)$$

which is called the line integral between P_0 and P.

The *vector product* (or *cross product*) of two vectors is a new vector, obtained by performing a mathematical operation on them. To illustrate it, let $\mathbf{A}$ and $\mathbf{B}$ be two vectors in a plane (Fig. 1.15). Decompose $\mathbf{B}$ into two other vectors, $\mathbf{B}_1$ and $\mathbf{B}_2$ (whose sum is $\mathbf{B}$). The vector $\mathbf{B}_1$ is in the direction of $\mathbf{A}$, while the vector $\mathbf{B}_2$ is perpendicular to $\mathbf{A}$. We now define a third vector that we call the vector product of $\mathbf{A}$ by $\mathbf{B}$, denoted by $\mathbf{A} \times \mathbf{B}$, whose characteristics are:

1. Its modulus is the product of the moduli of $\mathbf{A}$ and $\mathbf{B}_2$. In other words, it is equal to the product of the moduli of $\mathbf{A}$ and $\mathbf{B}$ with the sine of the angle between them, $AB \sin \alpha$;
2. Its direction is perpendicular to the plane spanned by $\mathbf{A}$ and $\mathbf{B}$ and is determined as follows. If the direction of rotation to superpose $\mathbf{A}$ on $\mathbf{B}$ is indicated by the index, middle, ring, and little fingers of the right hand (as shown in Fig. 1.15), then the thumb indicates the direction of $\mathbf{A} \times \mathbf{B}$ (assuming that the angle α between the vectors is smaller than $180°$).

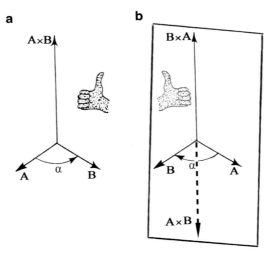

Fig. 1.15 a The vector product of two vectors $\mathbf{A}$ and $\mathbf{B}$ is a third vector, perpendicular to $\mathbf{A}$ and $\mathbf{B}$, whose modulus is the product of the moduli of $\mathbf{A}$ and $\mathbf{B}$ with the sine of the angle between them, or equivalently, the product of the modulus of one of them with the projection of the other on the direction perpendicular to the first. The direction of the vector product is given by the right-hand rule as shown in the figure. **b** The mirror image does not satisfy the definition for the vector product of two vectors, but obeys a left-hand rule, since the image of the right hand is the left hand.

Strictly speaking, the vector product of two vectors is not a true vector, but a *pseudovector*, since the mirror image does not satisfy the previous definition, but the *left-hand* rule, which is obviously not equivalent to it: the mirror image of the right hand is the *left* hand.

Consequently, the product $\mathbf{B} \times \mathbf{A}$ gives a vector of the same modulus but opposite direction to $\mathbf{A} \times \mathbf{B}$. This is an interesting result: the vector product is not commutative, but rather one can write $\mathbf{B} \times \mathbf{A} + \mathbf{A} \times \mathbf{B} = 0$, meaning that the vector product is *anticommutative*. In particular, $\mathbf{A} \times \mathbf{A} = 0 = \mathbf{B} \times \mathbf{B}$. This property can be generalized to higher dimensional spaces, and leads to the definition of exterior algebras or Grassmann algebras (see Chap. 7).

For the unit vectors, we have the properties:

$$\mathbf{i} \times \mathbf{i} = \mathbf{j} \times \mathbf{j} = \mathbf{k} \times \mathbf{k} = 0$$

and

$$\mathbf{i} \times \mathbf{j} = \mathbf{k}, \quad \mathbf{j} \times \mathbf{k} = \mathbf{i}, \quad \mathbf{k} \times \mathbf{i} = \mathbf{j}.$$

Since the product is anticommutative, if we exchange the pair on the left, the sign is changed on the right. In terms of components, we get

$$\mathbf{A} \times \mathbf{B} = (A_y B_z - A_z B_y)\mathbf{i} + (A_z B_x - A_x B_z)\mathbf{j} + (A_x B_y - A_y B_x)\mathbf{k}. \tag{1.9}$$

It is easily seen that the vector product vanishes if the vectors are parallel.

Transformations of vectors. Vector components transform like coordinates. For instance, under a rotation of the system of coordinates, the components A_x, A_y, A_z transform like the coordinates x, y, z. Under a positive (counterclockwise) rotation of angle θ around the z-axis, the position vector of a point P, expressed as $\mathbf{r} = x\mathbf{i} + y\mathbf{j} + z\mathbf{k}$ in the original system, is transformed in the rotated system to $\mathbf{r}' = x'\mathbf{i}' + y'\mathbf{j}' + z'\mathbf{k}$, where the new coordinates x', y', z' are given by the product of the rotation matrix $\mathbf{R}$ with the initial vector $\mathbf{r}$. The unit vectors in the rotated system are $\mathbf{i}', \mathbf{j}'$, whereas $\mathbf{k}$ does not change. The rotation matrix is an array of 3×3 numbers in three rows and three columns. The components of a matrix are labeled by two indices (i, j), where the first identifies the row and the second indicates the column. The rotated vector $\mathbf{r}'$ is the product of the rotation matrix $\mathbf{R}$ with the original vector $\mathbf{r}$. For the particular rotation of angle θ around the z-axis, we write this product as

$$\begin{pmatrix} x' \\ y' \\ z' \end{pmatrix} = \begin{pmatrix} \cos\theta & \sin\theta & 0 \\ -\sin\theta & \cos\theta & 0 \\ 0 & 0 & 1 \end{pmatrix} \begin{pmatrix} x \\ y \\ z \end{pmatrix}. \tag{1.10}$$

Under this rotation, the components of a vector $\mathbf{A}$ transform as

$$A'_x = A_x \cos\theta + A_y \sin\theta,$$
$$A'_y = -A_x \sin\theta + A_y \cos\theta,$$
$$A'_z = A_z.$$

Under an inversion of the coordinate axis, $(x, y, z) \rightarrow (x, y, -z)$, the vector **A** transforms as $(A_x, A_y, A_z) \rightarrow (A_x, A_y, -A_z)$. A pseudovector **P** transforms under rotations like the coordinates, but under an inversion, it remains the same, $(P_x, P_y, P_z) \rightarrow (P_x, P_y, P_z)$.

There is an alternative way of writing the previous 'vector' rotation. If we now denote the indices of components along x, y, z by $i = 1, 2, 3$, respectively, we may write the vector components as A_i. Further, we shall write the matrix **R** in terms of its elements as R_{ij} (row i and column j). Then, for instance,

$$A'_3 = \sum_j R_{3j} A_j = R_{31} A_1 + R_{32} A_2 + R_{33} A_3.$$

In what follows, we adopt Einstein's summation convention: if a term contains the same index twice, the summation over all values of that index is to be understood. Thus, $A'_3 = R_{3j} A_j$ means the sum over j, as j ranges over $1, 2, 3$. (From now on, we shall use the indices x, y, z as an alternative to $1, 2, 3$, understanding the correspondence $x \rightarrow 1, y \rightarrow 2, z \rightarrow 3$.)

Tensors. The *dyadic product* **AB** of two vectors **A** and **B** is a quantity with the property that

$$(\mathbf{AB}) \cdot \mathbf{C} = \mathbf{A}(\mathbf{B} \cdot \mathbf{C}). \tag{1.11}$$

The result is a vector in the direction **A**, since $\mathbf{B} \cdot \mathbf{C}$ is a scalar. As

$$(\mathbf{AB}) \cdot (c\mathbf{C}) = c[(\mathbf{AB}){\cdot}\mathbf{C})]$$

and

$$(\mathbf{AB}) \cdot (\mathbf{C} + \mathbf{D}) = (\mathbf{AB}) \cdot \mathbf{C} + (\mathbf{AB}) \cdot \mathbf{D},$$

the quantity **AB** is called a linear operator, or *tensor*, and (1.11) is a linear function of **C**. A tensor is a quantity whose components transform as a product of the coordinates. For instance, the component T_{xy} of a tensor **T** transforms as the product xy. The unit tensor is the dyadic $\mathbf{I} = \mathbf{ii} + \mathbf{jj} + \mathbf{kk}$. It is easy to check that $\mathbf{I} \cdot \mathbf{A} = \mathbf{A}$. In three-dimensional space, a second rank tensor can be written in the form

$$\begin{aligned} \mathbf{T} = {} & T_{xx}\mathbf{ii} + T_{xy}\mathbf{ij} + T_{xz}\mathbf{ik} \\ & + T_{yx}\mathbf{ji} + T_{yy}\mathbf{jj} + T_{yz}\mathbf{jk} \\ & + T_{zx}\mathbf{ki} + T_{zy}\mathbf{kj} + T_{zz}\mathbf{kk}. \end{aligned} \tag{1.12}$$

However, it is simpler to write it in the matrix form

$$\mathbf{T} = \begin{pmatrix} T_{xx} & T_{xy} & T_{xz} \\ T_{yx} & T_{yy} & T_{yz} \\ T_{zx} & T_{zy} & T_{zz} \end{pmatrix}. \tag{1.13}$$

By using the numerical indices $i, j = 1, 2, 3$, we may write the general component of $\mathbf{T}$ as T_{ij}. A tensor is symmetric if $T_{ij} = T_{ji}$, and antisymmetric if $T_{ij} = -T_{ji}$. An arbitrary tensor can be written as the sum of a symmetric and an antisymmetric tensor.

In a similar way we can define tensors of third rank as T_{ijk}, etc. For us, the most interesting third rank tensor is the completely antisymmetric unit tensor ϵ_{ijk}, called the Levi-Civita tensor. (Actually, it is a pseudotensor, because it behaves as a tensor except under the inversion of coordinates.) Its components are as follows: zero, if at least two indices are equal; $+1$, if the permutation of the (unequal) indices ijk is even (i.e. 123, 312, 231), and -1, if the permutation of the indices is odd (i.e., 213, 321, 132).

Let us consider two vectors represented by their components A_j and B_k. If we write the product of ϵ_{ijk} with these vectors, and sum over j and k, we get

$$C_i = \epsilon_{ijk} A_j B_k, \tag{1.14}$$

i.e.,

$$C_1 = \epsilon_{123} A_2 B_3 + \epsilon_{132} A_3 B_2,$$
$$C_2 = \epsilon_{231} A_3 B_1 + \epsilon_{213} A_1 B_3,$$
$$C_3 = \epsilon_{312} A_1 B_2 + \epsilon_{213} A_2 B_1.$$

Hence, $C_i = (A_2 B_3 - A_3 B_2, A_3 B_1 - A_1 B_3, A_1 B_2 - A_2 B_1)$; in other words, C_i with $i = 1, 2, 3$ are the components of the vector product $\mathbf{A} \times \mathbf{B}$.

Thus, the vector product $\mathbf{C} = \mathbf{A} \times \mathbf{B}$ can be written in components as $C_i = \frac{1}{2} \epsilon_{ijk} T_{jk}$, where T_{jk} are the components of the antisymmetric tensor $\mathbf{T}$:

$$\mathbf{T} = \begin{pmatrix} 0 & C_2 & -C_3 \\ -C_2 & 0 & C_1 \\ C_3 & -C_1 & 0 \end{pmatrix}. \tag{1.15}$$

The pseudovector $\mathbf{C}$ is called the dual pseudovector of the tensor $\mathbf{T}$.

Very important physical quantities are usually expressed as vector products. Examples are the angular momentum $\mathbf{L}$, or the magnetic field $\mathbf{B}$. The vector product will also be used in the expression (1.27) for the velocity written in a rotating system of coordinates, and it is useful to remember in these cases that it is a pseudovector, i.e., the dual of an antisymmetric tensor.

Fig. 1.16 Angular
momentum of a satellite S
that moves around the Earth
E. The angular momentum
is $\mathbf{L} = \mathbf{r} \times \mathbf{p}$.

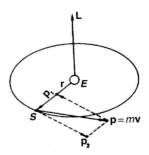

For a satellite of mass m that moves around the Earth, we can assume its velocity
at each instant to be the sum of two vectors: a component along the radius vector
and another perpendicular to it, contained in the plane of the orbit. The angular
momentum of the satellite around the Earth (Fig. 1.16) is given by the cross product
of the radius vector $\mathbf{r}$ of the satellite with respect to the Earth with the momentum
$\mathbf{p} = m\mathbf{v}$ of the satellite:

$$\mathbf{L} = \mathbf{r} \times \mathbf{p}. \tag{1.16}$$

1.4.3 Planetary Motion in Newton's Theory

It is instructive to analyze the motion of a planet around the Sun (or of the Moon
around the Earth) as a consequence of Newton's second law.

Assume that at a given instant the momentum of the planet is $\mathbf{p} = m\mathbf{v}$ around the
Sun. If the gravitational attraction could be switched off at that precise moment, the
planet would continue to move uniformly in a straight line, that is, with a constant
momentum $\mathbf{p}$. In the time interval Δt elapsed between two adjacent positions 1 and 2,
the planet suffers a change in its momentum due to the action of the Sun's attractive
force $\mathbf{F}$.

This change is represented by a vector $\Delta \mathbf{p} = \mathbf{F} \Delta t$ along the direction of the force
exerted by the Sun on the planet (Fig. 1.17). Then at the point 2, the momentum of
the planet becomes $\mathbf{p} + \Delta \mathbf{p}$. The process is reiterated at successive points so that the
resulting trajectory is a curve. This is due to the action at each instant of the force $\mathbf{F}$
that causes the planet to "fall" continuously toward the Sun.

As another example, assume that we let a stone fall freely, starting from a rest
position: its initial velocity is zero, but because of the Earth's gravitational attraction,
it acquires a momentum $\Delta \mathbf{p}$ that increases proportionally with time, and points in
the same direction as the force exerted by the Earth on the stone.

If the stone is thrown far away, it carries an initial momentum $\mathbf{p} = m\mathbf{v}$ (where m
is the mass of the stone and $\mathbf{v}$ its velocity). This initial momentum combines with the

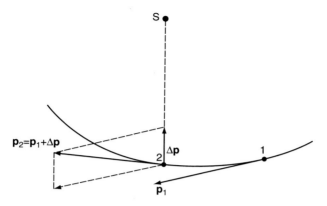

Fig. 1.17 The Sun exerts a continuous force on the planet producing an increase in its linear momentum by the amount $\Delta\mathbf{p}$ between two successive positions 1 and 2. This vector $\Delta\mathbf{p}$ is directed along the radius vector joining the planet to the Sun.

momentum $\Delta\mathbf{p}$ due to the Earth's gravitational attraction and results in an apparently parabolic trajectory (Figs. 1.18 and 1.19).

We would like to stress the important difference between the case of the object that is thrown vertically upward and the one which is put on a terrestrial orbit. In the first case, the initial momentum points in the same direction as the applied force and (unless it moves with a velocity greater than the escape velocity), the object will

Fig. 1.18 If the air resistance is neglected, the trajectory described by a stone thrown in the way shown in the figure is approximately a parabolic arc.

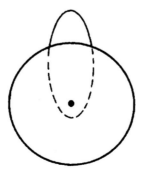

Fig. 1.19 Strictly speaking, the trajectory described by a body thrown like the one in the previous figure is an arc of a very eccentric ellipse, in which the centre of the Earth is one of the foci.

return to Earth, following the vertical. In the second case, the initial momentum **p** of the object (for example, a satellite) forms some angle (different from 0 or 2π) with the direction of the vertical, or with the radius vector **r** of the satellite with respect to the centre of the Earth. In our case, to carry out the motion of a satellite in orbit around the Earth, its angular momentum $\mathbf{L} = \mathbf{r} \times \mathbf{p}$ with respect to the centre of the Earth E must be different from zero (Fig. 1.16). For the case of motion along the vertical, the angular momentum is zero and there is no orbital motion.

An important property of the angular momentum for the case of a satellite orbiting the Earth, or a planet revolving around the Sun, is that its value is a non-vanishing constant, that is, it does not vary with time because the force acting here is a *central force*: its direction passes always through a fixed point, the *centre of forces*.

1.4.4 Newton's Third Law

Newton's third law is satisfied for the mutual gravitational attraction between the Sun and the planets, and for the planets among themselves. The force with which the Sun attracts the Earth is applied to the Earth, but a force equal and opposite is exerted by the Earth on the Sun, and both are directed along the line joining their centres. Action and reaction are always applied at different points.

Another example is provided by one body placed on top of another. A brick on a table exerts an action equal to its weight, and the table exerts on the brick an equal and opposite force. If the brick is put on a spring mattress, the mattress will be deformed, and partially flattened. Finally, if it is put on a stretched newspaper, held with both hands, this will probably break and the brick will fall to the floor. The newspaper would not be able to react with a force equal and opposite to the weight of the brick, and would therefore break. This would not occur if instead of the brick we placed a lighter body on the newspaper (for example, a coin). We see from these examples that, in general, the action produces some deformation of the body on which it acts, forcing it to give back an equal and opposite reaction, and that there exists a limiting value for the action force, when the body breaks.

In the two previous cases we have examples of action at a distance and of action by contact. The first is typical of the classical conception of the gravitational interaction between bodies, which was replaced later by the concept of *field*, as an intermediate agent for the interaction, propagating with finite speed. This will be discussed in more detail in later chapters.

Newton's third law should be interpreted with care in the atomic world, because of the finite velocity of propagation of the interactions. For instance, two charged particles in motion exert mutual forces of attraction or repulsion, but at a given instant the force exerted on one of the particles is determined by the position of the other at some previous instant, and the effect of its new position will be felt some time later.

Newton's third law can be considered as a manifestation of a much more general law. In fact, the expression action $\leftrightarrow$ reaction does not necessarily establish their equality, but rather the relation stimulus–response, in which the second is opposed to the first one. This can be found in all fields of physics. For instance, in electromagnetic theory, it is found that, when a magnetic field varies near a conductor, an electric current is induced on the latter (Faraday's law). But this electric current in turn creates a magnetic field which acts oppositely to the applied field (Lenz law). Furthermore, an electric charge in a medium creates an electric field, attracting charges of opposite sign, and the net effect is a screening of the charge and the field created by it.

In thermodynamics, that general law is expressed by Le Chatelier's principle: if some external actions are applied to a system in equilibrium, and if these tend to alter it, some reactions originate in the system which tend to compensate the external actions and take the system to a new state of equilibrium. For instance, if we heat a jar with a match at some point, the jar alters its state of equilibrium. However, the heat spreads across its mass cooling the hot point, and after some time, the jar reaches a new state of equilibrium at a uniform temperature higher than before because of the absorbed heat.

1.5   Conservation Laws

Starting from Newton's laws, and on the basis of a simple hypothesis about the interaction forces between the particles, it is possible to establish three conservation laws:

1. Conservation of linear momentum;
2. Conservation of angular momentum;
3. Conservation of energy.

The conservation of these quantities is usually accepted as valid in all fields of physics, and they can be derived as a consequence of the basic symmetry properties of space and time. Thus, the conservation of linear momentum is a consequence of the homogeneity of space, the conservation of the angular momentum is due to the isotropy of space (meaning that all the directions of space are equivalent for a given physical system, i.e., its properties do not vary when it is rotated as a whole), and the conservation of energy is a consequence of the homogeneity of time (in other words, the evolution of a system with respect to time, starting from an initial instant t_0, is the same for any value of t_0). This correspondence between symmetry properties and conservation laws is extremely general and crops up again in other theories, particularly in microscopic physics. The ultimate understanding of these relations was given by the German Jewish mathematician Emmy Noether (1882–1935), in the theorem which bears her name and which turned out to be one of the most influential works for the development of theoretical physics in the twentieth century.

1.5.1   *Conservation of Linear Momentum*

It is easy to demonstrate that for a system of particles under no external influences, the total linear momentum (the sum of the linear momenta of all the particles) is conserved when Newton's third law is satisfied. Put another way, the total linear momentum is conserved if the action and the reaction are equal in modulus but act in opposite directions.

It may happen that one or both interacting particles emit some radiation. In that case one must attribute some momentum to the radiation field in order that the linear momentum be conserved. When the radiation is assumed to be composed of quantum particles (for example, photons), the law of conservation of linear momentum is restored by including newly created particles carrying a certain amount of linear momentum.

We shall refer to an example from macroscopic physics. If we shoot a gun, the bullet, having a small mass, leaves the gun at a speed of several meters per second. The gun moves back in the opposite direction at a lower speed (at the moment of shooting we can neglect the action of the force exerted by the Earth's gravitational field). If m is the mass and $\mathbf{v}$ the speed of the bullet, and if M and $\mathbf{V}$ are the mass and speed of the gun, we find that $M\mathbf{V} = -m\mathbf{v}$. So the momentum acquired by the bullet is the same (but of opposite sign) as that acquired by the gun. The sum of two quantities equal in modulus but with opposite directions is zero, which was the initial value of the total linear momentum.

But what happens if we fix the gun to a solid wall? In this case the gun does not move back, the wall stops it. But now the conditions have changed. An external force is exerted on the gun, since it has been fixed to the wall, and this in turn is fixed to the Earth.

This means that the Earth should move back with a speed which, when multiplied by its mass, yields a momentum equal in modulus but opposite to that carried by the bullet. Let us suppose that the bullet has a mass of 100 g, and that its speed is $100 \, \text{m/s} = 10^4 \, \text{cm/s}$. The mass M of the Earth is 5.98×10^{27} g. From the equation $M\mathbf{V} = -m\mathbf{v}$, we find that, after the gun is fired, the Earth should recoil with a speed of the order of 10^{-20} cm/s. For all practical purposes, this is zero.

Something similar happens if we throw a rubber ball against a wall. The ball bounces and comes back with a velocity of approximately the same modulus, but in the opposite direction. Apparently, the linear momentum is not conserved, but the ball has subtracted a certain amount of momentum from the wall, or from the Earth, which recoils with insignificant velocity.

We should emphasize that, in the previous example, the velocity of the ball bouncing off the wall has opposite direction to what it had before the collision, but its modulus is actually somewhat smaller. The wall did not *give back* all the incident momentum, but absorbed a part of it. An extreme case occurs if we throw a ball of clay against the wall. In this case the ball does not rebound. All the linear momentum of the ball is transmitted to the wall, and as it is fixed to the Earth, its resulting change of motion is not perceptible. But if the wall were supported by wheels that could move without friction, it would start to move with the colliding ball of clay stuck on it. Its speed would be easily obtained: if M is its mass, and m and v are the mass and velocity of the ball of clay, we conclude that the modulus of the velocity of the wall V would be

$$V = v\frac{m}{m + M}.\qquad(1.17)$$

Conservation of Linear Momentum and the Mössbauer Effect. The previous example of the gun fixed to the Earth (that does not recoil) has an interesting analogy in nuclear physics, in the so-called Mössbauer effect. In this case, the gun is an atomic nucleus, and the bullet is the gamma radiation emitted by it. The gamma radiation emitted by a nucleus has a constant frequency, but when the nucleus is able to move, as happens in a gas, we have a case similar to the first example of the recoiling gun. The nucleus recoils when emitting the gamma radiation. This causes a range of frequencies to be observed, within a certain bandwidth $\Delta\omega$, that is, there are many values of the frequency in such an interval and a continuous set of frequencies is observed due to the different values of the energy lost by the recoil of the nucleus. The frequency no longer has a precise value, but lies in an interval of possible values, which we may call the imprecision or error.

However, in certain crystals (for example, iridium 197 and iron 57) phenomena occur as in the example of the gun fixed to the Earth, since the emitting nucleus is effectively fixed to the crystal (which does not recoil significantly). Then the frequency of the emitted radiation has an extraordinarily narrow band width $\Delta\omega$. In the case of iron 57, the band width divided by the frequency ω is of the order of $\Delta\omega/\omega \sim 3 \times 10^{-13}$. This is equivalent in units of time to an error of one second in an interval of 30,000 years.

As can be seen from this, the Mössbauer effect can be used to make very precise measurements of frequency.

1.5.2 Conservation of Angular Momentum

For the case of motion under the action of a central force (directed along the radius vector joining the planet with the Sun), angular momentum is conserved: it does not vary with time. Referring again to Fig. 1.16 of a satellite around the Earth, the change $\Delta\mathbf{p}$ in the linear momentum that the satellite acquires by the action of the terrestrial gravity force is always directed along $\mathbf{r}$. For that reason it does not contribute to the angular momentum, which is due only to the component $\mathbf{p}_2$ perpendicular to $\mathbf{r}$.

If $\mathbf{r}$ decreases, $\mathbf{p}_2$ increases so that the product $L = rp_2$ remains constant. This is equivalent to the statement of Kepler's second law, which is nothing but an expression of the conservation of angular momentum. The planets move faster when they approach the Sun (the radius vector diminishes) than when they are more distant.

In addition to the angular momentum due to the orbital motion around the Sun, the planets have an angular momentum due to the rotation around their axis. This creates a magnetic field due to the rotation of electric charges inside them.

Something similar takes place in the atomic world. An electron in an atom has some intrinsic angular momentum or spin that is retained even if it moves outside the atom, although it would not be correct to imagine the electron as a sphere that rotates around its axis.

The spin angular momentum is measured in terms of a unit $\hbar$ which is the Planck constant h divided by 2π, and whose value is 1.05×10^{-27} erg $\cdot$ s. Electrons, protons, neutrons, neutrinos, and other particles have spin equal to 1/2 of this unit. Photons have spin equal to 1 and π mesons have spin 0.

Particles with spin 1/2 (or any half-integer) are called *fermions*, in honour of the Italian physicist Enrico Fermi (1901–1954), and they obey Pauli's exclusion principle, formulated in 1925 by Wolfgang Pauli (1900–1958). Pauli's principle states that at most one fermion can exist in a given quantum state. On the other hand, if a particle has integer spin, it is called a *boson*, in honour of the Indian physicist Satyendra Nath Bose (1894–1974). Bosons do not obey the Pauli principle; consequently, in a given quantum state can exist an arbitrary number of bosons.

The angular momentum of an isolated system of particles is also conserved if the particles exert equal and opposite forces on one another.

As in the case of the linear momentum, it may happen that a particle loses a certain amount of angular momentum, which is carried by a newly created particle. This is the case of an electron in an atom: upon jumping from some level of energy to another one, it loses a certain amount of angular momentum, but the emitted photon carries precisely the missing angular momentum.

When neutron decay was investigated, it was observed that the resulting particles were a proton and an electron. Since the neutron had a spin angular momentum equal to 1/2, the same as the proton and the electron, it was a mystery why the total spin of the resulting particles was not 1/2. Furthermore, the energy was not conserved either. Then, in 1931, Pauli proposed the existence of a neutral particle that carries missing spin and energy. This particle was called neutrino and was assumed to have spin 1/2. Although it took more than 20 years, the existence of the neutrino was finally demonstrated in the laboratory. It took so long because neutrinos are particles whose interaction with matter is very weak. Neutrinos and weak interactions will be discussed in Chaps. 5, 9, 10, and 11.

1.5.3 Conservation of Energy

For a body of mass m that moves with speed $\mathbf{v}$, its kinetic energy is $\frac{1}{2}mv^2$. Unlike the linear momentum and the angular momentum, which are vectorial, the kinetic energy does not depend on the direction of motion.

Another form of energy is *potential* energy. If the same body of mass m is placed at a certain height h above the Earth, we say that it has a potential energy mgh with respect to the surface of the Earth (g is the acceleration due to the Earth's gravity, with the value 9.8 m/s^2), that is to say, this potential energy is equal to the product of the weight by the height. If the object falls freely, the potential energy diminishes due to the decreasing height with respect to the floor. But on the other hand, the body acquires an increasing speed: the decrease in potential energy produces an increase in kinetic energy, in such way that their sum is constant:

$$\frac{1}{2}mv^2 + mgz = \text{const.} = mgh,$$

where z is the height at any instant between the moment when the object was released and the moment when it touched the Earth.

For a planet of mass m that moves around the Sun, for example, its kinetic energy, denoted by T, is:

$$T = \frac{1}{2}m(v_r^2 + v_l^2)\,, \tag{1.18}$$

where v_r is its radial speed, directed along the radius that joins the planet with the Sun, and v_l is the velocity perpendicular to the radius vector. The potential energy (which is equal to the energy required to bring the planet from infinity to the point where it is located) is denoted by V and is equal to

$$V = -\frac{GMm}{r}. \tag{1.19}$$

Here, G is the universal constant of gravitation, M is the solar mass, r is the distance from the planet to the Sun (or from the centre of the planet to the centre of mass of the Sun–planet system, which is a point located in the Sun). The negative sign of the potential energy is due to the fact that the force between the planet and the Sun is attractive and the energy required to bring it from infinity to its present position is negative: it is not necessary to waste energy, because the Sun gives up this energy through its attractive force. The total energy would thus be

$$E = T + V = \frac{1}{2}m(v_r^2 + v_l^2) - \frac{GMm}{r}. \tag{1.20}$$

But as the angular momentum $L = mv_l r$ is constant, one can write

$$v_l = L/mr, \tag{1.21}$$

and the energy of the planet around the Sun is given by an expression that depends on the radius and the radial velocity. For a given value of the angular momentum, one has

$$E = m\frac{v_r^2}{2} + \frac{L^2}{2mr^2} - \frac{GMm}{r}, \tag{1.22}$$

and this is equivalent to the energy in case of the motion of a particle of mass m in one dimension, with a potential energy given by

$$U = \frac{L^2}{2mr^2} - \frac{GMm}{r}, \qquad (1.23)$$

called *effective potential energy*. If we represent graphically U as a function of r, we get the curve shown in Fig. 1.20. If the angular momentum $L = 0$, then $U = V$, and the graph is shown in Fig. 1.21.

Let us now analyze the case in which the angular momentum L is different from zero. If we put $U = E_1$ (E_1 is a straight line parallel to the r axis in Fig. 1.20), we get two intersection points, r_1 and r_2, between the curve U and the line E_1, if E_1 is negative. The points on the curve below the straight line correspond to the positions of the planet that are physically possible: the radial kinetic energy for them is positive, since $T_{rad} = E_1 - U$. This means that the distance of the planet from the Sun could take any value between r_1 and r_2, which are respectively the minimum (perihelion) and maximum (aphelion) distances from the Sun. This is the case of elliptical orbits, and the motion occurs in a finite region of space. That is, the radius oscillates between the values r_1 and r_2 as the planet moves around the Sun in its elliptical trajectory.

For the minimum energy value E_2, there is only one point of contact with the curve U, that is, only one value $r = r_0$. The planet is always at a fixed distance from the Sun, so the orbit is a circle.

In these examples, as already pointed out, the total energy is always negative. If the total energy is zero, that is, if the effective potential energy U is numerically equal

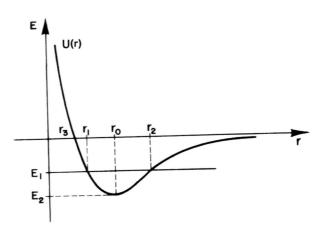

Fig. 1.20 The effective potential energy $U(r)$ of a planet in its motion around the Sun. For negative total energies E_1, motion is possible between the distances r_1 and r_2 (elliptical motion). The minimum value of the energy is E_2, corresponding to a circular orbit of radius r_0. For zero energy $E = 0$, the motion would be parabolic, between r_3 and $r = \infty$. For energies $E > 0$, the trajectories would be hyperbolas.

Fig. 1.21 If the angular
momentum is zero, the
potential energy, plotted as a
function of the distance r of
the planet to the Sun, has the
form shown in the figure. In
this case, only a linear
motion of the bodies is
possible, as in the case of
bodies that fall vertically.

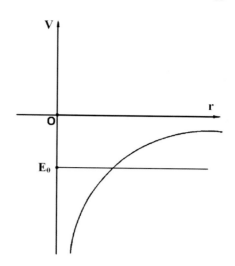

to the kinetic energy, there is only one point of intersection between the horizontal
axis and the curve U. The body could come from infinity and approach a minimum
distance r from the Sun, and then move away to infinity again. This motion corresponds to a parabolic orbit. This is valid for any body in free space, which being at
rest, or almost at rest, starts to feel the gravitational field of the Sun.

A similar situation occurs if the total energy E is positive, but in this case the orbit
is a hyperbola.

Some comets follow a parabolic or hyperbolic trajectory around the Sun, that is,
they approach it and later move away to infinity, never returning to the solar system.
A similar orbit will be described by any arbitrarily distant body with positive energy
as it approaches the Sun.

Even light does not circumvent the gravitational attraction law. Classical electrodynamics establishes that light has energy and momentum. Thus, light from a distant
star, when passing near the Sun, is expected to be deviated from the straight line,
describing a hyperbolic-like trajectory. Such light bending was suggested to occur
by Henry Cavendish (1731–1810), using Newton's corpuscular theory of light, and
was later calculated by Johann Georg von Soldner (1776–1833). Einstein's first calculations in 1911 (based on the gravitational time dilation) were in agreement with
Soldner's results, but in 1915 he proved that the total relativistic effect, which took
into account also the warping of space by massive bodies, was actually twice his
earlier calculated value (see Chap. 10).

Returning to our example, if the angular momentum is zero, $U = V$. This is the
case for a body thrown vertically upward. If r_0 is the Earth radius, the body could
reach a height r_1, and then fall back to the Earth's surface.

Let us imagine what would happen if a hypothetical hole were dug through the
Earth, along one of its diameters. Then the body could pass through the centre of
the Earth, where it would arrive with the maximum kinetic energy. After crossing

Fig. 1.22 If it were possible to make a hole right through the Earth, passing through its centre, a particle thrown down the hole would oscillate permanently between the two ends of the diameter.

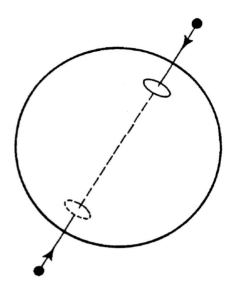

the centre, it would exit through the opposite end, reaching a position entirely symmetrical in the land of the antipodes. In principle, it would then come back to its starting point and thereafter oscillate indefinitely (Fig. 1.22). Its orbit would be a linear oscillator. At a point inside the Earth at a distance r from its centre, the force of gravity exerted on a body of mass m falling down the hole is the force produced by a sphere of radius r (due to Gauss's law). If ρ is the average Earth density, this mass is $M' = 4\pi r^3 \rho/3$, leading to a force $F = 4\pi Gm\rho r/3$.

In reality, it is not technically feasible to make such a hole through the centre of the Earth, due to the inner structure of the planet and its dynamical nature. Moreover, other factors must be considered, such as the friction between air and the moving body, which would heat up the body and damp the oscillations. The imagined experiment might be feasible inside an artificial satellite orbiting Earth in the vacuum of space.

If the body comes from infinity with an energy greater than or equal to zero (and $L = 0$), it will move toward the centre of forces (for example, the Earth) until it is stopped by the Earth's surface, at a distance r from the centre of forces.

What would happen in the case of a repulsive force? If the force is repulsive, the potential energy is positive, resulting in an effective potential that looks like the one depicted in Fig. 1.23. The total energy is always positive and the resulting orbits are hyperbolas.

A problem of this type occurs in the case of relative motion of electric charges of equal sign. This is interesting in connection with the famous experiment performed by Ernest Rutherford (1871–1937), in which a sheet of gold was bombarded with alpha particles (positively charged helium nuclei). By studying their deviations (assuming that the particles describe hyperbolic orbits), Rutherford proposed a *planetary model*

Fig. 1.23 For a positive potential, the total energy can be only positive. A positive potential occurs when particles repel each other, as in the case of an atomic nucleus interacting with alpha particles.

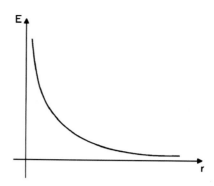

of the atom in which the nucleus was positively charged. We shall return to this point in Chap. 6.

If the sum of the kinetic and potential energies is constant, we say that the energy is conserved or that the system is conservative. This is the case for planets in their motion around the Sun, or for a falling object, until some instant when it hits the Earth. At the moment of impact, all the kinetic energy of the body is dissipated in the form of vibrations (for example, sound), elastic deformations, friction, and heat produced by friction.

1.6 Degrees of Freedom

A particle moving freely has three degrees of freedom—it can move independently in the three directions of space.

A pendulum oscillating in a plane has only one degree of freedom, which is the angle formed between the suspending cord and the vertical (Fig. 1.24).

Two free particles have six degrees of freedom, three for each. But if the particles are fixed to the ends of a bar, they lose one degree of freedom, and retain five: the three directions of space in which the bar can move, and the two angles which indicate its inclination say, around its midpoint (Fig. 1.25).

Fig. 1.24 A pendulum that oscillates in a plane has only one degree of freedom: the angle θ.

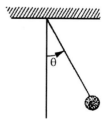

Fig. 1.25 Two particles joined by a rigid bar have five degrees of freedom: the three directions of space in which their centre of mass can move, and the two angles determining the positions of the particles with respect to it. This is almost the case for diatomic molecules, although for them the "bar" is not completely rigid, and can oscillate longitudinally. Therefore, diatomic molecules have six degrees of freedom.

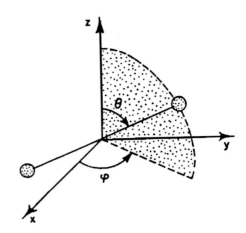

We shall discuss the mechanism of energy dissipation by using the example of the pendulum. We consider the pendulum as a mass hanging by a thread tied to a nail. When the pendulum oscillates, an enormous number of molecules of air (each of them having three degrees of freedom) collide with it. When the thread moves relative to the nail at the point of contact, it collides with a very large number of constituent particles of the nail (atoms and ions forming the lattice of the metal, and electrons). Energy dissipation in the pendulum (and in other physical systems) is related to the energy transfer from a system with very few degrees of freedom to other systems with a very large number of degrees of freedom, and the energy is in this case *disordered*.

The energy absorbed by a system with a very large number of degrees of freedom increases its *internal energy*. We shall consider this problem in more detail in the next chapter.

1.7 Inertial and Non-inertial Systems

As mentioned earlier, in order to describe the position and motion of a body, classical mechanics needs the concept of frames of reference. Such a frame could be a system of three perpendicular axes and a clock to measure the time. The origin O could be fixed to some body (Fig. 1.26).

For instance, in order to describe the Earth's motion around the Sun, the origin of the system of coordinates could be at the centre of the Sun. A frame of reference is said to be *inertial* if a free particle (on which no force acts) is at rest or moves with constant velocity along a straight line with respect to the frame (assuming that the other two laws of Newton are also valid). This definition is not free from difficulties, but it is very useful. Given an inertial frame of reference S, all the frames S', S'', etc. moving with respect to S with uniform motion along a straight line are also inertial.

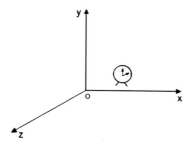

Fig. 1.26 A frame of reference is characterized by three perpendicular axes along which the three spatial coordinates x, y, z are measured, and a clock with which the time t is measured.

When we travel in a car and it accelerates abruptly, we feel a force pushing us back. If we brake, a force pushes us forward. If we follow a curved road in the car, a force pushes us outward (centrifugal force). All these are so-called *inertial forces* which appear in non-inertial frames of reference.

If the car accelerates, a pendulum will move away from the vertical line to some angle α. The same deviation α would occur in other cars if pendulums were placed in them when these cars move with respect to the first with the same acceleration, although different speeds. We thus conclude that the laws of mechanics (indicated by the verticality of the thread of the pendulum at rest with respect to the car) are not satisfied in a non-inertial frame, because fictitious forces, called *inertial forces*, appear. Furthermore, the laws of mechanics would not be valid in any frame of reference moving with constant velocity with respect to the first, non-inertial frame.

For such frames, some other set of mechanical laws is valid, modified by the inertial forces. Then the question arises: does a frame of reference exist in which the laws of mechanics are actually satisfied, if in fact one could have an enormous variety of frames of references?

In classical mechanics, we assume the existence of an *absolute frame* of reference in which Newton's laws are satisfied. They would also be satisfied for all the systems in uniform motion with regard to the absolute frame. Furthermore, an absolute time is assumed: in all the inertial frames the time is measured with the same universal clock.

Moreover, classical mechanics assumes that the interaction between particles takes place instantaneously. In such conditions, Galileo's relativity principle is satisfied. This states that the laws of mechanics are the same in all inertial systems. The principle implies that Galileo's transformations (1.24) are used when we want to describe the position of a particle with respect to two different inertial systems (Fig. 1.27).

If (x, y, z) are the numbers giving the position of a particle at some time t in the system S, and (x', y', z') the position at the same time $t = t'$ in the system S' moving with respect to S with the speed V along the x-axis, as shown in the figure, then the following equations hold between the two sets of numbers:

$$x' = x - Vt, \quad y' = y, \quad z' = z, \quad t' = t. \tag{1.24}$$

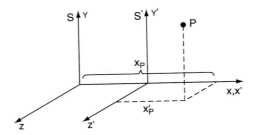

Fig. 1.27 Position of a particle P with respect to two systems of coordinates $S(x, y, z)$ and $S'(x', y', z')$. If S' moves along x with uniform velocity $\mathbf{V} = (V, 0, 0)$, the coordinates of a point P in the two systems are related by the Galilean transformation $\mathbf{r}' = \mathbf{r} - \mathbf{V}t$. Both are inertial systems.

This is a Galilean transformation. As we shall see in Chap. 5, these transformations are not satisfied by electromagnetic phenomena (in other words, Maxwell's equations are not covariant with respect to these transformations), nor indeed by any kind of physical phenomenon, although they are approximately valid for bodies moving at small velocities.

The need for a new principle of relativity in physics led to Einstein's principle of relativity, as we shall see in Chap. 5.

It is important to be able to compare two reference systems, the first S at rest and the second, non-inertial, S' rotating around an arbitrary axis at angular velocity ω radians per second (Fig. 1.28). The angular velocity is represented by a vector ω directed along the axis of rotation, as specified by the right-hand rule (Fig. 1.29). If we consider a vector $\mathbf{A}$ fixed to S' and forming an angle θ with the rotation axis, as shown in the figure, the change in $\mathbf{A}$ during the time interval from t to $t + \Delta t$ is given by

Fig. 1.28 Variation of a vector rotating around an axis with angular velocity ω in the time interval Δt.

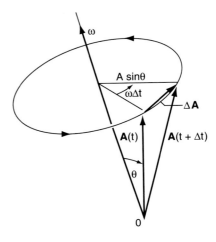

$$\Delta \mathbf{A} = \mathbf{A}(t + \Delta t) - \mathbf{A}(t) = A\omega \Delta t \sin \theta \, \mathbf{u} , \tag{1.25}$$

where $\mathbf{u}$ is a unit vector perpendicular to both ω and $\mathbf{A}$. When $\Delta t \to 0$, one can write the derivative of $\mathbf{A}$ with respect to t as the vector product

$$\frac{d\mathbf{A}}{dt} = \omega \times \mathbf{A}. \tag{1.26}$$

One can write the velocity of a particle indicated by the position vectors $\mathbf{r}$ and $\mathbf{r}'$ in S and S', respectively, considered as the time derivative of the position vector relative to each of the two systems, as

$$\frac{d\mathbf{r}}{dt} = \frac{d\mathbf{r}'}{dt} + \omega \times \mathbf{r}. \tag{1.27}$$

A similar expression for a rotated vector can be obtained from (1.10). If we assume an infinitesimal rotation of amplitude $d\theta$ such that $\cos d\theta \approx 1$ and $\sin d\theta \approx d\theta$, using (1.10) we obtain $d\mathbf{r}' = d\mathbf{r} - d\boldsymbol{\theta} \times \mathbf{r}$. Dividing by dt and defining $\omega = d\boldsymbol{\theta}/dt$, one gets (1.27). It follows that one can interpret the derivative operator with respect to time as transforming in agreement with the law (1.27). The second derivative or acceleration is

$$\frac{d^2\mathbf{r}}{dt^2} = \frac{d\mathbf{r}'^2}{dt^2} + 2\omega \times \frac{d\mathbf{r}}{dt} + \omega \times \omega \times \mathbf{r}. \tag{1.28}$$

From (1.28), for a body of mass m in S acted upon by a force $\mathbf{F}$, we can write in S'

$$\frac{d^2\mathbf{r}'}{dt^2} = \frac{d\mathbf{r}^2}{dt^2} - 2\omega \times \frac{d\mathbf{r}}{dt} - \omega \times \omega \times \mathbf{r}. \tag{1.29}$$

Two inertial forces appear, if one multiplies (1.29) by the mass m: the centrifugal force, $-m\omega \times \omega \times \mathbf{r}$, which acts in such a way as to move the body away from the axis of rotation, and the Coriolis force, $-2m\omega \times \frac{d\mathbf{r}}{dt}$.

A good example of a non-inertial system is the Earth. Because of its rotation, inertial forces appear, such as the centrifugal force which, together with the gravitational force, has produced the flattening of the Earth along the polar axis. The centrifugal force adds vectorially to the gravitational force, and the resultant is a vector which, at every point on the Earth, is directed along the vertical (if we neglect local irregularities), that is to say, it lies perpendicular to the Earth's surface.

The Coriolis force deflects the direction of bodies moving with respect to the Earth in a direction perpendicular to their velocities and to the Earth's axis of rotation. This force has a strong influence on the motion of large air masses; it is responsible for the fact that cyclones and tornados rotate counterclockwise in the northern hemisphere and clockwise in the southern hemisphere.

The local equivalence between inertial forces and gravitational field was a basic element in the formulation of a relativistic theory of gravitation, which was due to Albert Einstein, and is discussed in Chap. 10.

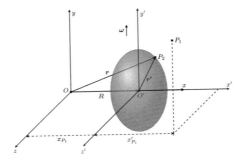

Fig. 1.29 A body in rotation with respect to the system of reference S. The system S' is fixed with respect to the body and is said to be non-inertial. The body rotates around the axis y'. The centre of the coordinate axes in S', denoted by O', is at distance $\mathbf{R}$ from O. The change in the position of any of its points in some time interval dt is given in S by the vector $d\mathbf{r}_i = d\mathbf{R} + d\boldsymbol{\theta} \times \mathbf{r}'_i$.

1.8 Rigid Bodies

A rigid body is a system of particles tightly bound in such a way that there is no relative motion among them. Under normal conditions, within certain pressure and temperature limits, bodies made out of metal, glass, stone, etc., can be considered rigid. Let us assume a system of coordinates centred at a point O' and fixed with respect to the rigid body. We shall call $M = \sum m_i$ the total mass of the body. The masses m_i are located at the points r'_i, where $i = 1, 2, \ldots, n$. The vector

$$Q = \frac{\sum_i m_i \mathbf{r}'_i}{M}, \tag{1.30}$$

with origin at O', determines a point called the *centre of mass of the rigid body*. If we consider the origin of the coordinate system S' to be placed at the centre of mass of the body, then $\mathbf{Q} = 0$. As a consequence, $\sum_i m_i \mathbf{r}'_i = 0$.

We assume that the body is moving along the x axis, and that at the same time it rotates with the angular velocity ω around the y' axis. As earlier, the time-dependent vector $\mathbf{R}$ fixes the position of O' with respect to O. An arbitrary point P of the rigid body is determined by the vector $\mathbf{r}$ with respect to O and by $\mathbf{r}'$ with respect to O'.

An equation similar to (1.27) can be derived in the case of the rigid body, for each particle composing it, seen from both the rest system of reference S and the moving one S'. The particles have a common angular velocity ω. Denoting $d\mathbf{r}_i/dt = \mathbf{v}_i$ and $d\mathbf{R}/dt = \mathbf{V}$, the velocity of a particle in a rigid body rotating around the y' axis can be written

$$\mathbf{v}_i = \mathbf{V} + \boldsymbol{\omega} \times \mathbf{r}'_i. \tag{1.31}$$

The velocity $\mathbf{v}_i$ is called the absolute velocity of the particle P. Recall that the system S' is fixed with respect to the rigid body, therefore the vectors $\mathbf{r}'_i$ are constant and $d\mathbf{r}'_i/dt = \mathbf{v}'_i = 0$.

Using the relation (1.31), we can find the expression for the kinetic energy of the rigid body as seen from the rest frame:

$$T = \sum m_i v_i^2 = \sum \frac{1}{2} m_1 V^2 + \sum m_i \mathbf{V} \cdot (\boldsymbol{\omega} \times \mathbf{r}_i') + \frac{1}{2} \sum m_i (\boldsymbol{\omega} \times \mathbf{r}_i')^2$$

$$= \frac{1}{2} M V^2 + \frac{1}{2} \sum m_i [\omega^2 r_i'^2 - (\boldsymbol{\omega} \cdot \mathbf{r}_i')^2], \tag{1.32}$$

where in the second term on the right we have used the equality $\sum m_i \mathbf{V} \cdot (\boldsymbol{\omega} \times \mathbf{r}_i') = (\mathbf{V} \times \boldsymbol{\omega}) \cdot \sum m_i \mathbf{r}_i' = 0$, since $\sum m_i \mathbf{r}_i' = 0$. In the third term, the squared vector product was expanded. The kinetic energy thus comprises two terms, the first being related to the translational motion of the body along the x-axis, and the second containing the kinetic energy of rotation. If we denote by θ the angle formed by $\boldsymbol{\omega}$ and $\mathbf{r}_i'$, the energy of rotation in (1.32) can be written as

$$\frac{1}{2} \sum m_i (\omega^2 r_i'^2 - (\boldsymbol{\omega} \cdot \mathbf{r}_i')^2) = \frac{1}{2} \sum m_i \omega^2 r_i'^2 \sin^2 \theta = \frac{1}{2} \sum m_i (x_i'^2 + z_i'^2) \omega^2,$$

where $I = \sum m_i (x_i'^2 + z_i'^2)$ is called the *moment of inertia* of the body around its axis of rotation. Actually, in the general case when the rigid body moves around a point, it can be shown that its axial moments of inertia are the diagonal components of a tensor, the *inertia tensor*, with the expression

$$I_{lk} = \sum_i m_i (\mathbf{r}_i'^2 \delta_{lk} - x_l' x_k'),$$

where $k, l = 1, 2, 3$. If I_{lk} is multiplied by the dyadic tensor $\omega_l \omega_k$, the kinetic energy of rotation can be written in the form

$$T_{rot} = \frac{1}{2} \boldsymbol{\omega} \cdot \mathbf{I} \cdot \boldsymbol{\omega}.$$

Returning to our body in motion around the y' axis, we see that the last term in (1.32) can be written as

$$\frac{1}{2} \sum m_i (\boldsymbol{\omega} \times \mathbf{r}_i') \cdot (\boldsymbol{\omega} \times \mathbf{r}_i') = \frac{1}{2} \boldsymbol{\omega} \cdot \mathbf{L}, \tag{1.33}$$

$$\frac{1}{2} \sum m_i (\boldsymbol{\omega} \times \mathbf{r}_i') \cdot (\boldsymbol{\omega} \times \mathbf{r}_i') = \frac{1}{2} \boldsymbol{\omega} \cdot \sum m_i \left(r_i'^2 \boldsymbol{\omega} - (\boldsymbol{\omega} \cdot \mathbf{r}_i') \mathbf{r}_i' \right) = \frac{1}{2} \boldsymbol{\omega} \cdot \mathbf{L}, \tag{1.34}$$

where we have used the permutation of the $\times$ and $\cdot$ products in the first term and

$$\mathbf{L} = \sum m_i \mathbf{r}_i' \times (\boldsymbol{\omega} \times \mathbf{r}_i')$$

is the rotating body's angular momentum around the y' axis. Let us consider the simple case in which the body has rotational symmetry around ω. The term of interest is the *moment of inertia* of the body around the y' axis, viz., $I_{22} = \sum m_i(x'^2 + z'^2)$. The moment of inertia with respect to the axis of rotation can also be defined as the angular momentum divided by the angular velocity, $I = L/\omega$, and conversely, given the moment of inertia tensor, the angular momentum can be written as $\mathbf{L} = \mathbf{I} \cdot \boldsymbol{\omega}$.

We define the torque applied to a rigid body composed of a set of particles labeled by i to be $\mathbf{N} = \sum_i \mathbf{r}_i' \times \mathbf{F}_i$, where $\mathbf{r}_i'$ are the position vectors with respect to the centre of mass and $\mathbf{F}_i$ are the external forces. Then the rotation of the body around an axis passing through it is described by the equation

$$d\mathbf{L}/dt = \mathbf{N}, \tag{1.35}$$

where $\mathbf{L} = \sum \mathbf{L}_i$ is the total angular momentum of the set of particles. The motion of the centre of mass is given by $M\dot{\mathbf{V}} = \mathbf{F}$, where $\mathbf{F} = \sum_i \mathbf{F}_i$ is the sum of the external forces acting on the particles. In the case of central forces, where $\mathbf{r}_i'$ is parallel to the force $\mathbf{F}_i$ exerted on the particle i, we have $d\mathbf{L}/dt = 0$, and the angular momentum is conserved. More details about rigid bodies can be found in the books mentioned in the bibliography.

1.9 The Principle of Least Action

The principle of least action states that, when a mechanical system evolves from an initial state 1 to a final state 2, it does so in such a way that some quantity, called the *action*, takes an extreme value when the system follows the actual or dynamic trajectory. This extreme value is in general a minimum (but it can be a maximum in some special cases). This principle is like a principle of economics in the field of mechanics.

But the action is not a directly observable quantity, like energy and momentum. To define the action, we first define the Lagrangian L of a system of particles as a function obtained by subtracting the potential energy from the kinetic energy:

$$L = T - V. \tag{1.36}$$

(The customary notation for the Lagrangian is L, like that for the angular momentum, so one has to be careful not to confound the two!) For a particle with mass m moving in a direction x with velocity $v \equiv \dot{x}$ in a potential $V(x)$, it is

$$L = \frac{1}{2}m\dot{x}^2 - V(x).$$

As $\dot{x}$ and x are functions of time, one can write $L = L(\dot{x}(t), x(t))$. Then the action S is defined as the integral of L evaluated along arbitrary trajectories $x(t)$ between two given times t_1 and t_2:

$$S[x(t)] = \int_{t_1}^{t_2} L[x(t), \dot{x}(t)]dt. \tag{1.37}$$

The action S is said to be a *functional*, since it takes as input a function of time, and returns a real number. The previous definition of action is easily extended to the case of N generalized coordinates $x(t) \rightarrow q_i(t)$, $i = 1, \ldots, N$, and to the case in which L depends explicitly on time. If the mathematical conditions are established for the action to be a minimum (in general, an extremum), a set of differential equations is obtained from the Lagrangian which determine the dynamical trajectory, and are called the Euler–Lagrange equations, in honour of Leonhard Euler (1707–1783) and Joseph-Louis Lagrange (1736–1813):

$$\frac{d}{dt}\frac{\partial L}{\partial \dot{q}_i} - \frac{\partial L}{\partial q_i} = 0. \tag{1.38}$$

Here, $\dot{q}_i$ are the velocities (or generalized velocities, like the angular speed $v_\theta = d\theta/dt$) and q_i are the coordinates (or generalized coordinates, like the angles θ and φ). These equations are the same as those obtained by starting out with Newton's laws.

For instance, for a free particle moving in one dimension x with the speed v, the Lagrangian is $L = \frac{1}{2}mv^2$. But as L does not depend explicitly on x, it follows from (1.38) that

$$m\frac{dv}{dt} = 0 , \tag{1.39}$$

with the solution $v = $ constant.

Consider two positions of a planet, P_1 and P_2, on its orbit around the Sun S. If the variation of the action is calculated between P_1 and P_2 when the planet follows the usual elliptical trajectory and when it is constrained to follow the straight line P_1P_2, one finds that the action is smaller for the elliptical trajectory P_1CP_2. Moreover, it remains smaller than any other value obtained by varying the trajectories between P_1 and P_2 (Fig. 1.30).

By calling $q(t)$ the solution for which S is an extremum, which we assume to be a minimum, this means that S increases when $q(t)$ is replaced by a function of the form $q(t) + \delta q(t)$, where $\delta q(t)$ is any function, called the *variation* of $q(t)$, satisfying the condition $\delta q(t_1) = \delta q(t_2) = 0$. The variation of a constant is obviously zero. The straight line trajectory in (Fig. 1.30) is of the form $q(t) + \delta q(t)$, where $q(t)$ is the dynamical trajectory. The variation vanishes at the extreme points, $\delta q(t(P_1)) = \delta q(t(P_2)) = 0$ (Fig. 1.30).

If two Lagrangians differ by a total derivative with respect to time of a function of the generalized coordinates and time $L'(q, \dot{q}, t) - L(q, \dot{q}, t) = \frac{d}{dt}f(q, t)$, then these Lagrangians describe the same system. This is so because the variation of the action is the same for both Lagrangians in the interval t_2, t_1, since it does not depend

Fig. 1.30 The action takes a
minimum value along the
dynamic trajectory $P_1 C P_2$.
For another trajectory, like
the straight line $P_1 P_2$, its
value would be larger.

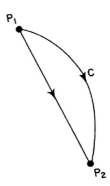

on the quantities $f(q(t_1), t_1)$ and $f(q(t_2), t_2)$, which are constants whose variation
is zero.

The principle of least action is of the utmost importance in all branches of physics,
and it was first formulated by Pierre-Louis Moreau de Maupertuis (1698–1759), and
in more complete form by William Rowan Hamilton (1805–1865).

A similar principle exists in optics for the trajectory followed by light when
it propagates through a medium. This is Fermat's principle, established by Pierre
Fermat (1601–1665). The principle states that, when light travels from one point to
another in a medium, it does it in such a way that the required *time* has an extremum
value, generally a minimum, although it could be a maximum.

Principles of least action have great importance in modern theoretical physics.

Lagrange Equations and Planetary Motion. Consider again the case of a planet
moving around the Sun. To describe its motion, it is convenient to use polar coordinates with the pole located at the Sun. Remember that the force exerted by the Sun
on the planet is a vector $\mathbf{F} = -GMm\mathbf{r}_0/r^2$. The polar coordinates are

$$x = r \cos \theta, \qquad y = r \sin \theta , \qquad (1.40)$$

where r, θ are the radius vector and the angle with respect to the polar axis, respectively. The velocities along and perpendicular to $\mathbf{r}$ are given by

$$v_r = \dot{r} = dr/dt, \qquad v_l = r\dot{\theta} = rd\theta/dt. \qquad (1.41)$$

The Lagrangian for the planet is the difference between the kinetic energy T and the
potential energy V, that is,

$$L = \frac{1}{2}m(\dot{r}^2 + r^2\dot{\theta}^2) + \frac{GMm}{r}. \qquad (1.42)$$

Since θ does not appear explicitly in L (only $\dot{\theta}$ appears), the first Lagrange equation,

$$\frac{d}{dt}\frac{\partial L}{\partial \dot{\theta}} - \frac{\partial L}{\partial \theta} = 0,$$

leads to the following equation:

$$\frac{d(mr^2\dot{\theta})}{dt} = 0 , \tag{1.43}$$

that is to say, $mr^2\dot{\theta} = C = $ const., which is the law of conservation of angular momentum (in this example only, we denote the value of the conserved angular momentum by C, not to confuse it with the Lagrangian). If $\dot{\theta} = C/mr^2$ is substituted into the expression for the total energy, $T + V$, we find that

$$E = \frac{1}{2}m\dot{r}^2 + \frac{C^2}{2mr^2} - \frac{GMm}{r}. \tag{1.44}$$

This in turn implies that

$$\dot{r} = \frac{dr}{dt} = \sqrt{\frac{2}{m}\left[E + \frac{GMm}{r} - \frac{C^2}{2mr^2}\right]}.$$

By combining $\dot{r}$ and $\dot{\theta}$, one obtains an equation for r as a function of θ, which is the parametric equation of the orbit:

$$d\theta = \frac{Cdr/r^2}{\sqrt{2m[E + GMm/r - C^2/2m\,r^2]}} ,$$

leading, upon integration, to

$$\theta = \arccos \frac{C/r - GMm^2/C}{\sqrt{2mE + G^2M^2\,m^4/C^2}}. \tag{1.45}$$

If we make the notations $d = C^2/GMm^2$, $\epsilon = \sqrt{1 + 2EC^2/G^2M^2\,m^3}$, we can finally write the equation for the orbit as the typical equation of a conic:

$$r = \frac{d}{1 + \epsilon\cos\theta} , \tag{1.46}$$

where ϵ is the eccentricity. If $E < 0$, then $\epsilon < 1$ and the orbit is elliptic. If $E = 0$, then $\epsilon = 1$ and the orbit is parabolic. If $E > 0$, then $\epsilon > 1$ and the orbit is hyperbolic.

1.10 Hamilton Equations

Instead of using generalized coordinates and velocities to describe the motion of a physical system with N degrees of freedom, it is sometimes easier to use coordinates and momenta, as an independent set of N pairs of *canonical coordinates*. If L is the Lagrangian of a system, considered as a function of the coordinates q_i, the velocities $\dot{q}_i$, and the time t, where $i = 1, 2, \ldots, N$, one can write the Lagrangian's total differential in terms of the generalized coordinates and velocities as

$$dL = \sum_{i=1}^{N} \frac{\partial L}{\partial q_i} dq_i + \sum_{i=1}^{N} \frac{\partial L}{\partial \dot{q}_i} d\dot{q}_i + \frac{\partial L}{\partial t} dt.$$

By definition,

$$p_i = \partial L / \partial \dot{q}_i$$

are the generalized momenta (also called canonically conjugated momenta). Recalling the Euler–Lagrange equations, it follows that

$$dL = \sum_{i=1}^{N} \dot{p}_i dq_i + \sum_{i=1}^{N} p_i d\dot{q}_i + \frac{\partial L}{\partial t} dt. \tag{1.47}$$

We can write $\sum p_i d\dot{q}_i = d\left(\sum p_i \dot{q}_i \right) - \sum \dot{q}_i dp_i$. Then, reorganizing the terms and defining

$$H = \sum_{i=1}^{N} p_i \dot{q}_i - L$$

as the *Hamiltonian* function, such that

$$dH = -\sum_{i=1}^{N} \dot{p}_i dq_i + \sum_{i=1}^{N} \dot{q}_i dp_i - \frac{\partial L}{\partial t} dt, \tag{1.48}$$

we conclude that

$$\dot{q}_i = \partial H / \partial p_i, \qquad \dot{p}_i = -\partial H / \partial q_i, \quad i = 1, 2, \ldots, N. \tag{1.49}$$

We also obtain the equation

$$\partial H / \partial t = -\partial L / \partial t.$$

If the Lagrangian does not depend explicitly on time, the Hamiltonian does not depend on it either. Then H is a constant of motion, similar to the total energy. Equations (1.49) are called Hamilton's equations. They constitute a set of $2N$ first order differential equations, equivalent to the set of N Euler–Lagrange equations of second order.

We consider as an example the harmonic oscillator of mass m and elastic constant k, described by the Lagrangian

$$L = \frac{1}{2}m\dot{q}^2 - \frac{1}{2}kq^2.$$

The velocity can be written as $\dot{q} = p/m$. The Hamiltonian is then

$$H = \frac{1}{2m}p^2 + \frac{1}{2}kq^2 \,,$$

which is the expression for the total energy. The Hamilton equations are

$$\dot{q} = p/m, \qquad \dot{p} = -kq.$$

Taking the derivative with respect to time of the first and substituting the result into the second, we get the equation

$$\ddot{q} = -\frac{k}{m}q \,,$$

whose general solution is

$$q = A\cos(\omega t + \varphi),$$

where $\omega = \sqrt{k/m}$, A is the amplitude, and φ is an arbitrary angle (initial conditions must be given for fixing the values of A and φ). The same equation and solution can be obtained using the Euler–Lagrange equation. We find

$$m\frac{d^2q}{dt^2} + kq = 0, \qquad (1.50)$$

which is the same equation as above. The harmonic oscillator is very important in all areas of physics. For instance, consider a massive particle whose potential energy $V(x)$ has a minimum x_0, where the particle is in a state of mechanical equilibrium. If the expansion around x_0 has the form $V(x) = V_0 + \frac{1}{2}V''(0)(x - x_0)^2 + ...$, and if the particle is pushed out of equilibrium and then allowed to move freely, it will behave as an oscillator.

The Hamiltonian formalism is of exceptional importance, mainly in connection with the transformation of a set of canonical coordinates p_i, q_i to another P_i, Q_i. We can consider the mechanical motion of a system as a canonical transformation of coordinates from some initial conditions to the set of canonical coordinates at some arbitrary instant t. It is possible to obtain a fundamental differential equation for the action S, the so-called Hamilton–Jacobi equation, whose solution allows us to find the equations of motion.

One can define a *phase space*, determined by the set p_i, q_i. The volume in phase space is invariant under canonical transformations. In this way, the phase space is an essential tool when dealing with systems having a very large number of particles, as it happens in statistical mechanics. The Hamiltonian formalism is also essential in quantum theory (see Chaps. 6 and 7).

Poisson Brackets. It is often necessary to define functions of the coordinates, the momenta, and the time, $f = f(q_i, p_i, t)$. The total derivative with respect to time is

$$\frac{df}{dt} = \frac{\partial f}{\partial t} + \sum_i \left(\frac{\partial f}{\partial q_i} \dot{q}_i + \frac{\partial f}{\partial p_i} \dot{p}_i \right). \tag{1.51}$$

Recalling the Hamilton equations, we can write

$$\frac{df}{dt} = \frac{\partial f}{\partial t} + \{H, f\}, \tag{1.52}$$

where

$$\{H, f\} = \sum_i \left(\frac{\partial H}{\partial p_i} \frac{\partial f}{\partial q_i} - \frac{\partial H}{\partial q_i} \frac{\partial f}{\partial p_i} \right)$$

is called the *Poisson bracket* of H and f. If f does not depend explicitly on time and $\{H, f\} = 0$, then f is called a constant of the motion. Among other reasons, the Poisson brackets are important because of their similarity with commutators in quantum mechanics, as we shall see in Chap. 6.

Let us consider as an example the Poisson bracket of the component of the angular momentum $L_z = xp_y - yp_x$ with the Hamiltonian $H = \frac{p^2}{2m} + V(x, y, z)$:

$$\{H, L_z\} = \{V, L_z\} = y\frac{\partial V}{\partial x} - x\frac{\partial V}{\partial y}. \tag{1.53}$$

Then L_z (like L_x and L_y) is not in general a constant of the motion. But if $V(\mathbf{r})$ is the potential associated with a central force, then L_x, L_y, and L_z, together with L^2, are all constants of the motion. The reader can check this by means of a simple calculation. If $V = V(z)$, that is, if the potential is independent of the x and y coordinates, whence there is rotational symmetry around the z axis, then $\{V, L_z\}=0$, and L_z is a constant of the motion.

1.11 Complements on Gravity and Planetary Motion

Gravity obeys the principle of superposition. The total gravitational force exerted on a body, say, the Moon, is the vector sum of the forces exerted by the Earth ($\mathbf{F}_E$), the Sun ($\mathbf{F}_S$), Venus ($\mathbf{F}_V$), and all other celestial bodies, so that the total force acting

Fig. 1.31 The gravitational potential created by a spherical shell of radius r at a point P taken on the x axis is the same as if the whole mass of the shell were concentrated at its centre O.

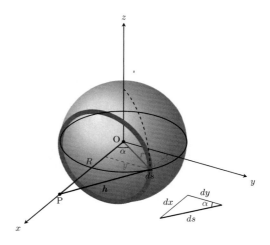

on the Moon is $\mathbf{F} = \mathbf{F}_E + \mathbf{F}_S + \mathbf{F}_V + \ldots$. These forces are, in general, functions of the coordinates and time. The same can be said with regard to other bodies, for instance, the Earth, on which the Sun and all other planets act. An example of the observable effects of the gravitational attraction of both the Sun and the Moon on the Earth are the tides. Tides are especially important for the oceans and atmosphere, and depend on the relative positions of the Sun and the Moon with respect to the Earth. Tides are maximal when the three bodies are aligned, as happens at full and new Moon, and minimal during the crescent and last quarter phases, when the Sun and the Moon lie in orthogonal directions with respect to the Earth.

Gravitational force produced by a spherical shell. We shall show that: (i) a spherical shell of uniform thickness and density attracts a point particle outside it as if its mass were concentrated in its centre; (ii) inside the spherical shell, the gravitational force is zero (Fig. 1.31). Suppose the spherical shell has radius r and thickness δr and consider a point P on the x-axis. Let us denote by R the distance from P to the centre of the sphere O.

We start by calculating the gravitational potential V at P, and from that, the gravitational force per unit mass, $g = -\partial V/\partial R$. To this end, we first calculate the gravitational potential created by a ring of radius y, generated by the rotation of a spherical arc of length ds around Ox.

We assume a constant volume density of mass ρ and surface density $\mu = \rho\delta r$. From the figure we note that the surface area of the ring on the sphere is $dA = 2\pi y ds$ and its mass is $\delta M = \delta\mu dA$. Assume also that all points of the ring are at the same distance h from P, where $h^2 = y^2 + (R - x)^2$, while $r^2 = x^2 + y^2$ and R are constants.

Finally, we express the total potential in terms of an integral over h between limits $R - r$ and $R + r$ for P outside the sphere and between limits $r - R$ and $r + R$, for P inside the spherical shell. All our reasoning will concern the plane (x, y), but by rotational symmetry it is valid for all planes passing through the x axis, that is, for all points of the ring. The gravitational potential created at P is

$$dV_s = -\frac{2\pi G\mu y ds}{h}. \tag{1.54}$$

Here $ds = \sqrt{dx^2 + dy^2}$, and the radius r joining the centre of the arc to the centre of the sphere O forms an angle α with the x axis, while the angle formed by ds and dy is also α, since ds is perpendicular to the radius r.

As $x = r\cos\alpha$ and $y = r\sin\alpha$, we have $dx = -yd\alpha$ and $dy = xd\alpha$, which implies $ds = rd\alpha$. Now, since $h^2 = y^2 + (R - x)^2$, we have

$$h^2 = y^2 + R^2 - 2Rx + x^2 = R^2 - 2Rx + r^2. \tag{1.55}$$

As r and R are constants, by differentiating (1.55) with respect to x, it follows that $hdh = -Rdx$. Notice also that $ds = rd\alpha$, and we get the chain of equalities

$$yds = r^2\sin\alpha d\alpha = -rdx = rhdh/R. \tag{1.56}$$

Thus, we write the potential created by the ring at P as

$$dV_s = -\frac{2\pi G\mu r dh}{R}. \tag{1.57}$$

When P is *outside* the spherical shell, as in case i), by integrating over h to get

$$\int_{R-r}^{R+r} dh = 2r,$$

we find

$$V_s = -\frac{4\pi\mu r^2 G}{R} = -\frac{GM_s}{R}, \tag{1.58}$$

where $M_s = 4\pi\mu r^2$ is the mass of the spherical shell, which behaves with respect to P as if all its mass were concentrated in its centre, and this is valid also for the gravitational field $g_s = -\partial V_s/\partial R = -GM_s/R^2$.

For a solid homogeneous sphere of radius r_0, substituting $\mu = \rho dr$ and integrating between $(0, r_0)$, (1.58) implies

$$V = -\frac{4\pi\rho r_0^3 G}{3h} = -\frac{GM}{R}, \tag{1.59}$$

where $M = 4\pi\rho r_0^3/3$ is the mass of the sphere. The gravitational field is $g = -\partial V/\partial R = -GM/R^2$.

For *P* *inside* the spherical shell, as in case ii), the integral over *h* must be taken between the limits $r - R \leq h \leq r + R$ and we get

$$\int_{r-R}^{r+R} dh = (r + R) - (r - R) = 2R. \tag{1.60}$$

This term cancels the denominator *R* in the potential, making the potential independent of the interior point where its value is calculated. Multiplying both the numerator and denominator of (1.57) by *r*, we obtain

$$V_s = -\frac{4\pi\mu r^2 G}{r} = -\frac{GM_s}{r}, \tag{1.61}$$

which is independent of the point *P* and thus a constant. Hence, the corresponding gravitational force per unit mass will be $g = -\partial V_s/\partial R = 0$.

Deviation of bodies and light by the gravitational force. Consider a free particle of mass *m* and velocity *v* very far from the Sun. Its momentum is $p = mv$. The distance between the direction of the momentum and the centre of the Sun is denoted by *d* and called the impact parameter. The kinetic energy of the particle is $E = mv^2/2 > 0$ and its angular momentum is $L = mvd$. As the particle approaches the Sun, it is deviated from the straight line, and having positive energy, it must describe a hyperbolic trajectory. We would like to calculate the angle of deviation of the particle from the straight line in terms of *E* and *L*. The expression obtained will be applied to a proton (of mass $m = 1.6 \times 10^{-24}$ g) traveling at 1/3 the speed of light, and to a massless particle (a photon), traveling at the speed of light *c*, passing by the Sun's limb (or edge), where $d = 6.96 \times 10^{10}$ cm is the Sun's radius and $M_\odot \approx 1.98 \times 10^{33}$ g is its mass.

We start from the expression for the angle of a hyperbolic motion in polar coordinates. We consider the branch located in the Cartesian halfplane $x < 0$, which is symmetric with respect to the *x*-axis. The equation for the asymptotes is given by taking $r \to \infty$ in (1.56) for a planetary orbit (the pole is taken at the focus *S* in Fig. 1.32), yielding the angle

$$\Delta\theta = \arccos \frac{-GMm^2/L}{\sqrt{2mE + G^2M^2 m^4/L^2}}. \tag{1.62}$$

The total deviation from the straight line is $2\delta\phi = \pi - 2\Delta\theta$, leading to

$$\delta\phi = \frac{\pi}{2} - \Delta\theta = \arcsin \frac{GMm^2/L}{\sqrt{2mE + G^2M^2 m^4/L^2}}. \tag{1.63}$$

We assume that *m* is small enough to make the second term in the denominator negligible compared to the first. Then, we replace *m* by $2E/v^2$ in the remaining terms to obtain

Fig. 1.32 Deviation of the
trajectory of a particle
having positive energy by the
action of a gravitational field,
according to Newtonian
theory.

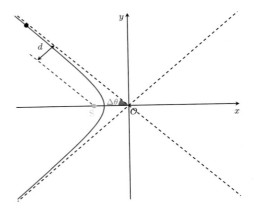

$$\delta\phi = \arcsin \frac{2GM}{v^3\, L/E}.$$ (1.64)

As ϕ is usually very small, we can approximate $\sin\phi \approx \phi$. Finally, the total deviation
is

$$2\delta\phi = \frac{4GM}{v^3\, L/E}.$$ (1.65)

Recall that, at large distances from the Sun, $E = \frac{1}{2}mv^2$ and $L = mvd$. Thus,

$$v^3\, L/E = 2v^2 d.$$ (1.66)

Note that, after substituting in (1.65), the result is that $2\delta\phi$ is *independent* of the
value of the particle mass.

By substituting values for the proton case, we get

$$2\delta\phi = \frac{2GM}{v^2 d} = \frac{2 \times 6.67 \times 1.98 \times 10^{25}}{6.96 \times 10^{30}} = 3.79 \times 10^{-5} \text{rad.}$$ (1.67)

As $1° = 3600''$, using the conversion of units

$$\frac{\text{rad}}{\pi} = \frac{\text{deg}}{180},$$

we get finally $2\delta\phi = 7.8''$. Formally, for the photon, in the limit $m \to 0$, we take
$v = c$ in (1.67), and find $2\delta\phi = 7.8''/9 = 0.87''$. This is in accord with von Soldner's
(1776–1833) calculation of 1804, but is exactly half the angle predicted by general
relativity for the deflection of the light coming from a distant star and passing by the
limb of the Sun, viz., $2\delta\phi = 1.75''$ (see Sect. 10.2).

Velocity and acceleration in plane polar coordinates. In plane polar coordinates,
the position of a particle is given by a vector **r** and the angle θ, where **r** is expressed

in terms of the unit vectors in Cartesian coordinates $\mathbf{i}, \mathbf{j}$ as

$$\mathbf{r} = x\mathbf{i} + y\mathbf{j}, \tag{1.68}$$

with $x = r \cos \theta$, $y = r \sin \theta$, and $r = \sqrt{x^2 + y^2}$, $\theta = \arctan y/x$. Let us find the expressions for the velocity and acceleration expressed in terms of the unit vectors $\mathbf{r}_0$, $\boldsymbol{\theta}_0$ in polar coordinates and show that the components of the velocity are $v_r = \dot{r}$ and $v_\theta = r\dot{\theta}$, respectively, while those of the acceleration are $a_r = (\ddot{r} - r\dot{\theta}^2)$ and $a_\theta = (r\ddot{\theta} + 2\dot{r}\dot{\theta})$.

We shall write (1.68) as

$$\mathbf{r} = r(\cos \theta \mathbf{i} + \sin \theta \mathbf{j}) = r\mathbf{r}_0, \tag{1.69}$$

where $\mathbf{r}_0 = \cos \theta \mathbf{i} + \sin \theta \mathbf{j}$. Recall that $\boldsymbol{\theta}_0 = -\sin \theta \mathbf{i} + \cos \theta \mathbf{j}$ is the unit vector in the direction θ, and the relation

$$\boldsymbol{\theta}_0 = \frac{d}{d\theta}\mathbf{r}_0, \qquad d\boldsymbol{\theta}_0/d\theta = -\mathbf{r}_0.$$

Differentiating $\mathbf{r}$ with respect to time, we get the velocity as $\mathbf{v} = d\mathbf{r}/dt$, and the acceleration as $\mathbf{a} = d\mathbf{v}/dt$, and we have finally

$$\frac{d\mathbf{r}}{dt} = \mathbf{v} = \dot{r}\mathbf{r}_0 + r(d\mathbf{r}_0/d\theta)\dot{\theta} = \dot{r}\mathbf{r}_0 + r\dot{\theta}\boldsymbol{\theta}_0. \tag{1.70}$$

Differentiating $\mathbf{v}$ with respect to time, the acceleration is found to be

$$\mathbf{a} = (\ddot{r} - r\dot{\theta}^2)\mathbf{r}_0 + (r\ddot{\theta} + 2\dot{r}\dot{\theta})\boldsymbol{\theta}_0, \tag{1.71}$$

Note that the radial acceleration has two terms, $a_r = \ddot{r} - v_\theta^2/r$, where the first is the acceleration of the modulus of $\mathbf{r}$ and the second, written in terms of v_θ, is the centripetal acceleration, which is directed along $\mathbf{r}_0$, and is due to the existence of a nonzero angular velocity of rotation. Multiplied by m, the centripetal acceleration yields the centripetal force. If the accelerated particle is acted on by an external force $\mathbf{F}$ along $\mathbf{r}$, it is also parallel to the vector $\mathbf{r}_0$, and its acceleration is also along $\mathbf{r}_0$. The acceleration along $\boldsymbol{\theta}_0$ vanishes, that is, it is $r\ddot{\theta} + 2\dot{r}\dot{\theta} = 0$, which is equivalent to saying that

$$\frac{d}{dt}(r^2\dot{\theta}) = 0. \tag{1.72}$$

This implies that the so-called areolar velocity $r^2\dot{\theta} = const$, leading to the second Kepler law, which is equivalent to the conservation of angular momentum $L = mr^2\dot{\theta} = const$. It should be emphasized that, if the component of the acceleration vector along $\boldsymbol{\theta}_0$ vanishes, this does not imply that the angular acceleration $\ddot{\theta}$ vanishes. This occurs only if the orbit is circular.

Two bodies in motion due to gravitational force. Consider two bodies of masses m_1 and m_2 and position vectors $\mathbf{r}_1$ and $\mathbf{r}_2$, respectively, with respect to a given system of coordinates. Let us assume that the two bodies attract each other through a gravitational force obtained from a potential $V(r) = -Gm_1m_2/r$, where $\mathbf{r} = \mathbf{r}_2 - \mathbf{r}_1$ is the position vector of the second body with respect to the first. We shall write down the equations of motion and the total energy and discuss the motion of the system. This is a complement to the discussion of the Kepler problem earlier in Sects. 1.5 and 1.9.

We call the total mass $m_1 + m_2 = M$. The centre of mass of the system is given by the vector

$$\mathbf{R} = \frac{(m_1\mathbf{r}_1 + m_2\mathbf{r}_2)}{M}. \tag{1.73}$$

We wish to express the initial vectors $\mathbf{r}_1$ and $\mathbf{r}_2$ in terms of $\mathbf{R}$ and $\mathbf{r}$.

From the expressions for $\mathbf{r}$ and $\mathbf{R}$ in terms of $\mathbf{r}_1$, $\mathbf{r}_2$, it is easy to obtain

$$\mathbf{r}_1 = (\mathbf{R} - m_2\mathbf{r}/M), \qquad \mathbf{r}_2 = (\mathbf{R} + m_1\mathbf{r}/M). \tag{1.74}$$

The force is $F = -\partial U(r)/\partial \mathbf{r} = -Gm_1m_2\mathbf{r}_0/r^2$, where $\mathbf{r}_0 = \mathbf{r}/r$ is a unit vector. As this does not depend on $\mathbf{R}$, the acceleration of the centre of mass $\ddot{R} = 0$. Thus, the centre of mass behaves as a free particle, and by Newton's first law, there is an inertial system in which it is at rest. We can take it as the origin of the coordinates, namely take $\mathbf{R} = 0$, leading to

$$\mathbf{r}_1 = -(m_2/M)\mathbf{r}, \qquad \mathbf{r}_2 = (m_1/M)\mathbf{r}. \tag{1.75}$$

The equations of motion can be written

$$m_1\ddot{\mathbf{r}}_1 = -\mathbf{F}, \qquad m_2\ddot{\mathbf{r}}_2 = \mathbf{F}. \tag{1.76}$$

The position vectors and also the accelerations of the two bodies are taken in opposite directions with respect to the centre of mass. Inserting (1.75) in either of the equations (1.76) and defining $\mu = m_1m_2/M$ as the reduced mass, we get

$$\mu\ddot{\mathbf{r}} = F. \tag{1.77}$$

The solution is (1.46), expressed in terms of the masses μ and M as follows. The angular momentum is $L = \mu\dot{r}^2\dot{\theta}$ and the energy is $E = T + V$, which is negative for elliptic (in particular, circular) motion, and we assume this is so in the present case. Here, the kinetic energy is $T = \mu\dot{r}^2 = T_1 + T_2$, where

$$T_1 = \frac{1}{2}(m_1\dot{\mathbf{r}}_1^2), \qquad T_2 = \frac{1}{2}(m_2\dot{\mathbf{r}}_2^2). \tag{1.78}$$

That is, by taking $d = L^2/GM\mu^2$ and $\epsilon = \sqrt{1 + 2EL^2/G^2M^2\mu^3}$, we obtain

$$r(t) = \frac{d}{1 + \epsilon \cos \theta(t)}. \tag{1.79}$$

Thus, $r_1 = -(m_2/M)r(t)$ and $r_2 = (m_1/M)r(t)$ define two ellipses (in particular, circles) of different sizes, the two bodies rotating one around the other, in such a way that along their rotation motion their position vectors with regard to the centre of mass are aligned in opposite directions. The explicit dependence of $\mathbf{r}$, θ on time would require additional equations.

In the problems proposed for this chapter, we apply the above results to obtain the location of the centre of mass of the Earth–Sun and Jupiter–Sun systems. In the second case, it is seen that the Sun "wobbles" significantly. Observed in other stars, such a "wobble" could lead to the discovery of extrasolar planets.

A clue about the formation of planetary rings. A ring is the result of the natural evolution of a cloud of particles (like rocks and ice) in the gravitational field of a central body like a planet or a star, when it loses energy in inelastic interparticle collisions while angular momentum is conserved. The particles here are assumed to be located within the Roche limit, an approximate distance limiting a region inside which any satellite rock may break up under the tidal effects exerted by the planet or star. It is assumed that a planet of mass M and radius r_0 is surrounded by a cloud of particles having a total mass $m \ll M$, at an average distance $r \gg r_0$. The cloud has a total energy E and angular momentum $\mathbf{L}$, due to its rotation around an axis, which we will take to be the z-axis. It continuously loses internal energy due to inelastic collisions, so that, although its angular momentum L is conserved, its total energy decreases. In a simplified version of the problem, it can be shown that there is a maximum energy which can be lost by the system, and that once this has been lost, the material must lie in a circular ring around the star (although not necessarily uniformly distributed).

One must consider two symmetries: the central symmetry due to the gravitational force exerted by the planet or star, and directed toward its centre, plus the axial symmetry of the body due to its rotation around its z-axis. Consider two particles having masses, positions, and velocities m_j, $\mathbf{r}_j$, $\mathbf{v}_j$, respectively, with $j = 1, 2$, and angular momenta $\mathbf{L}_1 = \mathbf{h} + \mathbf{n}$ and $\mathbf{L}_2 = \mathbf{h} - \mathbf{n}$, where $\mathbf{h} \parallel \mathbf{L}$ and $\mathbf{n} \perp \mathbf{L}$. We assume that to a good approximation almost all particles can be paired off in such a way as to have the properties discussed above, this exhausting most of the particles contained in the initial cloud. The remaining unpaired particles should lose their $\mathbf{n}$ angular momentum in collisions with the formed ring. We also assume that, when not involved in a collision, each particle follows a Keplerian orbit as discussed in Sect. 1.5.3 and that their total angular momentum is $2\mathbf{h} = \mathbf{L}_1 + \mathbf{L}_2$. Note that the motion along L is responsible for the fact that $\mathbf{n} \neq 0$. As a consequence, the energy lost by friction, conserving angular momentum, is due to a decrease in the kinetic energy along L. The motions are finally confined to a ring in the x, y plane (the axial symmetry becomes dominant). As the mass is not necessarily uniformly distributed in the ring, some parts of it may become centres of force, leading afterwards to the formation of satellites (or planets).

One can write

$$\mathbf{L}_i = m_i (\mathbf{r}_{i\perp} + \mathbf{r}_{i\parallel}) \times (\dot{\mathbf{r}}_{i\perp} + \dot{\mathbf{r}}_{i\parallel}) \tag{1.80}$$

$$= m_i [(\mathbf{r}_{i\perp} \times \dot{\mathbf{r}}_{i\perp}) + (\mathbf{r}_{i\perp} \times \dot{\mathbf{r}}_{i\parallel} + \mathbf{r}_{i\parallel} \times \dot{\mathbf{r}}_{i\perp})]. \tag{1.81}$$

Obviously, the term $\mathbf{r}_{i\parallel} \times \dot{\mathbf{r}}_{i\parallel} = 0$, but $m_i (\mathbf{r}_{i\perp} \times \dot{\mathbf{r}}_{i\perp}) = \mathbf{h} \neq 0$, and assuming $\mathbf{r}_{2\parallel} = -\mathbf{r}_{1\parallel}$ as well as $\dot{\mathbf{r}}_{2\parallel} = -\dot{\mathbf{r}}_{1\parallel}$, the term $(\mathbf{r}_{i\perp} \times \dot{\mathbf{r}}_{i\parallel} + \mathbf{r}_{i\parallel} \times \dot{\mathbf{r}}_{i\perp}) = \pm \mathbf{n}$, respectively, for $i = 1, 2$.

The initial kinetic energy of the particle i is $T_i = \frac{1}{2} m_i (\dot{r}_{i\perp}^2 + \dot{r}_{i\parallel}^2)$, and due to friction, the parallel components decrease, and finally, $\dot{r}_{i\parallel} = 0$, as well as $\mathbf{r}_{i\parallel} = 0$. In consequence, the two particles describe elliptic orbits in the plane orthogonal to L containing the centre of forces (the centre of the star). A further decrease in the energy is achieved when the velocity vector $\dot{r}_{i\perp}$, which lies in the plane of the orbit, loses its component parallel to $\mathbf{r}_{i\perp}$, and is reduced to the perpendicular component, whence $\mathbf{r}_{i\perp} \cdot \dot{r}_{i\perp} = 0$: the orbit becomes a circle. What is valid for the case of two particles can be extended to all the rest of the particles.

For circular motion of a planet in a gravitational field, as the effective potential U has a mimimum (see Sect. 1.5.3), it leads to the equality between the moduli of the centripetal and gravitational forces:

$$\frac{mv^2}{r} = \frac{GMm}{r^2}.$$

This in turn implies

$$\frac{1}{2} mv^2 = \frac{GMm}{2r}.$$

Thus, the total energy $E = T + U$ is

$$E = -\frac{GMm}{2r}. \tag{1.82}$$

Assuming that the total initial energy is $E^I = \sum_i m_i (\dot{r}_{i\perp}^2 + \dot{r}_{i\parallel}^2) - \frac{GMm_i}{r_i} (< 0)$, the final energy is $E^F = -\sum_i \frac{GMm_i}{2r'_{i\perp}}$, which corresponds to circular particle trajectories. Thus, the maximum energy loss is

$$E^I - E^F = \sum_i \left(m_i (\dot{r}_{i\perp}^2 + \dot{r}_{i\parallel}^2) - \frac{GMm_i}{r_i} \right) + \sum_i \frac{GMm_i}{2r'_{i\perp}},$$

where $E^F < E^I$, since the system has lost energy and both quantities are negative. Obviously, $\left|E^F\right| > \left|E^I\right|$, because the modulus of the energy for the circular orbits is greater than the modulus of the negative energy corresponding to elliptical orbits (See Fig. 1.20). It should be pointed out that, when the system lies in a plane, friction may continue, and loss of energy will lead to an increase in the radius of the orbit around the star or planet, since the angular momentum is conserved.

1.12 Advice for Solving Problems

Dimensionality. It is essential in solving problems to consider dimensionality, which is the relation between a physical quantity and a set of fundamental quantities called basic dimensions. These are the mass (M), length (L), and time (T). The dimensionality is usually represented by denoting the quantity in brackets. For instance, the dimension of force is $[F] = MLT^{-2}$. Then we can find the dimensions of the gravity constant G. We have for the force of gravity $[F_g] = [G]M^2L^{-2}$. Then $[G] = M^{-1}L^3T^{-2}$. In the CGS system it is $6.674 \cdot 10^{-8}$ cm^3/g · s^2, and in the IS system, it is $6.674 \cdot 10^{-11}$ m^3/kg · s^2. Throughout our book, we use mainly CGS units, but units in another systems are sometimes used as well.

Symmetry and conservation laws. The laws of symmetry determine which quantities are conserved, and this often allows one to find a simpler way for solving a problem. For instance, as pointed out in Sect. 1.5, the conservation of linear and angular momentum arise from the space translation and rotation symmetries of a physical system, respectively, and the conservation of energy, from homogeneity under time translation.

A simplified problem. Sometimes, we encounter a problem that is difficult to solve, but where there is a similar problem that is simpler, for instance, because there is an additional symmetry, and can be easily solved. In that case, one can get a lot of information by solving the simpler problem. This is the case for instance, when the closed orbit of a body around its centre of forces is an ellipse of small eccentricity. The discussion of the problem for a circular orbit is considerably simpler and provides important information. In this case, the radial kinetic energy is much less than the centripetal energy, namely $m\dot{r}^2 \ll L^2/2mr^2$. This can be seen from the effective potential (1.23), which we rewrite here

$$U(r) = L^2/2mr^2 - GM_\odot m/r.$$

It reaches its minimum for a value of r given by $\partial U(r)/\partial r = 0$. This leads to the equality between the centripetal and gravitational forces for circular orbits, viz.,

$$mv^2/r = GMm/r^2. \tag{1.83}$$

The small eccentricity of most planets justifies assuming circular orbits as a first approximation. Apart from Mercury, whose eccentricity is $\epsilon = 0.2056$, other planets have smaller values. The highest among the other planets is Mars with $\epsilon = 0.0934$, followed by Saturn with $\epsilon = 0.0541$. For Jupiter and Uranus they are respectively $\epsilon = 0.0484$ and $\epsilon = 0.0472$. The Earth has $\epsilon = 0.0167$, and Neptune and Venus, the smallest values, viz., $\epsilon = 0.0086$ and $\epsilon = 0.0068$, respectively. Due to mutual interactions, these values change slowly with time.

Similarity. Recall that, if the Lagrangian is multiplied by an arbitrary factor, the equations of motion are not altered. Let us assume that the potential energy is a homogeneous function of the coordinates. This happens in several cases. Suppose now that we multiply all coordinates by the same constant α. We have for the potential energy

$$U(\alpha\mathbf{r}_1, \alpha\mathbf{r}_2, \ldots, \alpha\mathbf{r}_n) = \alpha^k U(\mathbf{r}_1, \mathbf{r}_2, \ldots, \mathbf{r}_n), \tag{1.84}$$

where k is the degree of homogeneity of the function U. If all coordinates are multiplied by α and time is multiplied by another constant β, the kinetic energy term is multiplied by the constant α^2/β^2. To have a unique factor multiplying the Lagrangian and unaltered equations of motion, we must have $\alpha^2/\beta^2 = \alpha^k$. This implies that $\beta = \alpha^{1-\frac{1}{2}k}$.

The scales of time and space satisfy the relationship

$$t'/t = (l'/l)^{1-\frac{1}{2}k} \tag{1.85}$$

Other quantities give rise to similar relationships. For example, for velocities, energies, and angular momenta,

$$v'/v = (l'/l)^{\frac{1}{2}k}, \qquad E'/E = (l'/l)^k, \qquad L'/L = (l'/l)^{1+\frac{1}{2}k}. \tag{1.86}$$

For a uniform field, $k = 1$ and then $t'/t = \sqrt{l'/l}$. For the oscillator we have $k = 2$, so the period does not depend on the oscillation amplitude. In the case of the Newtonian gravitational potential, $k = -1$, with the result $t'/t = (l'/l)^{3/2}$, which implies Kepler's third law: the squares of the periods of revolution of the planets vary as the cubes of their mean distances from the Sun.

Problems

Problem 1.1 The total energy E of the harmonic oscillator is the sum of its kinetic energy $K = \frac{1}{2}m\dot{x}^2$ and its potential energy $V = \frac{1}{2}kx^2$. Using the general solution for the coordinate (see Sect. 1.11), show that the total energy can be written as

$$E = K + V = \frac{1}{2}m\dot{x}^2 + \frac{1}{2}kx^2 = \frac{1}{2}kA^2, \tag{1.87}$$

where the amplitude $A = \sqrt{2E/k}$. Then find out at which points the potential and kinetic energies obtain their maximal and minimal values.

Problem 1.2 Circular orbits: oscillatory projection and angular momentum. We have already found that, for circular orbits, planets move at constant speeds $v = \sqrt{GM_\odot/r}$, where r is the distance between the centres of the planet and the Sun, G is the gravitational constant, and $M_\odot$ is the solar mass. As we have seen, the total energy of the moving planet is $E = GM_\odot m/2r$. (a) Show that the projection of the position of the planet (treated as a point) on a diameter of the orbit behaves as an oscillator of frequency $\omega = \sqrt{GM_\odot/r^3}$. (b) Show that the angular momentum of the planet can be written $L = m\sqrt{GM_\odot mr} = GM_\odot m^{3/2}/(2|E|)^{1/2}$.

Problem 1.3 The Venusian year. The semimajor axis of the Earth's orbit is approximately 1.496×10^{11} m. For Venus the semimajor axis is 1.0800×10^{11} m. Calculate the period of rotation T_V of Venus around the Sun in years (for the Earth it is $T_E = 1$). Hint: Use Kepler's third law.

Problem 1.4 Locate the centre of mass of (a) the Sun–Earth system, (b) the Sun–Jupiter system. (a) For the Sun–Earth system, take $m_1 = M_\odot = 1.988 \times 10^{33}$ g as the Sun's mass, $m_2 = 5.98 \times 10^{27}$ g as the Earth's mass, and $r = 1$ AU, where the astronomical unit is 1 AU $\sim 150 \times 10^6$ km, i.e., the mean distance between the centres of the Earth and Sun. (b) For the Sun–Jupiter system, take $m_J = 1.8 \times 10^{30}$ g as the mass of Jupiter and 5.2 AU as the mean distance of Jupiter from the Sun. Using (1.75), find $\alpha = m_2/M_\odot$ and the reduced mass m, and from it the location of the centre of mass of each system relative to the centre of the Sun.

Problem 1.5 The Sun's wobble due to Jupiter. One way of detecting planets orbiting other stars is by observing the way those stars wobble due to their motion around the star–planet centre of mass. This can be done because the wobble causes a shift in the frequency of its light. The idea was first proposed in 1952 by Otto Struve (1897–1963), but the technology was not yet refined enough. In more recent times, advances in Doppler spectroscopy are such that an increase in the velocity of the emitting body of $\delta v \sim 1$ ms^{-1}, and even less, in the direction of the observer on Earth, is enough to be detected. Consider the system Sun–Jupiter, and find the velocity of rotation around its centre of mass of (i) the Sun and (ii) Jupiter.

Problem 1.6 Simple estimate of the galactic mass. The Sun is at an approximate distance of $R = 27\,200$ light-years (lyr) from the centre of the Milky Way and it takes a time $T = 170\,000$ yr to complete a closed orbit. We approximate the orbit by a circle of radius R. The mean distance of the Earth from the Sun is 1 AU $= 149.6 \times 10^6$ km, or approximately $= 150 \times 10^6$ km. Find: (a) the mass of the Milky Way contained inside the sphere circumscribed by the solar orbit, in units of solar mass, and (b) the total mass of the Milky Way, assuming that the galaxy is spherically shaped and has a uniform density, and that the Sun is located at a distance of $R = 0.55 R_G$, where R_G is the average galactic radius. Note that one year is approximately 3.15×10^7 s.

Problem 1.7 A train pulled by gravity? Two cities are connected by a tunnel of length 320 km which can be interpreted geometrically as the chord of a great circle through the Earth's surface. A train can move due to the gravitational force, and by assuming a constant Earth density ρ, it moves as a harmonic oscillator, as discussed earlier. (We are ignoring friction forces.) Calculate the speed of the train at the point where it reaches its maximum value. Let M and m_T be the masses of the Earth and the train, respectively, and assume the Earth's radius to be $R = 6350$ km.

Problem 1.8 Due to the decrease in the frequency of rotation of the Earth produced by tidal friction effects, the length of the day has increased and the spin angular momentum S_E, due to rotation around its axis, has slightly decreased. Let the Moon's angular momentum on its orbit around the Earth be L_M. One expects $S_E + L_M = L_T$ to be conserved. a) Calculate S_E by assuming the Earth to be a homogeneous sphere of constant density ρ and radius R. b) The decrease in the spin S_E of the Earth is compensated by the increase in L_M so that L_T remains constant. Thus, the Moon is receding. Assume circular motion and estimate how much the Moon recedes each year by taking the average increase in the length of the day every year to be $\delta T = 1.75 \times 10^{-5}$s.

Literature

1. J.D. Bernal, *Science in History* (MIT, Cambridge, Massachusetts, 1969). Contains a detailed discussion of the work of Copernicus, Kepler, Galileo, and Newton
2. J.R. Newman (ed.), *The World of Mathematics* (Dover, New York, 2003). Interesting details about those who discovered the laws of planetary motion
3. A. Koestler, *The Sleepwalkers. The history of man's changing vision of the Universe* (Hutchinson and Co., London, 1959). An essay of exceptional value and a tremendous piece of historical research on the topic of the people who discovered the laws of planetary motion
4. K.R. Symon, *Mechanics*, 3rd edn. (Addison Wesley, Reading, Massachusetts, 1971). An excellent introductory text to classical mechanics
5. H. Goldstein, C. Poole, J. Safko, *Classical Mechanics*, 3rd edn. (Addison Wesley, New York, 2002). An advanced text, but pedagogically very clear
6. L.D. Landau, E.M. Lifshitz, *Mechanics*, 3rd edn. (Butterworth-Heinemann, Oxford, 2000). A classical text on classical mechanics
7. M. Chaichian, I. Merches, A. Tureanu, *Mechanics* (Springer, Berlin Heidelberg, 2012). A monograph on analytical mechanics, with applications to various branches of physics

8. C. Sagan, *Cosmos* (Random House, UK, 1980). An exceptional book that synthesizes the author's television series, in which he deals with a broad range of topics, among them the early stages of planetary exploration and the search for extraterrestrial life

9. M.S. Longair, *Theoretical Concepts in Physics* (Cambridge University Press, London, 1984). An original approach to theoretical physics, through a series of case studies, intended to deepen understanding and appreciation of the incredible achievements of theorists in creating the structure of modern physics

10. D. Sobel, *Galileo's Daughter* (Penguin, USA, 2000). A biographical work about Galileo, mainly supported by the letters between Galileo and his daughter, the nun Maria Celeste

11. Y.-k. Lim (ed.), *Problems and Solutions on Mechanics* (World Scientific, 1990). Some problems were adapted from this excellent book

Chapter 2
Entropy, Statistical Physics, and Information

What is entropy? This term is used with several meanings in science and technology. For a chemist, "*it is a function of the state of a thermodynamic system*", whereas a physicist would say that "*it is a measure of the disorder of a given system*", and a communications engineer would come with a very different idea, since for him entropy would be "*the average information transmitted or received in a series of messages*".

The concept of entropy was introduced in thermodynamics in 1864 by Rudolf (1822–1888) as a quantity relating to a thermodynamic system that does not vary in a reversible cyclic process, but always increases in every irreversible process, being a function of the state of the system.

In 1872, Ludwig Boltzmann (1844–1906), in his work on the kinetic theory of gases, started out by assuming the molecular structure of matter and then applied the laws of classical mechanics and the methods of probability theory to give a statistical interpretation of entropy. Through his famous H theorem, he introduced a microscopic interpretation of macroscopic irreversibility. In essence, the function H defined by Boltzmann was identifiable as the negative of the entropy of Clausius, i.e., $H = -S$.

Thus, the concept of entropy, subtle from a thermodynamic point of view, acquired a new life in the theory of Boltzmann. One would expect that, after Boltzmann's work, the understanding of the thermodynamic properties of matter would be clearer in the minds of his scientific contemporaries, but very frequently new ideas are not widely accepted without resistance, and Boltzmann's work on the molecular kinetic theory came under heavy attack by some of his contemporaries, like the outstanding physicists Ernst Mach (1838–1916) and Wilhelm Ostwald (1853–1932). They objected to the atomic–molecular theory and argued that a physical theory should deal only with macroscopically observable quantities, whence the concept of atoms should be rejected.

In 1898, Boltzmann wrote: "*I am conscious of being only an individual struggling weakly against the stream of time.*" Subject to increasing depressions, aggravated also by an apparent loss of memory, Boltzmann committed suicide in 1906.

© The Author(s), under exclusive license to Springer-Verlag GmbH, DE, part of Springer Nature 2021
M. Chaichian et al., *Basic Concepts in Physics*, Undergraduate Lecture Notes in Physics, https://doi.org/10.1007/978-3-662-62313-8_2

The statistical entropy introduced by Boltzmann served as a basis for later developments of this concept within the framework of statistical physics, and it must be said that, following the invention of quantum mechanics, it fitted the new quantum formalism admirably.

In 1948, Claude Shannon (1916–2001), when studying the transmission of information through a channel in the presence of noise, introduced a quantity which, in line with Boltzmann's ideas, formally reproduced the expression for the entropy in statistical mechanics.

Shannon planned to baptize this new quantity with the name *information*, but as this word was already widely used, he preferred *uncertainty*. However, John von Neumann (1903–1957) suggested to Shannon to use the name entropy:

> *You must call it entropy for two reasons: In the first place, your uncertainty function has already been introduced in statistical mechanics with the denomination of entropy, so that it has already such a name; in the second place, and even more importantly, nobody knows what the entropy actually is, so in a debate you always will have the advantage.*

Some time later, Léon Brillouin (1889–1969) found the relation between Shannon's entropy and the entropy of Clausius and Boltzmann. He thus contributed to clarifying an old and widely discussed problem that dated from the previous century: the famous paradox of Maxwell's Demon.

2.1 Thermodynamic Approach

Let us see some useful definitions. In any system one can define *macroscopic parameters* which characterize it, such as the pressure, density, volume, etc. Those determined by the positions of bodies external to the system are called *external parameters*. These include the volume, or magnetic, electric, or gravitational fields created by external sources. Those depending on the spatial distribution or the motion of the particles making up the system are called *internal parameters*. These include the density, pressure, magnetization, etc.

The set of independent macroscopic parameters determines the *state of the system*, and the quantities which, independently of the previous history of the system, are determined by the state of the system at a given instant are called *state parameters*. If the state parameters are constant in time and there are no stationary currents whatsoever owing to the action of external sources, then the system is said to be in a state of *thermodynamic equilibrium*. Thermodynamic parameters are defined to be those that characterize a system in thermodynamic equilibrium.

If the thermodynamic parameters are independent of mass or the number of particles, they are called *intensive parameters* (for instance, pressure, temperature, etc.); those proportional to the mass and the number of particles are said to be additive or *extensive parameters* (for instance, volume, energy, entropy, etc.).

A system is said to be *isolated* if it does not exchange energy or matter with external bodies. Then a principle is usually postulated which establishes that, in the course of time, an isolated system always reaches a state of thermodynamical equilibrium and can never depart spontaneously from it. (This principle is not considered to be valid for the Universe as a whole.)

Another principle (sometimes called the zeroth principle), establishes the existence of temperature as a parameter characterizing the state of thermodynamic equilibrium of a system, by the reason that the temperature is the same for all its subsystems. Suppose that two bodies A and B in thermodynamic equilibrium are put in thermal contact (they can exchange energy without changing the external parameters). Then there are two possibilities: either they remain the same without changing their thermodynamic states, or some alterations are produced in their equilibrium states, leading finally to a new state of equilibrium. Then it is said that, the temperature was the same between the two bodies in the first case, and that it was different in the second. Moreover, by transitivity, if two systems are in thermal equilibrium with a third system, then they are in thermal equilibrium with each other.

Thermodynamics is then based on three laws which we now state.

2.1.1 First Law of Thermodynamics

This law establishes the conservation of energy in thermodynamic systems. It states that every increase in the internal energy of a system is produced in one of two ways (or both): (a) by means of work performed on the system and/or (b) by means of heat absorbed by it.

In order to give or take work from a system, its external parameters, such as the volume or applied external electric or magnetic fields, must be varied. In order to give or take heat from a body, thermal contact must be established with another body, without necessarily varying its external parameters.

The first principle of thermodynamics is expressed by the relation:

$$dU = \delta Q + \delta W, \tag{2.1}$$

where dU is the infinitesimal variation of the internal energy and δQ and δW are respectively infinitesimal increments of heat and work given to the system.

The different notation for the increments is due to the fact that dU is mathematically an exact differential. Thermodynamically speaking, this means that the internal energy U is a function of the specific conditions of the thermodynamic state at a given instant and is independent of the history of the system.

On the other hand, it does not make sense to ask how much heat or how much work a system *has* under given conditions. Heat and work are actually forms of *energy exchange* between a body and its environment, and δQ and δW are not exact differentials: they depend on the history of the body, that is, on the previous states of the body.

2.1.2 Second Law of Thermodynamics

The second law deals with an asymmetry of the transformation of heat into work and vice versa. In contrast to the work performed on a system, which may be transformed directly into other forms of energy (electric, elastic, potential in the presence of a field), heat is transferred only to increase the internal energy of a system, unless it is first converted into work.

A system that converts heat into work is shown in Fig. 2.1. The subsystem M subtracts heat from a source A at temperature T and transfers work W to a receiver B, in such a way that M performs a cyclic process, that is, M starts from an initial state and returns to it at the end, after subtracting the heat Q and transferring the work W. This can only happen if M delivers a part Q' of the absorbed heat Q to a thermal system C whose temperature is $T' < T$. In other words,

$$W = Q - Q', \tag{2.2}$$

where Q' is referred to as the compensation heat. Then one can state that it is impossible to transform heat into work in a cyclic process without compensation. Having two sources at different temperatures $T > T'$ is a necessary condition.

This leads to the idea of an irreversible process as a transition from a state A to another state B such that the reverse process $B \to A$ involves the transformation of heat into work without compensation. Since the latter is forbidden, such reverse processes are not found in Nature, although we will see later that, from the statistical point of view it would be more exact to say that such reverse processes have an extraordinarily small probability of occurrence.

In contrast, a reversible process would be one whose reverse does not involve the transformation of heat into work without compensation. Actually, reversible processes occur in Nature only approximately. The principle previously stated can be included in the formulation of the second law of thermodynamics, which establishes

Fig. 2.1 Outline of a heat engine. One must have $T > T'$.

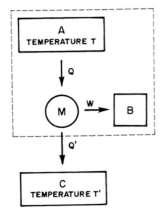

the impossibility of completely converting some amount of heat into work by means of a cyclic process by extracting it from a source of heat at uniform temperature. The process requires two sources at different temperatures, and a transfer of heat of compensation from the hot one to the cold one.

But one could also formulate the second law by establishing the existence of a new function of state, the entropy, which increases to reach a maximum in any isolated system. The entropy variation in an infinitesimal process in which a certain amount of heat δQ is absorbed at an absolute temperature T is given by

$$dS \geq \frac{\delta Q}{T}, \tag{2.3}$$

where the equality sign is valid if the process is quasi-static, and the strict inequality sign is valid if it is non-static. (A quasi-static process occurs so slowly that the system can be considered to be in equilibrium at each instant. Otherwise it is said to be non-static.)

When we speak about *absolute temperature*, we refer to the Kelvin scale of temperature, established by William Thomson, Lord Kelvin (1824–1907), which can be justified on thermodynamic grounds, and whose zero is at $-273.15\,°\text{C}$. Then the absolute temperature is equal to the temperature in centigrade increased by 273.15.

2.1.3 Third Law of Thermodynamics

Finally, we state the third law of thermodynamics, which establishes that, if the absolute temperature of a system tends to zero, its entropy has a minimum value, which can be zero. (Actually, zero entropy cannot be attained in practice, since this is the entropy of the ideal crystal which is an abstract model.) As a consequence of the third law, it can be demonstrated that the absolute zero of temperature cannot be reached.

2.1.4 Thermodynamic Potentials

In addition to the internal energy U and the entropy S, there are some other quantities which are functions of the thermodynamical state. Among these are the thermodynamic potentials.

The internal energy U is a thermodynamic potential. It is a function of the entropy and volume, $U = U(S, V)$. As $\delta Q = T dS$ and $\delta W = -p dV$, we have

$$dU = T dS - p dV.$$

Other thermodynamic potentials are:

- The Helmholtz free energy,

$$F = U - TS.$$

It is a function of the temperature and the volume, $F = F(T, V)$, such that $dF = -SdT - pdV$. The work done on a body in an isothermal irreversible process is equal to the change in its free energy.

- The enthalpy or heat function,

$$H = U + pV.$$

The quantity of heat gained by a body in processes occurring at constant pressure is equal to the change in the heat function. The enthalpy H is a function of the entropy and the pressure, $H = H(S, p)$ and we have $dH = TdS + Vdp$.

- The Gibbs free energy (sometimes called free enthalpy, especially in chemistry), thus called in honour of Josiah Willard Gibbs (1839–1903),

$$G = U - TS + pV.$$

This is a thermodynamic function $G = G(T, p)$ of the temperature and the pressure, and $dG = -SdT + Vdp$. If there are other external parameters, say, a magnetic field $\mathbf{B}$, the conjugated internal parameter is the magnetic moment of the system $\mathbf{M}$, and G will also depend on that, i.e., $G = G(T, p, M)$. Then one has $dG = -SdT + Vdp + \mathbf{B} \cdot d\mathbf{M}$.

These quantities are called thermodynamic potentials because they reach their extrema in the state of thermodynamical equilibrium. This is similar with the mechanical equilibrium, which is achieved for extremum values of the potential energy.

2.2 Statistical Approach

First, we must point out that any macroscopic body is composed of an enormous number of molecules, atoms, or ions. In fact, one mole of any gas contains 6.023×10^{23} molecules. This gigantic quantity is known as Avogadro's number, in honour of the Italian physicist Amedeo Avogadro (1776–1856). Secondly, we must recall that the dimensions of the molecules in simple gases are of the order of one angstrom ($1\,\text{Å} = 10^{-8}$ cm).

For the following discussion, we consider as a model a box that contains a gas made up of the same kind of molecules (e.g., O_2 or N_2). We say that the gas is *ideal* if the molecules interact weakly among themselves and with the walls of the box. (The meaning of 'weakly' is relative here.) We assume further that the gas has low density and that the mean free path, or average distance traveled by a molecule between two collisions, is large as compared with the sizes of the molecules.

Fig. 2.2 The molecules are distributed approximately uniformly in the two halves of the box, but the numbers n and n' fluctuate continually.

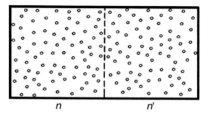

n n'

State of Equilibrium and Fluctuations. For further simplification, we assume that our gas is isolated, which means that the external world does not influence the motion of the molecules. This gas is a system in thermodynamic equilibrium when its macroscopic properties (pressure, density) do not vary in time. Otherwise the system is in a non-equilibrium state.

More rigorously, it would be better to say that, in equilibrium, the *average values* of these quantities do not vary in time, because precise measurements would show that all these quantities do actually fluctuate due to the chaotic motions of the molecules. For instance, if there is a total of N molecules inside the box, and if we assume our box is divided into two equal parts by an imaginary wall or partition, we might expect there to be $N/2$ molecules in each half. However, what we would actually find is that, if n and n' are the numbers of molecules in each half box, these numbers will change in time, fluctuating continually around the value $N/2$, but always satisfying the obvious condition (see Fig. 2.2):

$$n + n' = N. \tag{2.4}$$

A simple example serves to illustrate these ideas. Assume that there are $N = 4$ molecules in the box. Actually, when we consider a small number of molecules, it has no meaning to speak about a thermodynamic system, equilibrium, etc. The reader should accept these examples only as oversimplified models to help understand the behaviour of the system when the number of particles is increased. Let us see in how many ways the four molecules can be distributed in the two halves, and evaluate the corresponding probabilities.

Let us count the molecules from one to four and construct Table 2.1 with the various possibilities. We indicate with an R the fact that a molecule is on the right, and with an L that it is on the left. That is, for four molecules there are two situations in which all of them are concentrated in one of the two halves of the box, eight situations in which there are three in one half and one in the other, and six in which they are distributed symmetrically.

It is easy to see that the numbers of distributions of molecules in the box is obtained from the binomial coefficients (combinations). The possible situations for $N = 6$ and $N = 8$ are shown in Tables 2.2 and 2.3.

Table 2.1 Counting possible distributions of four molecules in a box with partition.

Molecules				Possible distributions		
1	2	3	4	n	n'	Total
R	R	R	R	4	0	1
R	R	R	L			
R	R	L	R			
R	L	R	R	3	1	4
L	R	R	R			
R	R	L	L			
R	L	R	L			
R	L	L	R	2	2	6
L	R	L	R			
L	R	R	L			
L	L	R	R			
R	L	L	L			
L	R	L	L			
L	L	R	L	1	3	4
L	L	L	R			
L	L	L	L	0	4	1
					Total	16

Table 2.2 Counting possible distributions of six molecules in a box with partition.

$N = 6$

Possible distributions		
n	n'	Total
6	0	1
5	1	6
4	2	15
3	3	20
2	4	15
1	5	6
0	6	1
	Total	64

From these tables, one can see that, as N increases:

1. The situations in which all the molecules accumulate in one half of the box are less frequent. The relative frequencies, or equivalently, the probabilities are, respectively, for $N = 4, 6, 8$: 1/8, 1/32, 1/128, ...
2. The cases for which the molecules distribute symmetrically are also less frequent. The probabilities are, respectively: 3/8, 5/16, 35/128, ...; however, the probabilities of situations in which the numbers of molecules in both halves are the same

Table 2.3 Counting possible distributions of eight molecules in a box with partition.

$N = 8$			
Possible distributions			
n		n'	Total
8		0	1
7		1	8
6		2	28
5		3	56
4		4	70
3		5	56
2		6	28
1		7	8
0		8	1
		Total	256

or differ relatively little from the average $N/2$ increases with increasing N. Thus, for example, for $N = 4$, we have:

- 6 cases of symmetric distribution,
- 8 cases in which n or n' differ from $N/2$ (2 molecules) by $N/4$ (1 molecule),
- Total 14 cases.

To this group of 14 cases, there corresponds a probability of $14/16 = 0.875$. But for $N = 8$, we find that there are:

- 70 situations with symmetric distribution,
- 112 situations in which n or n' differ from $N/2$ by less than $N/4$ (2 molecules),
- 56 situations in which n or n' differ from $N/2$ (4 molecules) by $N/4$ (2 molecules),
- Total 238 cases.

The probability of occurrence of one of these 238 cases is: $238/256 = 0.930$.

If similar calculations are carried out for $N = 12$, we obtain a probability of $987/1024 = 0.964$. It can be shown that, as N increases, the quantity q increases according to

$$\frac{n - N/2}{N} = 1/2^q, \quad q = 1, 2, 3 \ldots, \tag{2.5}$$

making the relative fluctuations smaller and smaller.

For N extremely large, we have

$$\left(n - \frac{N}{2}\right)/N \approx N^{-1/2}. \tag{2.6}$$

This means that the number n will fluctuate closely around $N/2$. Thus, if N is of the order of 10^{18} molecules, the fluctuations in n are of the order 10^9 molecules, and the relative quotient is

$$10^9/10^{18} = 10^{-9}.\tag{2.7}$$

In our previous analysis, we calculated the probability as the quotient of the number of favourable cases divided by the total number of possible cases. But one could proceed directly, using the elementary methods of probability theory.

Actually, if the probability of finding a molecule on the right is p, while on the left it is q, and if V is the volume of the box, we have (since $p + q = 1$)

$$p = \frac{(V/2)}{V} = \frac{1}{2} = q.\tag{2.8}$$

Then, according to the binomial theorem, the probability of having n molecules on the right and n' on the left is

$$P_{n,n'} = \frac{N!}{n!(N-n)!}p^n q^{N-n}.\tag{2.9}$$

In particular, the probability that they all accumulate on the right is

$$P_{N,0} = 2^{-N}.\tag{2.10}$$

For $N = 100$, $P_{N,0} \sim 10^{-30}$. To get an idea of how unbelievably small this number is, we introduce the hypothesis that each of our molecular situations has a duration of 10^{-7} s. The accumulation of all the molecules on one side would occur, on average, once every

$$10^{30} \times 10^{-7} = 10^{23} \text{ s}.\tag{2.11}$$

There are 3.15×10^7 s in one year. This means that the order of magnitude of the period of time between two such accumulations (recurrence) is 10^{15} years. The calculated age of the Universe is five orders of magnitude smaller, since it is estimated to be of order 10^{10} years!

It must be emphasized that we have chosen a modest number of molecules for our calculation. The reader may repeat the calculations for the case of 1 cm^3 of nitrogen gas under standard environmental conditions (temperature of 25 °C and pressure of 100 kPa), in which case $N = 2.5 \times 10^{19}$ molecules. The number then obtained for the probability according to (2.10) is considerably smaller. In fact, for all practical purposes, it is zero.

Assume now that the gas is prepared as shown in Fig. 2.3, that is, all the molecules are concentrated in the right half, while the left half is empty. Although these conditions are impossible to achieve exactly in the laboratory, one could achieve a reasonable approximation. If at a given instant the partition gate opens, both the right

Fig. 2.3 **a** The molecules are concentrated in the left half. **b** After removal of the partition, they are allowed to move throughout the whole volume. The probability that they spontaneously return to occupy only the left half decreases exponentially with the number of molecules.

Fig. 2.4 A sequence of events whose probability of occurrence is extraordinarily small. It would be equal to the probability of the reverse of an irreversible process. Those who seek the *perpetuum mobile* have a similar probability of success.

and left volumes will become available to all the molecules, and in a short time the gas will be distributed uniformly in the box, reaching equilibrium.

We have seen that the probability for the molecules to return spontaneously to their original position is a number so extraordinarily small that, in order to observe that phenomenon, such a long interval of time would be required that the present calculated age of the Universe could be neglected in comparison. A phenomenon or process like the one in the previous example, where the probability of the reverse process is negligibly small, is said to be *irreversible* (Fig. 2.4).

The macroscopic world has the special feature that all the processes that occur in it are irreversible; for instance, the conduction of heat and electric current, diffusion, etc. In the systems accessible to human experience and observation, equilibrium is only attained approximately in some cases, and then only for short intervals of time.

2.3 Entropy and Statistical Physics

To give a more exact description of the behaviour of a macroscopic system like the ideal gas from our previous example, we need to know the probability of finding the system in its available microscopic states under given conditions. The microscopic states are quantum states and have to be characterized accordingly, in which case one usually takes the energy as the fundamental quantity (and not the position, which was the quantity used in the classical approach of our previous example). As a matter of fact, in a more rigorous treatment of classical statistical mechanics, the energy is also utilized as a fundamental quantity.

Thus, instead of characterizing the microscopic configuration of the system by giving the number of molecules that have a given speed and position (within certain intervals), a set of possible molecular states $E_1, E_2, E_3, \ldots, E_p, \ldots$ is specified. A given energy corresponds to each state, and may be common to several states (in this case, we say the that these states are degenerate with respect to the energy). Then, a microscopic state of the system could be described by giving the number of molecules in each molecular state, e.g., $(2, 3, 0, \ldots, 5, \ldots)$.

A more detailed analysis of this problem would require us, among other things, to consider the so-called property of *indistinguishability* attributed to identical particles in the atomic and subatomic world. In addition, depending on the value of the spin quantum number (or intrinsic angular momentum), which is either an integer or a half-integer, particles manifest completely different properties, and they are said to obey Bose–Einstein or Fermi–Dirac statistics, respectively.

Any number of particles of integer spin can be simultaneously in a given quantum state, but for particles of half integer spin, each quantum state can be occupied by one particle or none at all, in accordance with Pauli's exclusion principle.

At high temperatures and low pressures the particles can be described by means of the so-called Maxwell–Boltzmann statistics, which is almost equivalent to considering them as distinguishable, leading to a limiting case of the Bose–Einstein and Fermi–Dirac statistics. We say almost, because some other factors must be introduced which can be justified only within the framework of quantum theory. This is because in classical statistics, based on a model of distinguishable particles, the *additivity* of thermodynamic quantities like the entropy must be conserved. (For example, under the same conditions, if the energy, the volume, and the number of particles is multiplied by some constant number, the resulting entropy is expected to be multiplied by the same number.) Otherwise, it leads to the famous Gibbs paradox, which goes as follows: when the expression for the entropy is written in classical statistics, for a system of N identical particles, it is necessary to divide by $N!$, which is the number of permutations of the N particles, in order to satisfy the above-mentioned property of additivity. But this factor $N!$ can be justified rigorously only within the theory of identical particles in quantum mechanics (see Chap. 6) and it has no place in classical mechanics.

But let us return to our original purpose of introducing the concept of entropy in quantum statistics. According to the different distributions of molecules in the molecular states, or cells, there is an enormous number of possible microstates for the system. To each of these microstates we could assign a probability P_i. The entropy of the system is then defined by means of the expression

$$S = -k \sum_i P_i \ln P_i, \tag{2.12}$$

where k is Boltzmann's constant, which is approximately 1.38×10^{-16} erg/K. In (2.12) the sum runs over all possible microstates allowed to the system and compatible with the given macroscopic conditions. It can be shown that this expression can be identified with the entropy as defined in thermodynamics.

When the system is in a specified state i with certainty ($P_i = 1$, $P_{k \neq i} = 0$), it is natural to assume that it has maximum order or minimum disorder. Such a case would occur in the absence of thermal motion, in other words, at zero absolute temperature.

Actually, it is impossible to isolate a system from all external influence, and absolute zero is unattainable. One could, however, say that, if T is very near zero, the system will tend to occupy just a few states close to one state called the ground state, and as a consequence, S will tend to zero (because $P_i = 1$ and $\ln 1 = 0$), which justifies the third law of thermodynamics. As a matter of fact, in some cases there are reasons to expect the entropy to tend to a constant value S_0 when T tends to zero.

For an isolated system, equilibrium is reached when the system has equal probability of being found in each of its accessible states. If there are N allowed states, the probability of finding the system in any one of them is $1/N$. Then (2.12) leads to

$$S = -kN \frac{1}{N} \ln \frac{1}{N} = k \ln N, \tag{2.13}$$

and the entropy is proportional to the logarithm of the number of accessible states.

Since the probability of finding the system in each of the N accessible states is the same, one could say that the disorder of the system is maximum. Under such conditions, the entropy is a maximum.

One could take the entropy as a measure of the disorder of the system. If there were some states with probability greater than others, it is not difficult to see that the entropy and the disorder would be lower than in equilibrium.

As an example, let us choose a system with two accessible states, A and B. When the probabilities of finding the system in these two states are equal ($P_A = P_B = 1/2$), the entropy is given by

$$S_e = k \ln 2 = 0.693k. \tag{2.14}$$

If the probabilities are not equal, for instance, $P_A = 1/4$ and $P_B = 3/4$, its entropy is

$$S_{ne} = -k(1/4 \ln 1/4 + 3/4 \ln 3/4) = 0.612k. \tag{2.15}$$

As a consequence,

$$S_e > S_{ne}. \tag{2.16}$$

When the probabilities P_i are not equal and vary in time, it is possible to prove that the entropy S defined in (2.12) also varies in time, but in such way that it never decreases for an isolated system. This result is well known in statistical physics and goes by the name of the Boltzmann H theorem. It was demonstrated by Ludwig Boltzmann for a gas, starting from his famous kinetic equation, and it gives a microscopic model for irreversibility, based on a probabilistic description of the system.

2.4 Temperature and Chemical Potential

The system in equilibrium we shall consider here is a gas in a box which has been partitioned as in the discussion above. The total entropy S must be a maximum, with the total energy E and the number of particles N being constant. We then have: $S = S_1 + S_2$, $E = E_1 + E_2$, $N = N_1 + N_2$, where S_1, E_1, N_2 and S_2, E_2, N_2 are the entropy, energy, and number of particles of subsystems 1 and 2, respectively. If there is an energy exchange between the subsystems 1 and 2, the relation $\delta E_1 + \delta E_2 = 0$ must be satisfied. Furthermore, since S reaches a maximum upon the energy exchange, we can write

$$dS = \frac{\partial S_1}{\partial E_1} \delta E_1 + \frac{\partial S_2}{\partial E_2} \delta E_2 = 0. \tag{2.17}$$

For constant energy, $\delta E_1 = -\delta E_2$, consequently

$$\frac{\partial S_1}{\partial E_1} = \frac{\partial S_2}{\partial E_2}. \tag{2.18}$$

The quantity $\partial S/\partial E = \beta = 1/T$ characterizes the equilibrium when there is an energy exchange between two bodies, and T represents the common absolute temperature of the two bodies.

Similarly, we can define the quantity $\partial S/\partial N = \mu/T$ to characterize the equilibrium when there is an exchange of particles. The quantity μ is called the *chemical potential*. It can be shown that the chemical potential is the Gibbs free energy per particle, $\mu = G/N$.

In this case, the thermodynamic potentials contain an additional term dependent on dN. We write

$$dF = -SdT - pdV + \mu dN, \qquad dG = -SdT + Vdp + \mu dN.$$

Note that, as V and S are extensive variables, we can write them as $V = Nv$ and $S = Ns$, where v and s are the volume and entropy per particle, respectively. We then have

$$dG = N(-sdT + vdp) + \mu dN.$$

But as $d\mu = -sdT + vdp$, we have $dG = d(\mu N)$, or $G = \mu N$.

2.5 Statistical Mechanics

When we think of a thermodynamic system like the one described above, it is impossible to find a mechanical interpretation of its behaviour because it has such an enormous number of degrees of freedom. Let us consider the case of an ideal gas of N molecules. The system's phase space has $3N$ coordinates and $3N$ momenta. At each instant, the configuration of the gas can be represented by a point in phase space. The point is continuously moving around and maps out all the available phase space (Fig. 2.5).

For example, let us consider the simple case of an isolated "gas" with two molecules of equal mass m and total energy E (see Fig. 2.6). Then, we can write

$$E = \frac{p_1^2 + p_2^2}{2m}, \tag{2.19}$$

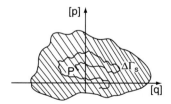

Fig. 2.5 Schematic representation of phase space. Along the horizontal axis we have the coordinates (configuration space) and along the vertical axis, the momenta (momentum space). The region of phase space allowed to the system is denoted by $\Delta\Gamma_S$. The time evolution of the system is represented by a point describing a curve inside $\Delta\Gamma_S$. After a large enough time the curve passes at an arbitrarily small distance of any point inside $\Delta\Gamma_S$. This is the content of the so-called *ergodic hypothesis*.

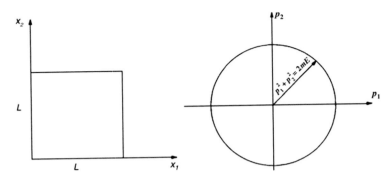

Fig. 2.6 Two molecules of mass m move in one dimension along a line of length L with momenta p_1, p_2 respectively. The allowed phase space is the product of the configuration space L^2 by the space of momenta, which is a circle bounded by the circumference $p_1^2 + p_2^2 = 2mE$.

where $p_1^2 = p_{1x}^2 + p_{1y}^2 + p_{1z}^2$ and $p_2^2 = p_{2x}^2 + p_{2y}^2 + p_{2z}^2$. In other words, the momentum space has six dimensions and the energy defines a hypersphere of radius $\sqrt{2mE}$. As the energy is constant, this last requirement constrains the point representing the system to move only on the hyperspherical surface in momentum subspace. The six-dimensional coordinate space, or configuration space as it is usually called, is determined by the volume. If we assume that the molecules are inside a cubic container of side L, then as the molecules can move freely, the available space is the whole volume of the container. Therefore, the total available volume for the two molecules is represented by all the interior points of a six-dimensional box of side L. The total phase space has twelve dimensions, and the system is represented by a point whose projection in configuration space lies inside the six-dimensional box, while its projection in momentum space lies on the hypersphere mentioned above.

For N molecules there is a similar situation with the corresponding generalization in phase space. Let us assume an ideal gas of volume V, temperature T, and pressure p. In this case, we can imagine a gigantic set made of an infinite number of similar systems, all of them in the macroscopic state of the originally given system. In a sense, this set describes our ignorance about the (microscopic) state of the given system. Following Gibbs, we refer to this set as the *ensemble* representation of the gas. The idea of an ensemble is as follows: consider the simple case of throwing a die, and imagine that we wish to know the probability of getting a given face, for example a 3. One way to find it out is to throw the die a large number of times and calculate the probability over a long period of time (determined by the number of throws). Alternatively, we can throw a large number of identical dice *at the same time*, and calculate the probabilities on that set at any instant t. This set of dice would be the ensemble of the original die. If instead of a die, we have a given physical system, we can think of a huge number of imaginary copies of that system such that all the copies replicate the macroscopic characteristics of the original system but differ in the microscopic configuration compatible with the macroscopic one. In other words, the components of the ensemble have the same characteristics regarding the external parameters (volume, applied fields, etc.) and internal parameters (pressure, temperature, etc.) but differ in the microstates occupied at any given instant. We define a phase space, given by a hypervolume with $3N$ dimensions in the coordinate or configuration space, and by a region in momentum hyperspace with $3N$ dimensions. This region in momentum space is defined by a series of spherical shells corresponding to the different possible values of the energy. For a member of the ensemble of given energy E, its representative point lies on the hypersurface of energy E.

As the ensemble has a large set of representative points, it also covers a very large region in phase space. We thus define the density $\rho = \rho(E)$ characterizing the distribution of representative points in phase space.

2.5.1 Canonical Ensemble

Of particular importance is the canonical ensemble, which describes a system which exchanges energy with a large "heat bath", such that the temperature remains constant in the process. The corresponding density is expressed as

$$\rho(p, q) = \frac{1}{Z} e^{-E/kT}, \tag{2.20}$$

where Z is a constant called *partition function*. The energy $E = E(p, q)$ varies due to the exchange with the other system at constant temperature. The mean energy U, that is, the average of E over the ensemble:

$$U = \frac{1}{Z} \int E e^{-E/kT} d\Gamma_S, \tag{2.21}$$

is constant. In (2.21), the single integral symbol actually denotes a multiple integral over all $6N$ degrees of freedom, and $d\Gamma_S = \prod_i dp_i dq_i$. The constant Z is defined such that the integral of the density $\rho(E)$ over all phase space is unity:

$$\int e^{-E/kT} d\Gamma_S = Z. \tag{2.22}$$

Then ρ can be interpreted as a probability density. In other words, $\rho d\Gamma_S$ is the probability of finding elements of an ensemble inside a region of phase space $d\Gamma_S$. The volume in phase space Γ_S has dimensions of angular momentum to the power $3N$. The integral in (2.21) can be written as an integral over the energy by performing a coordinate transformation and replacing $d\Gamma_S = \prod_i dp_i dq_i$ by the integral over $(d\Gamma_S/dE)dE$. Then $W(E) = \rho(E)d\Gamma_S/dE$ gives us the distribution of elements in the ensemble with energies between E and $E + dE$. Based on the reasoning presented in this chapter, when N is large, the value of the energy does not vary appreciably around U, and $W(E)$ has a sharp maximum around U. From the mean value theorem of integral calculus, it can be shown that there is an energy interval ΔE around U such that

$$W(U)\Delta E = 1. \tag{2.23}$$

This implies

$$\frac{1}{Z}e^{-U/kT}\Delta\Gamma_S = 1, \tag{2.24}$$

where $\Delta\Gamma_S = (d\Gamma_S/dE)\Delta E$ is the phase space volume available to the ensemble representing the system (Fig. 2.5) in the energy interval ΔE. It is proportional to the number of quantum states in that interval of energy.

Equation (2.23) shows that for large N, the distribution of states with energy around U is approximately constant, meaning that each of the possible microstates, whose total number is $\Delta\Gamma_S$, is equally probable. Consequently, we can write the entropy:

$$S = k \ln \Delta\Gamma_S. \tag{2.25}$$

Using (2.24) and (2.25), we get

$$S = k \ln Z + \frac{U}{T}. \tag{2.26}$$

From (2.26) we get $F = U - TS$, where $F = -kT \ln Z$ is the Helmholtz free energy. Tracing back the expressions of Z and U in terms of $\rho(E)$, i.e. using (2.21) and (2.22), we re-write the entropy as

$$S = -k\frac{1}{N!}\int \rho \ln \rho d\Gamma_S. \tag{2.27}$$

We must recall that the factor $1/N!$ in the classical formula is required to guarantee the additive property of the entropy, and it appears in this form only in the description of systems whose particles do not interact among themselves (like the ideal gas). If the particles interact, this factor is replaced by other expressions, sometimes very complicated.

Let us illustrate the above with a simple example of an ideal gas. The total energy is

$$E = \frac{1}{2m}\left[p_{1x}^2 + p_{1y}^2 + p_{1z}^2 + p_{2x}^2 + \cdots + p_{Nz}^2\right], \tag{2.28}$$

and we wish to calculate the entropy using formula (2.27)

We can simplify the calculation by using the Gaussian integral

$$\int_{-\infty}^{\infty} e^{-ax^2} dx = \sqrt{\frac{\pi}{a}}.$$

From the normalization condition (2.22), we find that the integral becomes a product of $6N$ integrals, of which $3N$ correspond to the coordinates. Upon integration, each molecule contributes a factor V in configuration space, resulting in the end in an overall factor V^N. In momentum space we have $3N$ Gaussian integrals, with each molecule contributing a factor $(2\pi m k T)^{3/2}$, resulting in a total of $(2\pi m k T)^{3N/2}$. To guarantee the additivity property mentioned above, we must include the factor $1/N!$. In quantum mechanics, due to Heisenberg's uncertainty relation, a particle in the phase space is not described by one point, but by a cell of minimum volume $(2\pi\hbar)^3$. This is due to the fact that we cannot in principle know with perfect accuracy both the position, and the momentum of the microparticle (we shall elaborate more on this issue in Chap. 6). For our present purpose, in order to establish the correspondence with the number of quantum mechanical states we have to divide $d\Gamma_S$ in (2.22) by $(2\pi\hbar)^{3N}$. Thus,

$$\frac{V^N}{\lambda^{3N} N!} = Z. \tag{2.29}$$

The quantity $\lambda = 2\pi\hbar/(2\pi m k T)^{1/2}$ has the dimension of length and is called *de Broglie thermal wavelength*. Let us estimate its value, for instance, for the hydrogen molecule of mass $m \sim 10^{-24}$ g. We recall that $h = 2\pi\hbar = 6.63 \times 10^{-27}$ erg·s, Boltzmann's constant is equal to 1.38×10^{-16} erg/K, and consider the value of λ at room temperature, i.e., $T = 300$ K. Substituting in these values, we find that

$$\lambda \simeq 10^{-8} \text{cm} = 1 \text{ Å}. \tag{2.30}$$

Another characteristic length for a gas is the average distance between molecules $d = (V/N)^{1/3}$. For an ideal gas with 3×10^{19} molecules per cm^3, it is approximately $d = 30$ Å. When $d \gg \lambda$, the gas behaves classically, as in the previous example. Quantum properties arise when $d \simeq \lambda$. This happens if T is taken to be 100 times smaller, for instance $T = 1\text{–}3$ K. At such low temperatures dilute gases may exhibit Bose–Einstein condensation, while more dense matter may display superfluidity or superconductivity (see Chaps. 3 and 8).

The average energy U can be found using the second Gaussian integral,

$$\int_{-\infty}^{\infty} x^2 e^{-ax^2} dx = \frac{1}{2}\sqrt{\frac{\pi}{a^3}}. \tag{2.31}$$

From (2.21) and (2.29), we get the contribution of N integrals ($3kT/2$ for each). Finally, we obtain

$$U = \frac{3}{2} N k T \tag{2.32}$$

as the caloric equation of state for the ideal gas.

If we divide U by $3N$, we get an average energy of $kT/2$ per degree of freedom. Actually, this average energy corresponds to each coordinate or momentum component contributing a quadratic term to the total energy of the system, and it is named the *theorem of equipartition of energy*, valid for classical thermodynamics. In the quantum regime it is not valid.

Let us introduce the heat capacity at constant volume as a thermodynamical quantity which measures the change of internal energy with regard to temperature. From (2.32), for the ideal gas, it has the expression $C_V = \frac{\partial U}{\partial T} = \frac{3}{2}Nk$. The internal energy and the heat capacity are proportional to mass, and U and C_V are assumed implicitly to refer to some amount of mass. It is called specific heat the heat capacity per unit mass, using the same symbol C_V. If we use as unit of mass the mass of one mole (the molecular weight expressed in grams), which contains N_A molecules, where N_A is Avogadro's number, we denote the specific heat by C_v. We call $kN_A = R$.

We must emphasize that the model of a gas as a set of free point-like molecules is too restrictive, and in most real gases extra degrees of freedom must be introduced since they play a significant role. Thus, for air in normal conditions of pressure and temperature, $C_v = \frac{5}{2}R$, which is due to the fact that air is composed of diatomic molecules, having 5 degrees of freedom (the kinetic energy of rotation of molecules contributes also to the specific heat).

For simple atomic solids, there are three extra oscillatory degrees of freedom, leading at room temperature to $C_v = 3R$, which is the so-called Dulong–Petit law, due to Pierre L. Dulong (1785–1838) and Alexis T. Petit (1791–1820). At low temperatures, quantum phenomena play a fundamental role, and for solids, as $T \to 0$, $C_v \to 0$ proportionally to T^3.

Let us discuss in more detail the integration procedure in phase space and the probability density as a function of energy. To this end, let us consider first a problem of N-dimensional geometry. We want to show that in a sphere with an arbitrary number of dimensions N, most of the volume tends to concentrate in a surface layer with decreasing thickness (in fact, approaching zero as N increases!). The volume of a sphere in N dimensions can be expressed in general as $V_N = K_N R^N$, where R is the radius and $K_N = \pi^{N/2}/\Gamma(N/2 + 1)$, where for N an integer, the Gamma function is defined as $\Gamma(N) = (N - 1)!$. Let us consider the volume of the spherical shell $V_c = K_N R^N - K_N(xR)^N$, where x is close to but strictly less than 1, i.e., $x < 1$. The relative volume is

$$\frac{V_c}{V_N} = \frac{K_N R^N - K_N(xR)^N}{K_N R^N} = 1 - x^N. \tag{2.33}$$

Assuming $x = 0.9$ and taking $N = 32$, we get $x^N < 0.04$. For N of the order of 10^{23}, we can take x extremely close to 1, and still get $x^N \sim 0$.

Let us come back to our problem above. For a given value of the energy E, the expression (2.28) defines a hypersphere of radius $(2mE)^{1/2}$ in the momentum space with $3N$ dimensions. Integrating with the help of hyperspherical coordinates, we have $3N - 1$ angles and one radial coordinate. The $3N - 1$ angles can be integrated to give a constant $C_{3N} = (2\pi)^{3N/2}/\Gamma(3N/2)$, where for large N, $3N/2$ can always be approximated by an integer. The radial coordinate contribution is a factor $E^{\frac{3N}{2}-1}dE$. For example, in two-dimensional polar coordinates, there is a polar angle θ and the radial coordinate r. The latter contributes the factor $r\,dr$, while $C_2 = 2\pi$. In spherical coordinates, the polar angle θ and the azimuthal angle φ contribute the factor $C_3 = 4\pi$, while the radial part is $r^2 dr$ (in $3N$ dimensions the radial part is $r^{3N-1}dr$). Then $\frac{d\Gamma_s}{dE} = DC_{3N}E^{\frac{3N}{2}-1}$, where $D = \frac{1}{N!}\left(\frac{(2m)^{3/2}V}{(2\pi\hbar)^3}\right)^N$. The probability density as a function of energy is

$$W(E) = C'_{3N}E^{\frac{3N}{2}-1}e^{-E/kT}, \tag{2.34}$$

where $C'_{3N} = C_{3N}D/Z$. The maximum of $W(E)$ is at the point

$$E_{\max} = \left(\frac{3N}{2} - 1\right)kT. \tag{2.35}$$

As N increases, the difference between the mean value of the energy given by (2.32) and the maximum energy can be neglected.

To study the behaviour of $W(E)$ around $E_{\max}$, let $E = x E_{\max}$. We shall study the relative distribution

$$\frac{W_N(E)}{W_N(E_{\max})} = f_N(x), \tag{2.36}$$

where $f_N(x) = x^{\frac{3}{2}N-1}e^{-(\frac{3}{2}N-1)(x-1)}$. The function $f_N(x)$ vanishes for both small and large values of x, and it has a maximum at $x = 1$. This maximum becomes sharper as N increases, as shown in Fig. 2.7. This is expected, considering the behaviour of the volume of a hypersphere as N increases, as mentioned above.

Let us obtain expressions for $\Delta\Gamma_S$ and ΔE. We can proceed from (2.24) and neglect unity as compared to $3N/2$, whence

$$\frac{\lambda^{3N}}{V^N}N!e^{-3N/2}\Delta\Gamma_S = 1. \tag{2.37}$$

Using Stirling's formula, $\ln N! \simeq N \ln N - N$, we can write the entropy of the ideal gas as $S = k \ln \Delta\Gamma_S$, which yields

Fig. 2.7 The function $f_N(x)$ has a peak that becomes sharper as N increases. This means that the probability of values different values from $E_{\max}$ decreases with increasing N.

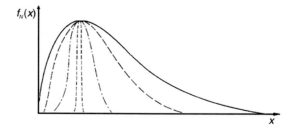

$$S = \frac{3}{2}Nk(1 - \ln \lambda^2) + kN \ln \frac{eV}{N}. \tag{2.38}$$

From (2.23), by taking the average energy to be $U = \left(\frac{3N}{2} - 1\right) kT$ and neglecting unity subtracted from $3N/2$ once again, we obtain with the help of (2.29) and (2.34),

$$\frac{2e^{-3N/2}(3N/2)^{3N/2}\Delta E}{\Gamma(3N/2)kT} = 1. \tag{2.39}$$

Using Stirling's formula, $N! \sim \left(\frac{N}{e}\right)^N \frac{1}{\sqrt{2\pi N}}$, we have

$$\Delta E = kT \frac{(3N/2)!e^{3N/2}}{2(3N/2)^{3N/2}} \simeq \frac{kT}{N^{1/2}}. \tag{2.40}$$

In other words, the energy curve width is $\Delta E \simeq \frac{kT}{N^{1/2}}$ and, as shown below, $\Delta E/kT$ is of the order of magnitude of the relative energy fluctuations $\Delta U/U$ (where $\Delta U = \sqrt{\langle (E - U)^2 \rangle} = \sqrt{\langle E^2 \rangle - U^2}$). The last result is understood better if we assume the system as composed by N subsystems. In particular, each subsystem can be a molecule. The subsystems are considered as independent from the statistical point of view, and we use the labels i, j, m for denoting them, where i, j, $m = 1, 2, \ldots, N$. Notice that, due to the statistical independence, the energy of a molecule minus its mean energy, $E_i - U_i = \delta E_i$, is independent of the same quantity for other molecules, $E_j - U_j = \delta E_j$. If we fix one of them, say δE_i, and multiply by all the possible positive and negative values of the other, δE_j, for each positive product $\delta E_i \delta E_j$ there will exist a negative one, $-\delta E_i \delta E_j$, and the sum of all of them is zero. Thus, we may state that $\langle \delta E_i \delta E_j \rangle = 0$ for $i \neq j$. However, $\langle \delta E_i^2 \rangle > 0$.

We have

$$\langle (E - U)^2 \rangle = \left\langle \sum_{ij} (E_i - U_i)(E_j - U_j) \right\rangle$$

$$= \left\langle \sum_i (E_i^2 - 2E_i U_i + U_i^2) \right\rangle$$

$$= N(\langle E_m^2 - U_m^2 \rangle).$$

Thus, $\Delta U/U = \sqrt{(\langle E_m^2 \rangle - U_m^2)}/N^{1/2}U_m$, since $U = NU_m$. It can be shown by using the Maxwell distribution below, for instance, that $E_m^2 = 15k^2T^2/4$ and $U_m = 3kT/2$ are respectively the average square energy and mean energy per molecule. Thus, $\Delta U/U = \sqrt{2/3}N^{-1/2}$. For 1 cm^3 of an ideal gas at room temperature $T \sim 300$ K, $N \sim 10^{19}$, we have $\Delta E \sim 10^{-23}$ erg, which is 10^{17} times smaller than the electron rest energy.

As we have seen, the ideal gas is an excellent example from which we may learn about many thermodynamical facts. For it is valid the so-called equation of state. For an ideal gas in equilibrium, if p is its pressure, V is its volume, N the number of molecules and T its temperature, the equation of state is

$$pV = NkT.$$

However, for real gases we must take into account the interactions between molecules, and the fact that they have actually a finite volume. This would lead to more exact equations of state, for example,

$$\left(p + \frac{a}{V^2}\right)(V - b) = NkT,$$

where the quantity a accounts for the molecular cohesive forces, and b is due to the molecular volume. This equation bears the name of Johannes van der Waals (1837–1923).

2.5.2 Maxwell Distribution

If the system under study is *one* molecule in equilibrium with the rest of the system, the energy is now $E = (p_x^2 + p_y^2 + p_z^2)/2m$, and the volume element in phase space is $d\Gamma_S = dp_x dp_y dp_z dx dy dz$. The probability of finding the representative point of the system in this element of volume $d\Gamma_S$ is

$$\rho d\Gamma_S = \frac{e^{-E/kT} d\Gamma_S}{\int e^{-E/kT} d\Gamma_S}. \tag{2.41}$$

From this we get the density

$$\rho(\mathbf{p}, \mathbf{r}) = \left(\frac{1}{2\pi mkT}\right)^{3/2} \frac{1}{V} e^{-\frac{E}{kT}}. \tag{2.42}$$

The mean energy per molecule is

$$u = \int E \, \rho(E) d\Gamma_S = \frac{3kT}{2}, \tag{2.43}$$

and for N molecules, this yields $U = Nu = \frac{3NkT}{2}$.

If we wish to express the probability in terms of the modulus of the velocity $v = \sqrt{p_x^2 + p_y^2 + p_z^2}/m$, we find the probability density

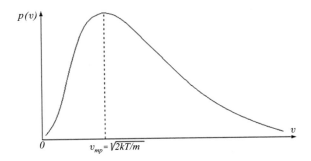

Fig. 2.8 The Maxwell velocity distribution.

$$p(v) = \left(\frac{m}{2\pi kT}\right)^{3/2} 4\pi v^2 e^{-mv^2/2kT}, \qquad (2.44)$$

which is represented in Fig. 2.8. The maximum of $p(v)$ gives the most probable velocity as

$$v_{mp} = \sqrt{2kT/m},$$

whereas the root mean square (RMS) velocity is

$$v_{RMS} = \sqrt{\overline{v^2}} = \left(\int_0^\infty v^2 p(v)\,dv\right)^{1/2} = \sqrt{3kT/m}.$$

The various characteristic velocities of molecules grow with temperature. To get an idea about orders of magnitude, for nitrogen (N_2) at room temperature $T = 300\,K$, with the mass of one molecule about $5 \cdot 10^{-23}$ g, one finds $v_{RMS} \approx 500$ m/s, which is of the order of the sound speed in the gas.

2.5.3 Grand Canonical Ensemble

If we assume that the system under study also exchanges particles with a large system, we must use the so-called *grand canonical ensemble*, whose density is:

$$\rho(p, q) = \frac{1}{\mathcal{Z}} e^{-(E-\mu N)/kT}. \qquad (2.45)$$

where μ is the chemical potential and $\mathcal{Z} = \sum_{N=0}^\infty e^{\mu N/kT} Z(N)$ is named grand partition function. Here, $Z(N)$ is the partition function for a system of N particles. Usually $\Omega = -kT \ln \mathcal{Z}$ is called the grand potential, thermodynamically defined as $\Omega = U - TS - \mu N$. In the case of the grand canonical ensemble, the mean number of particles is constant. The mean energy U, which is the average of E over the ensemble, is now

$$U = \frac{1}{\mathcal{Z}} \sum_{N_j} \frac{1}{N_j!} \int E e^{-(E-\mu N_j)/kT} d\Gamma_S, \tag{2.46}$$

where we sum over the number of particles N_j, which varies in the different systems of the ensemble. The grand canonical ensemble is particularly important in the study of the Fermi–Dirac and Bose–Einstein distributions, which we shall obtain in Chap. 8 by using a different method.

2.6 Entropy and Information

For a source which is able to send N messages or symbols with the probabilities $P_1, \ldots, P_N$, Claude Shannon had the idea of introducing a measure of the information gained (or lost through uncertainty) when one of the messages is received, by the function he called entropy. We represent this function by the symbol I, because it corresponds to the function that we now call information:

$$I = -k \sum_{i=1}^{N} P_i \ln P_i, \tag{2.47}$$

where k is an arbitrary constant, which may be the Boltzmann constant or ln 2 depending on the units used to measure the entropy (either thermodynamic units or bits).

If the N messages are equally probable, $P_i = 1/N$ and one can write

$$I = k \ln N. \tag{2.48}$$

This formula allows us to understand the notion of information as a lack of uncertainty, when we consider the occurrence of a certain number of possible events, whether or not they are messages.

Initially there is no information, $I = 0$, and the uncertainty is large, depending on N. That is, the larger N, the larger the number of possible alternatives. When receiving a message or event, the information is $I = k \ln N$. This quantity could be interpreted as a measure of the decrease in uncertainty with respect to the initial conditions after choosing one message or event out of N possibilities.

If the final situation does not lead to the knowledge of one specific event, but to some set of them N_I, which implies the reduction of the possible outcomes from N to $N_I < N$, the information gained or uncertainty lost is $I = k \ln N/N_I$.

John von Neumann made the observation that Shannon's ideas, were rooted in the assertion made by Boltzmann in 1894, that the entropy is related to *information loss*, since it is determined by the number of possible alternatives of microstates allowed to a physical system after all the macroscopic observable information about such a system has been obtained. But the relation between Shannon's entropy or information, and the entropy in statistical physics or thermodynamics was clarified by Léon Brillouin, in a way we discuss below.

Assume a physical system is in an initial state with N_0 equally probable available states such its information and its entropy are, respectively,

$$I_0 = 0, \quad S_0 = k \ln N_0.$$

If now we get information from the system, e.g., by reducing the number of available states to N_1, the obtained information and the entropy are:

$$I_I = k \ln N_0 - k \ln N_I, \quad S_I = k \ln N_I.$$

If the states available to the system are not equally probable (which occurs, for example, if the system is in equilibrium with a thermal bath at constant temperature T), its initial entropy is

$$S_0 = -k \sum P_i \ln P_i. \tag{2.49}$$

After getting information about the system, the probability of the accessible states changes to $q_1, q_2, q_3, \ldots$, whence the information obtained is

$$I_i = -k \sum P_i \ln P_i + k \sum q_i \ln q_i, \tag{2.50}$$

and its final entropy is

$$S_f = -k \sum q_i \ln q_i. \tag{2.51}$$

In any event, it follows that the information obtained is equal to the decrease in entropy

$$I_i = S_0 - S_f, \tag{2.52}$$

or

$$S_f = S_0 - I_i. \tag{2.53}$$

In other words, any increase in information we have from a given system, starting from a given state, implies a decrease in its entropy.

For example, for a gas that contains N molecules, it can be shown that its entropy depends on its volume V according to the expression

$$S_0 = kN \ln V + S(T), \tag{2.54}$$

where $S(T)$ is a function of the temperature T. If T is fixed and the volume is reduced by one half, its entropy becomes

$$S_f = kN \ln V/2 + S(T). \tag{2.55}$$

The decrease in entropy is

$$S_0 - S_f = kN \ln 2. \tag{2.56}$$

It is immediately clear that, by reducing the volume by one half, more information (or less uncertainty) is obtained about the position of the molecules. This information can be readily measured assuming once again that the molecules are allowed to be in either of the two halves in 2^N different ways.

Knowing that all the molecules are accumulated in one of the two halves, the occurrence of one of the 2^N possible situations gives then

$$I_l = k \ln 2^N = kN \ln 2. \tag{2.57}$$

2.7 Maxwell's Demon and Perpetuum Mobile

At the end of his book *Theory of Heat*, published in 1871, Maxwell wrote that, if there were a *creature* or *demon* within a box containing air (for example, the box divided in two portions A and B by a partition with a small hole and a gate), and if the demon had the ability to see the molecules, then by opening and closing the gate he could allow the slowest molecules to pass from B to A and the fastest from A to B (Fig. 2.9). This would raise the temperature in B and decrease it in A, without expending any work, in contradiction with the second law of thermodynamics.

Fig. 2.9 Maxwell's demon.

Fig. 2.10 Trajectory of a
Brownian particle.

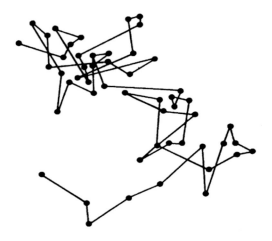

   This is the paradox of Maxwell's demon, considered as a mechanical or electronic system, capable of bringing about some specific increase in order by selecting from several possible options based on some previously obtained information. It attracted the attention of several generations of physicists and mathematicians, among whom one must mention in particular the names of Marian Smoluchowski (1872–1917), Leó Szilárd (1898–1964), Norbert Wiener (1894–1964), John Clarke Slater (1900–1976), and Léon Brillouin (1889–1969).

   The latter gave the following solution for the paradox: since the system is assumed to be in equilibrium, black body radiation is present inside the gas. This makes it impossible to distinguish any object distributed isotropically inside the box. Then, in order to be able to *see* a molecule, the demon needs to illuminate it with radiation at a temperature higher than that of the gas. The demon is informed about the speed of the molecule and opens or closes the gate accordingly. In the process, there is a decrease in the entropy of the gas, but the radiation at higher temperature, after being absorbed by the *eye* of the demon (a photodetector), at lower temperature, increases its entropy. The net result is not a decrease but an increase in the entropy of the system as a whole (gas + radiation + demon).

   We see that Maxwell's demon cannot achieve its objective of creating a difference of temperatures in the two halves of the box, that would permit the operation of a thermal motor, starting from a system in equilibrium. But it is possible to conceive of other systems which, not operating exactly at the molecular level, would take advantage of fluctuations. As we have seen, such fluctuations are a consequence of the thermal motion of the molecules.

   For instance, if we observe grains of pollen suspended in water with a powerful microscope, they are seen to describe chaotic trajectories. This is called Brownian motion (Fig. 2.10), after its discoverer, the botanist Robert Brown (1773–1858), and it is due to the collisions of the water molecules with the suspended particles.

The Brownian particle is bigger than the molecules. Thus, at each time t, it suffers a large number of random collisions, leading to some average velocity, but moving in a random direction. If the velocity of the particle is denoted by $\mathbf{v} = \dot{\mathbf{r}}$, where $\mathbf{r}$ is the position vector with regard to some coordinate system, the dynamical description of its motion is given by the so-called Langevin equation, in honour of Paul Langevin (1872–1946):

$$\dot{\mathbf{v}} = -\gamma\mathbf{v} + \mathbf{A}(t), \tag{2.58}$$

where the first term at the right is the friction force per unit mass and γ is a constant. The quantity $\mathbf{A}(t)$ is the random force (also per unit mass) whose average, in magnitude and direction, is zero. Equation (2.58) is a *stochastic equation*. In contrast to the dynamical systems discussed in Chap. 1, which had well-defined trajectories, here the random nature of the force acting on some mass m leads to a *stochastic motion* (random variables are used to describe *stochastic processes*). For instance, one can only predict the average squared distance. For long periods of time, one finds $\langle r^2 \rangle = 6kTt/m\gamma$.

Diffusion describes the spread of particles by means of random motion (like Brownian motion). A typical example is found in the spread of droplets of ink in water. Particles move stochastically from regions of higher density to regions of lower density. If the density of particles is $n(\mathbf{r}, t)$, it obeys the diffusion equation:

$$\frac{\partial n}{\partial t} = D\nabla^2 n, \tag{2.59}$$

where $\nabla^2 = \partial^2/\partial x^2 + \partial^2/\partial y^2 + \partial^2/\partial z^2$ (in one dimension it can be taken simply as $\nabla^2 = \partial^2/\partial x^2$) and D is the diffusion coefficient.

The solution of (2.59) is

$$n(\mathbf{r}, t) = \frac{1}{\sqrt{4\pi Dt}} e^{-\frac{r^2}{4Dt}}, \tag{2.60}$$

which is a Gaussian describing the spread of particles in space as time goes by. With this distribution, we find the average of the squared distance as

$$\langle r^2 \rangle = \int r^2 n(\mathbf{r}, t) d\mathbf{r} = 6Dt. \tag{2.61}$$

Since (2.58) and (2.59) describe the same process, we obtain the diffusion coefficient as $D = kT/m\gamma$.

Brownian motion and other stochastic processes can be described by path integrals, which are multiple integrals defined over an infinite number of variables. However, this goes beyond the scope of the present book.

Similarly to Brownian motion, in a highly sensitive mirror galvanometer in the absence of a current, the position of the reflected ray oscillates on the scale because the plane of the mirror also oscillates chaotically as a consequence of molecular collisions, and this leads to fluctuating pressure (Fig. 2.11).

Fig. 2.11 Brownian motion
in a mirror galvanometer.

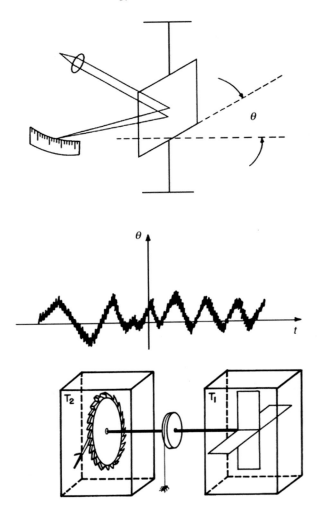

Fig. 2.12 The
Smoluchowski–Feynman
motor.

Consider now the mechanical rectifier depicted in Fig. 2.12, as suggested by
Marian Smoluchowski and Richard Feynman. It consists of a shaft with a paddle wheel at one end and a ratchet wheel at the other. A pawl allows it to rotate only
in one direction. A thread is being wound around the axis with a small weight, for
instance, an ant, hanging from it.

We assume that the paddle and the ratchet wheels are at the same temperature.
Due to the fluctuations, one would expect the paddle wheel to be able to acquire
enough energy to cause the ratchet wheel to rotate, lifting the weight. But then also
with the same frequency, due to the fluctuations, the pawl would get enough energy
to rise, allowing opposite rotations of the ratchet wheel. Since at a given temperature
the probability that the paddle wheel acquires some rotation energy is equal to the

probability that the pawl acquires the same amount of elastic energy to rise and allow the ratchet wheel to rotate in the opposite direction, the net result is that no rotation takes place. The machine cannot operate.

Now, if the temperature of the paddle wheel is greater than that of the ratchet wheel, the probability that the paddle wheel will acquire the necessary energy to move and raise the pawl (according to statistical physics), is greater than the probability that the pawl will rise spontaneously to allow free motion. The net result is that there is rectified motion, i.e., rotation. The machine works, but now its operation is allowed by the second law of thermodynamics.

There is a very interesting electronic analog. The characteristic curve intensity–voltage of an ideal diode is given by two half lines, one horizontal and another forming some angle with the vertical, meeting at the origin. Such an ideal diode would be able to rectify the thermal noise, or fluctuating voltage, produced between the extremes of a resistance, whose mean squared variation is given by the Nyquist formula:

$$V_R^2 = 4RkT\Delta f. \tag{2.62}$$

Here R is the resistance, k is Boltzmann's constant, T is the absolute temperature, and Δf is the bandwidth over which the fluctuating voltage is observed. The quantity V_R^2 gives a measure of the amplitude of fluctuations, and obviously increases with temperature.

But the above-mentioned characteristic curve does not correspond to any real semiconductor diode. The curve (Fig. 2.13) for a real diode can only be used to rectify the voltage above a certain threshold. Below that threshold, there is current in both directions. The thermal voltage of a resistance at the same temperature as the diode would give on average zero current, in analogy with the mechanical rectifier machine which on the average produces zero rotation, due to the Brownian motion of the paddle wheel. Again the perpetual motor fails.

But if the source of thermal noise warms up, the amplitude of the fluctuations in the voltage could exceed the thermodynamic threshold of the diode, and noise would be rectified, but this time with two different temperatures, in agreement with the second law of thermodynamics.

Fig. 2.13 Characteristic curve of a semiconductor diode. A similar picture corresponds to the Smoluchowski–Feynman mechanical rectifier, by plotting the angular frequency as function of the applied torque.

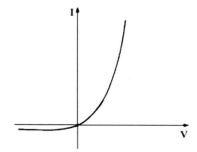

The presumed perpetual motors or *perpetuum mobile* considered above started out from conditions at equilibrium. In non-equilibrium systems, such motors can operate, for example, by using two thermal sources at different temperatures. One can conceive still of other systems in which the non-equilibrium is less evident, as in the case of a liquid and its vapour.

If we have a liquid in the presence of its vapour, there are vapour molecules colliding continuously with the liquid phase, and some of them remain in it. At the same time, within the liquid, some molecules reach energies high enough to leave it, and pass into the vapour phase. The equilibrium settles down when, on average, the number of condensing molecules is the same as the number of evaporating ones. Otherwise there is no equilibrium.

If there are more evaporating molecules than condensing ones, we have an interesting situation. If some liquid is put on a dry surface, it rapidly begins to evaporate, and in the process the liquid cools. This is because the molecules subtract energy when evaporating, in an amount greater than the average energy of the liquid that remains without evaporating.

There is a very interesting toy, which appears at first glance to be a perpetual motor, based on this property. It is the so-called Chinese duck, which consists of a glass bottle of suitable shape, as shown in Fig. 2.14, fixed to a metal stand in such a way that it can rotate in a vertical plane. Inside the toy there is a volatile liquid. The neck is inclined at several degrees with respect to the vertical in the position of equilibrium.

The head of the duck is covered with cotton wool. If this wool becomes wet, by submerging the beak in a glass of water, the duck begins to move continuously, lifting its neck up to the normal position, then returning to drink water, and doing so indefinitely, apparently behaving as a perpetual motor.

The explanation of the motion of the duck goes as follows: when evaporating, the water from the cotton wool in the head cools down the head and the vapour pressure of the volatile liquid inside the duck decreases, raising the level of the liquid in the neck. This leads to a displacement of the centre of gravity, and the duck thus leans

Fig. 2.14 Scheme of the Chinese duck.

forward until it sinks its beak into the water. In the process, the lower end of the neck has risen on the surface of the liquid in the bottle, and the pressure is equalized in the head and body. This allows the liquid to flow back to the body, leading to a backward displacement of the centre of gravity, and the duck straightens itself once more.

In environmental conditions of high humidity the motion slows down, and may even stop. If the toy is put under a glass bell jar it stops completely, but as soon as the bell jar is removed, it starts to move.

The Smoluchowsky–Feynman motor as well as the Chinese duck suggest that a Maxwell demon can operate in a non-equilibrium system. Such is the case for biological systems. The cell membranes select some substances to be absorbed, and reject others, acting like a true demon; but one can justify this, because biological processes take place essentially in out-of-equilibrium conditions, and this allows living organisms to get information about the environment around them by physical or chemical means.

Does this mean that the second law of thermodynamics is universal? Not necessarily. What one can say is that, on the terrestrial scale and even at the scale of the visible Universe, we have not found any contradiction. But of course, this does not validate the hypothesis of the so-called *thermal death of the Universe*, which assumes that increasing entropy will lead the whole Universe to a final state of equilibrium after a long enough interval of time. The present cosmology is much more dynamic (see Chaps. 10 and 11) and does not support this idea.

On the basis of the discussion above, we would advise those who seek perpetual motion to abandon their task. Although the possibility of success is not denied statistically, its probability is so low that, in order to produce a favourable fluctuation allowing them to achieve their aim, they would have to wait an interval of time given by a figure so incredibly large that it escapes the realms of human understanding.

Landauer's principle and Maxwell's demon. Closely related to Maxwell's demon and the erasure of information is Landauer's principle, first suggested in 1961 by Rolf Landauer (1927–1999). This states that there is a minimum possible amount of energy required to erase one bit of information, known as the Landauer limit $kT \ln 2$. Some physicists consider Landauer's principle to give compelling reason why Maxwell's demon cannot work. The demon would need to erase (or forget) the information it used to select the molecules after each operation, and this would release heat and increase the entropy, more than counterbalancing the entropy lost by the demon, getting information in each observation.

Another way of presenting Landauer's principle is to state that, if an observer loses information about a physical system, the observer loses the ability to extract work from that system. If no information is erased, computation may in principle be achieved in a thermodynamically reversible form, and require no release of heat. This is of importance in the study of reversible computing.

2.8   First Order Phase Transitions

The chemical potentials and the temperature play an important role in the study of two phases in equilibrium through a first order phase transition, characterized by a discontinuous change in the macroscopic parameters, like volume. Examples are the liquid–solid or liquid–gas transitions. Under some special conditions, when the two-phase curves intersect, the three phases, solid, liquid and gas, coexist in equilibrium. This occurs for water: the triple point for water is at $0.01\,°C$. The zero of the Kelvin scale is taken at $-273.16\,°C$ from this point.

Let us consider the case of two phases, labeled 1 and 2. Equilibrium requires that they have equal temperatures $T_1 = T_2$ and chemical potentials $\mu_1 = \mu_2$. Since the chemical potential is the Gibbs free energy per particle $\mu = G/N$ and $d\mu = -s\,dT + v\,dp$, where s and v are the entropy and volume per particle, respectively, the equality of the chemical potentials implies

$$- (s_1 - s_2)dT + (v_1 - v_2)dp = 0. \tag{2.63}$$

Note that $\Delta h = h_1 - h_2 = T(s_1 - s_2)$ is the latent heat or enthalpy change (per particle) in the phase transition (the change in its internal energy) and $\Delta v = v_1 - v_2$ is the change in its volume per particle. For a first order phase transition, in a $p - T$ diagram, the curve separating the two phases is known as the coexistence curve. We shall find the differential equation giving the slope of that curve in terms of ΔH and ΔV. This is the so-called Clausius–Clapeyron equation, that is, the equation $p = f(\Delta H, \Delta V, T)$.

First, we multiply (2.63) by T. For a given amount of substance of N units (for instance, N moles), we multiply by N and write $TN(s_1 - s_2) = T(S_1 - S_2) = T\Delta S = \Delta H$, and also $TN(v_1 - v_2) = T(V_1 - V_2) = T\Delta V$. Finally, we write

$$\frac{dp}{dT} = \frac{\Delta H}{T\Delta V}, \tag{2.64}$$

which is the Clausius–Clapeyron equation.

We integrate (2.64) in the particular case of a liquid–gas transition. Call their volumes V_L, V_G, respectively, so that $V_G \gg V_L$ and $V_G = RT/p$ per mol (assume the ideal gas equation of state $pV = RT$). The pressure is obtained as

$$p = const. \, e^{-\Delta H/RT}. \tag{2.65}$$

Notice from (2.65) that the pressure increases with increasing T and vice versa. In high mountains, where the atmospheric pressure is significantly lower than 100 kPa, water boils below $100\,°C$. For high pressures, the boiling temperature of water is increased (see Problem 2.5 below).

We would like at this point to mention some physical reasons which would make it unrealistic to dig a hole right through the Earth, passing through its centre, as mentioned in Chap. 1. If we drop a stone through such a hypothetic hole, for example, down to an average of 40 Km, it will fall through the rigid outer crust (which is thicker for continents and thinner for oceans). After that it will encounter the hot mantle, made primarily of solid but plastically flowing matter, the temperature increasing up to hundreds of degrees centigrade. Deeper still, it will find a liquid outer core, and a solid inner core, with temperatures increasing from a few thousand degrees in the liquid outer core, up to the extremely hot solid core at 5700 K, of the same order as the temperature of the Sun's photosphere. This core is conjectured to be composed mainly of an iron–nickel alloy around 70 and 20%, respectively (and also probably gold, platinum, and other heavy elements). Studies of the solid–liquid phase transition show that iron can solidify at such high temperatures, which occur at the high pressures in the Earth's core, estimated as being in the range 339–360 gigapascals (from 3.3 to 3.6 million atmospheres). The solid inner core is slowly growing at the expense of the liquid outer core at the boundary. This is due to the gradual cooling of the Earth's interior (which is estimated as 100 °C every billion years). We can conclude that the high pressures and temperatures in the mantle and core would prevent any possibility of digging a stable hole passing through the centre of the Earth.

The meaning of $\delta Q, \delta W$ not being exact differentials. Starting from the expressions for the internal energy $U = nC_V T$ and the equation of state for an ideal gas $pV = nRT$ (where n is the number of moles, $R = N_A k$, and $k = 1, 38 \cdot 10^{-16}$ erg/K is the Boltzmann constant), it will be shown that the quantities ΔQ, ΔW in (2.1), for a finite process, depend on the path followed by the system when it passes from the initial state 1 with internal energy U_1 to the final state 2 with internal energy U_2.

The change in internal energy is $\Delta U = U_2 - U_1$. As mentioned earlier, since Q and W are not generally functions of the thermodynamic state, the quantities $\delta Q, \delta W$ are not exact differentials.

Assume that the transition $1 \rightarrow 2$ occurs in two steps: first an isobaric process from the initial state p_1, V_1, T_1 to an intermediate state characterized by values p_1, V_2, T', then an isochoric process (constant volume) from the intermediate state, to the final state p_2, V_2, T_2. We want to show that, whereas ΔQ, ΔW are dependent on the intermediate step, the thermodynamic variables U and S do not depend on it.

The work done on the system in the isobaric step is

$$\delta W_1 = -p_1 \int_{V_1}^{V_2} dV = -p_1(V_2 - V_1) = -nR(T' - T_1) \qquad (2.66)$$

and the heat received is

$$\delta Q_1 = nC_p(T' - T_1). \qquad (2.67)$$

The second process is isochoric from the intermediate state, characterized by the values p_1, V_2, T', to the final state, p_2, V_2, T_2, where $T_2 > T'$. The amount of heat given to the system is equal to its increase in internal energy:

$$\delta Q_2 = nC_V(T_2 - T'). \tag{2.68}$$

The total heat supplied to the system is

$$\delta Q_T = \delta Q_1 + \delta Q_2 = nC_p(T' - T_1) + nC_V(T_2 - T'), \tag{2.69}$$

where C_p is the specific heat at constant pressure. The total work done on the system is

$$\delta W_T = -nR(T' - T_1), \tag{2.70}$$

which is negative, since the work was actually done by the system. Note that both δQ_T and δW_T depend on the intermediate state. But, since $C_p - C_V = R$, the increase in internal energy is

$$\begin{aligned} \Delta U &= \delta W_T + \delta Q_T \\ &= -nR(T' - T_1) + nC_p(T' - T_1) + nC_V(T_2 - T') \\ &= nC_V(T_2 - T_1), \end{aligned} \tag{2.71}$$

and depends only on the initial and final states. The increase in entropy is

$$\Delta S = \int_1^2 \frac{dU + pdV}{T} = nC_p \ln\frac{T'}{T_1} + nC_V \ln\frac{T_2}{T'} = \ln\left(\frac{T_2}{T_1}\right)^{nC_V} + \ln\left(\frac{V_2}{V_1}\right)^{nR}. \tag{2.72}$$

The last expression follows by using the proportionality $T'/T_1 = V_2/V_1$. We also see that ΔS does not depend on the intermediate state, which is what we expected.

Problems

Problem 2.1 The temperature of an ideal gas expanding freely in a partially empty recipient. An insulated chamber is divided into two boxes, as shown in Fig. 2.3a. The left half, whose volume is V_1, contains an ideal gas at temperature T_0, while the right half, of volume V_2, is empty. After removal of the partition, the gas flows through it, and the system comes to equilibrium. No heat is exchanged with the walls. (a) What is the final temperature of the gas? (b) Show that the gas expansion process is irreversible. Hint: Calculate the change in entropy.

Problem 2.2 Show that, at constant temperature T, the work done by a system is equal to the change in its Helmholtz free energy.

Problem 2.3 Show that, at constant pressure p, the heat exchange $\delta Q = T dS$ of a system is equal to the change in enthalpy.

Problem 2.4 One mol of a monoatomic ideal gas initially at temperature T_0 expands from volume V_0 to volume $2V_0$, (a) at constant temperature, (b) at constant pressure. Calculate the work of expansion and the heat absorbed by the gas in each case.

Problem 2.5 Where would sea water boil at 400 °C? Assuming that the temperature at which water boils at the ocean surface is 100 °C, where the pressure is one atmosphere, find how deep in the ocean we must go for raising the boiling temperature of water to 400 °C. The latent heat of vaporization for water is $\Delta H = 40\,700$ J/mol and $R = 8.31$ J/mol·K.

Problem 2.6 Calculate the free energy of the ideal gas starting from (2.29). Use this to obtain the entropy and the internal energy.

Literature

1. I.P. Bazarov, *Thermodynamics* (Pergamon Press, London, 1964). Contains an excellent discussion of the laws of thermodynamics
2. L. Brillouin, *Science and Information Theory* (Academic Press, New York, 1962). One of the first books on information theory and the relation between thermodynamic entropy and information. The problem of Maxwell's demon is discussed
3. R.P. Feynman, *The Feynman Lectures on Physics*, vol. 1, 2 (Addison Wesley, Reading, Massachusetts, 1969). Heat engines, irreversibility, and other topics are discussed in Feynman's usual enlightening style
4. F. Reif, *Statistical Physics, Berkeley Physics Course*, vol. 5. (McGraw-Hill, New York, 1965). The foundations of statistical physics are discussed in a clear and pedagogical way
5. G.H. Wannier, *Statistical Physics* (Dover, New York, 1987). A comprehensive book, containing excellent discussions on the basic concepts of thermodynamics and statistical physics
6. L.D. Landau, E.M. Lifshitz, *Statistical Physics*, 3rd edn. (Pergamon, London, 1981). Like all the Landau–Lifshitz books, this is characterized by its originality in dealing with the concepts of thermodynamics and statistical physics
7. R.K. Pathria, *Statistical Mechanics*, 2nd edn. (Elsevier, Oxford, 2006). An excellent treatise on statistical physics, containing a very good treatment of a broad range of topics
8. M. Chaichian, A. Demichev, *Path Integrals in Physics. Vol. 1: Stochastic Processes and Quantum Mechanics* (IOP, Bristol, UK, 2001). The interested reader will find in this book the description of stochastic processes (and quantum mechanics) by means of path integrals
9. Y.-k. Lim (ed.), *Problems and Solutions on Thermodynamics and Statistical Mechanics* (World Scientific, 1990). Some problems were adapted from this excellent book

Chapter 3
Electromagnetism and Maxwell's Equations

It is said that, around 600 BCE, Thales of Miletus already knew that amber, when rubbed, attracts pieces of straw. Our word for electricity comes from the Greek word *elektron*, which means amber.

The study of electricity and magnetism began at the beginning of the seventeenth century, starting with William Gilbert (1544–1603), Queen Elizabeth's physician, Otto von Guericke (1602–1686), Stephen Gray (1666–1736), who discovered the conduction of electricity, Charles François Dufay (1698–1739), who was the first to mention two kinds of electricity, Ewald Georg von Kleist (1700–1748) and Pieter van Musschenbroeck (1692–1761), inventors of the Leyden jar. The American Benjamin Franklin (1706–1790) brought various contributors, especially the invention of the lightning-rod. Charles Coulomb (1736–1806) established that electrical forces obey an inverse square law, similar to the gravitational force discovered by Newton. The same law was also discovered by Henry Cavendish, a contemporary of Coulomb. Luigi Galvani (1737–1798) discovered the reaction of frog muscles to electrical stimulation, and Alessandro Volta (1745–1827) made the first battery.

The relation between electricity and magnetism was found for the first time by Hans Christian Oersted (1777–1851) in Copenhagen. He discovered that an electric current affected the direction of a magnetized needle. An immediate consequence was the invention of the electromagnet by William Sturgeon (1783–1850) in 1823 and its improvement by Joseph Henry (1797–1878) in 1831, which led to the realization of the telegraph and the electric motor.

From the theoretical point of view, the discovery made by Oersted provided the background for the work done by André-Marie Ampère (1775–1836), Carl Friedrich Gauss (1777–1855), and Georg Simon Ohm (1787–1854). Until that time only central forces were known, and the appearance of new forces with directional properties (the magnetic force was perpendicular to the line joining the magnetic pole with a wire carrying a current) opened the way for consideration of a vectorial physical theory, where direction, as well as distance, played an essential role (Figs. 3.1 and 3.2).

M. Chaichian et al., *Basic Concepts in Physics*, Undergraduate Lecture Notes in Physics, https://doi.org/10.1007/978-3-662-62313-8_3

Fig. 3.1 Michael Faraday, English physicist and chemist. Among his various contributions to science are the introduction of the concept of lines of force, the laws of electrolysis, and the discovery of electromagnetic induction.

In 1831, the complementary relation between electricity and magnetism was discovered by Michael Faraday (1791–1867), not accidentally as it happened to Oersted but by carefully planned experiments. Faraday found that, by moving a conducting wire in a suitable way in a magnetic field, it was possible to generate an electric current.

In this way, electricity and magnetism were found to be interrelated, and the science of electromagnetism was born. The discovery made by Faraday had transcendent applications many years later with the invention of generators. This started the electrical utilities industry. However, Faraday was hardly interested in those problems. He wanted to find a relation between the forces or physical agents known at that time: electricity, magnetism, heat, and light. The physical ideas of field and lines of force were established by Faraday in intuitive form, and subsequently put onto a more formal foundation by Maxwell.

James Clerk Maxwell (1831–1879) was one of the most outstanding physicists of the nineteenth century. He developed the mathematical theory of Saturn's rings at the age of 24, and also made fundamental contributions to the kinetic theory of gases and thermodynamics. In his most important work, *A Treatise on Electricity and Magnetism*, he completed the program begun by Faraday, devising a system of equations that describes electromagnetic phenomena in mathematical language.

In particular, Maxwell's theory shows that light is an electromagnetic wave, in full agreement, at least for a certain time, with Thomas Young (1773–1829), Augustin Jean Fresnel (1788–1827), and other followers of Christiaan Huygens (1629–1695),

Fig. 3.2 It was the Scottish physicist James Clerk Maxwell who established electromagnetism as a science. Not only did he unify magnetism and electricity, but he also predicted the existence of electromagnetic waves. Apart from that, Maxwell made other important contributions to physics, especially to the kinetic theory of gases.

and against those who acknowledged the Newtonian corpuscular theory. Maxwell's theory predicted the existence of electromagnetic waves, and these were discovered by Heinrich Hertz (1857–1894) in his laboratory in 1888, providing the foundations for future radio telecommunication, due to Alexander Popov (1859–1905) and Guglielmo Marconi (1874–1937).

As mentioned above, many experiments in electricity made it clear that there were two kinds of electric charges, referred to as positive and negative. This fact led to the formulation of Coulomb's law.

3.1 Coulomb's Law

Two electric charges are subjected to mutual forces, which could be attractive (if the charges have opposite signs, one + and one −) or repulsive (if the charges have the same sign, both + or both −). For two electrically charged bodies with charges q_1 and q_2, such that the distance between their centres is r, the mutual force is proportional to the product of the charges q_1 and q_2 and inversely proportional to the square of the distance r between them:

$$\mathbf{F} = \frac{Cq_1q_2}{r^2}\mathbf{r}_0, \tag{3.1}$$

where C is a constant of proportionality depending on the system of units used and $\mathbf{r}_0$ is the unit vector along $\mathbf{r}$. We shall use the CGS electrostatic system, or esu, in which $C = 1$.

The direction of the electric forces is depicted in Fig. 3.3. The mutual force is attractive if q_1 and q_2 have opposite signs and repulsive if they have the same sign. If the charges are at rest, or if they move at very small velocities (compared with the speed of light), $\mathbf{F}_1$ and $\mathbf{F}_2$ have opposite directions. At each moment, they satisfy Newton's third law. If the charges move at velocities near the velocity of light, the mutual forces are no longer given by the expression (3.1).

The electron carries the smallest measured quantity of negative electric charge (equal and opposite in sign to that of the proton), $e = -4.803 \times 10^{-10}$ esu. However, there is increasing evidence that protons, neutrons, π mesons, and other particles which interact strongly (hadrons) are composed of other subparticles called quarks, whose electric charge is $\pm 2e/3$ or $\pm e/3$. Since quarks have never been observed isolated, it is believed that they must always form part of the above-mentioned hadrons. They are said to be confined to the regions of space occupied by the hadrons. We shall discuss this issue in some detail in Chap. 11.

Assume we have a body with an electric charge q, and we place another charge q' at any point in the space around q (Fig. 3.4). Then the charge q' is subjected to a force of attraction or repulsion, depending on the relative signs of q and q'. Obviously, q' will also exert a force on q. If we assume that the mass of the body with the charge q' is very small compared to the mass of the body charged with q, then q' will not influence q appreciably. However, in this case, the force that q exerts on q', divided

Fig. 3.3 The force exerted between two charges is attractive if they have opposite sign and repulsive if they have the same sign.

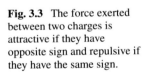

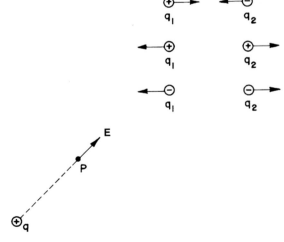

Fig. 3.4 The field created by a positive charge q at a point P is a vector whose direction is indicated in the figure and whose magnitude is equal to the charge divided by the square of its distance to the point P. The field constitutes a physically measurable entity, and the interaction of the charge q' with the field $\mathbf{E}$ produced by q leads to the resulting force $\mathbf{F} = q'\mathbf{E}$.

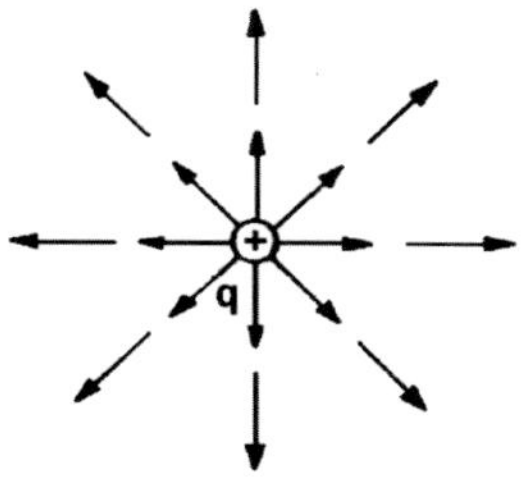

Fig. 3.5 At each point in
space around a charge q
there exists an electric field
produced by this charge.

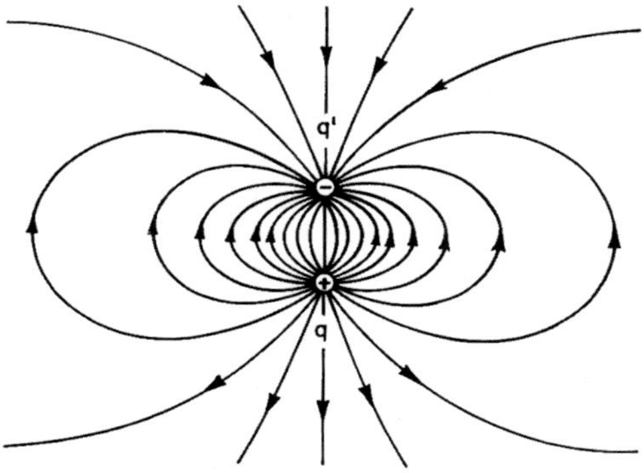

Fig. 3.6 Lines of force
between two electric charges
q, q' equal in modulus but
with opposite signs.

by q' is a very important quantity:

$$\mathbf{E} = \frac{\mathbf{F}}{q'} = \frac{q}{r^2}\mathbf{r}_0. \tag{3.2}$$

$\mathbf{E}$ is called the electric field generated by the charge q and has the same or opposite
direction to $\mathbf{F}$, depending on the sign of q. For a positive charge q, the field at
a point P is represented as in Fig. 3.4, outward from the charge q. If the charge
were $-q$, the field would point toward the charge.

At each point in space around a charge q, one could imagine a vector representing
the electric field applied at that point (Fig. 3.5). Multiplying the intensity of the field
at a point by the charge located there, we obtain the force exerted on the charge.

For two equal charges of opposite signs, the resultant field is indicated in Fig. 3.6.
The field is oriented at each point along the tangents to certain curves. These curves
are called *lines of force*, and their density increases where the field is stronger (the
number of lines per unit area is proportional to the strength of $\mathbf{E}$) (Figs. 3.5 and 3.6).

In Fig. 3.6, for example, if the charge q' is not kept fixed, it will move toward q by
the action of the attractive force exerted by q. We could say that the charge q', when
held fixed, has a certain amount of potential energy with respect to q (similarly to
the case of a mass m near the Earth, which has some potential energy with respect
to it).

If we divide the potential energy by q', we obtain a quantity called the electric potential, which characterizes the electrical properties of the medium around it.

3.2   Electrostatic and Gravitational Fields

Up to this point, we have mentioned electric fields of charges either at rest or moving very slowly. If we refer to charges at rest, it is more appropriate to talk about electrostatic fields. For moving charges, especially when their velocity is comparable to that of light, some other considerations will be required.

It is interesting to compare the electrostatic and gravitational forces. There are important analogies between them, the most important being that both forces decrease as the inverse of the square of the distance. In modern physics, such forces are said to be *long-range forces*.

However, there are also three essential differences between them:

1. The gravitational *charge* coincides with the mass of the body, while the electric charge is not related to the mass. For instance, the positron and the proton have equal electric charges, but the proton is about 1,840 times more massive than the positron;
2. There are electric charges of opposite polarity which attract each other, but when they have the same sign they repel each other. In contrast, all masses have the same sign for all bodies, and the gravitational force is always attractive. An electric charge placed in a medium attracts charges of opposite sign which tend to distribute around it, producing a screening effect which decreases its action on other charges. The gravitational force, on the other hand, cannot be screened;
3. The electric charge is conserved in any process, while the mass is not necessarily conserved.

We shall illustrate this with an example depicted in Fig. 3.7. When an electron and a positron of equal mass and opposite charge collide, they may annihilate mutually, giving rise to two photons. Although the photons have zero charge, the total electric charge has been conserved, because initially the net charge was also zero $e + (-e) = 0$. The total mass was $2m$ at the beginning and zero at the end (since the photon's mass is zero). However, the total energy is conserved. If the process occurs inside a closed box where the two photons are kept moving around, for instance, by continuously reflecting them in mirrors, then for an external observer the mass does not change (it is the same before and after the annihilation). This is due to the fact that, although the mass of the photons is zero, their energy E is different from zero. According to special and general relativity (see Chaps. 5 and 10), if the photons are kept moving around within the box, their external effect is equivalent to that of a mass $2E/c^2 = 2m$, from the formula $E = mc^2$.

Fig. 3.7   An electron e^- and a positron e^+, with masses different from zero, collide and annihilate to produce two photons of zero mass. In the process, the electric charge is conserved but the mass is not. However, if the photons are kept reflecting by suitable mirrors inside a body, e.g., an artificial satellite, then for an external observer the mass does not change, since the energy of the photons divided by c^2 is equivalent to the mass of the initial particles.

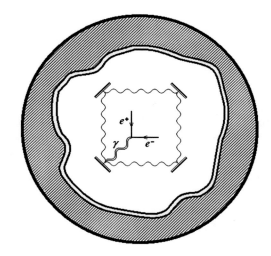

3.3   Conductors, Semiconductors, and Insulators

Almost everybody knows by experience that metals are good conductors of electricity, which means to say that the electric charges move easily through them. For this reason, they are called conductors.

Other substances like glass, rubber, and plastics are poor conductors, and we call them insulators. There is a set of substances, like germanium and silicon, that have intermediate properties, and these are called semiconductors.

In order for charges inside a conductor to move permanently in some direction producing an electric current, it is necessary to apply an electric field, or equivalently, to establish a potential difference between two points. When connecting the + and − terminals of a dry battery with a wire, a current of intensity I flows from the positive to the negative pole. The electrons, which are the charge carriers in a metal, move in exactly the opposite direction (Fig. 3.8).

Due to their motion, the electrons collide with one another and with the ionic lattice inside the metal. Friction, and the corresponding increase in metal temperature, will develop a resistance to the current flow. Under similar conditions, some metals (for example copper and silver) are characterized by having a low resistance to an electric current, while other metals, like tungsten, have a high resistance. For this reason, when a current flows through a tungsten wire, its temperature increases, and the wire can become incandescent. The high melting point of tungsten and its chemical inertness allow it to be used for lamp filaments in industry.

Fig. 3.8 By convention it is assumed that the current flows through a conductor from the positive to the negative terminal of the battery. However, the electrons, as charge carriers, move in the opposite direction, from the negative toward the positive terminal.

3.4   Magnetic Fields

Magnetism is familiar to everybody, and we know that magnets have two poles, called north and south. A magnetized needle suspended by a thread orients itself from north to south, following the direction of the Earth's magnetic field (Fig. 3.9). When two magnets are brought together, we observe that like poles repel each other, while unlike poles attract.

The Earth is a giant magnet with a magnetic south pole in the northern hemisphere, and a magnetic north pole in the southern hemisphere. These attract the north and south poles of a compass needle, respectively. The Earth's magnetic field is produced by the rotation of charges in its core. In fact, the magnetic north and south have oscillated and swapped places several times during the evolution of the Earth. This magnetic field, whose strength lies in the range 0.25–0.65 G, protects the Earth's surface from incoming particles in the solar wind, which is a stream of electrically charged particles released from the upper atmosphere of the Sun. The solar wind is a plasma consisting mainly of electrons, protons, and alpha particles, and their interaction with the Earth's magnetic field produces what is called the magnetosphere. In the inner magnetosphere, there are two belts called the Van Allen radiation belts (Fig. 3.9) which consist of charged particles surrounding the Earth in doughnut-shaped regions, trapped by the Earth's magnetic field. They spiral along the magnetic lines of force from pole to pole and may produce auroras. Most of the particles are thought to come from the solar wind but some may also be from cosmic rays. The inner belt contains electrons and some protons, the outer mainly protons. Ions and antiprotons may also be found in the belts.

When a magnet is divided into two parts, two magnets are obtained. In other words, any attempts to isolate the north and south poles of a magnet are unsuccessful. However small the piece removed from the initial magnet, a new magnet will always be obtained with both north and south poles.

Analogous to the case of an electric charge with a surrounding electric field, there exists a magnetic field around a magnet. If another magnet is placed at any point in space, a force will be exerted which may be attractive or repulsive, depending on the position of the magnet's poles relative to the field.

One can also speak of magnetic lines of force, which emerge from the north pole and end in the south pole of the magnet. Does an electric current have an effect on

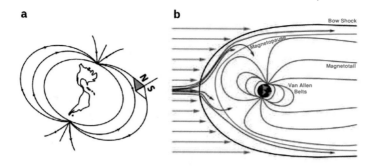

Fig. 3.9 a A magnetized needle lies along the north–south direction following the lines of force created by the Earth's magnetic field. **b** The solar wind is also deflected by it, creating the magnetosphere. In its inner region are located the Van Allen belts, where incoming particles from the Sun are trapped. They spiral along the magnetic lines of force from pole to pole. The magnetopause is the boundary between the Earth's magnetic field and the solar wind, where the pressures of the two systems are equal, whereas the magnetotail is the region of the magnetosphere swept back by the solar wind in the direction away from the Sun. Curved arrows indicate cusps with small magnetic fields through which solar wind particles can enter the Earth's magnetosphere.

Fig. 3.10 An electric current creates a magnetic field around it which orients a needle in the way shown here.

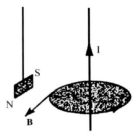

a magnet? The answer is affirmative. This phenomenon discovered by Oersted is easily observed by placing a needle near a wire carrying a direct current (Fig. 3.10).

Circular lines of force appear around the wire, with the magnetic field **B** oriented along their tangents. The magnetic field grows stronger if the current increases, and for a given current, **B** weakens with increasing distance from the wire.

Current-carrying solenoids behave like magnets, with definite north and south poles. In fact, the magnetic properties of magnets are due to small currents produced by the motion of electrons in the atoms or molecules of the magnetic substance, or due to the intrinsic magnetic moment of the electrons. Having an intrinsic angular momentum or spin, each electron behaves like a small magnet, and the interaction between these "microscopic magnets" may lead to a macroscopic magnetic field.

3.5 Magnetic Flux

Magnetic lines of force are represented like electric lines of force, so that the number of lines per unit area is proportional to the strength of the magnetic field. In regions where the field is stronger, the density of lines of force will be higher, and conversely. The magnetic flux is a useful concept, which we shall introduce now.

Consider a point P in space where the magnetic field is $\mathbf{B}$, and a small surface S which contains that point. Decompose $\mathbf{B}$ into one component tangent to S and another perpendicular to it, denoted by B_n. The latter is the one which is of interest here. The magnetic flux Φ is equal to the product of B_n and S (Fig. 3.11).

For the loop depicted in Fig. 3.12, we can calculate the flux by dividing the sub-tended surface into smaller surfaces. When the flux through each of the small surfaces is calculated and added up, we get the flux through the loop. An exact procedure for calculating this flux is to consider the so-called surface integral. If ΔS_i is the area of each of the N surfaces into which the initial surface has been divided, the flux is

$$\Phi = \sum_{i=1}^{N} B_{ni} \Delta S_i, \tag{3.3}$$

where B_{ni} is the value of the normal component of $\mathbf{B}$ in the region ΔS_i. In the limit as the area ΔS_i approaches zero in (3.3), the flux is represented by the surface integral of $\mathbf{B}$ on S:

$$\Phi = \int_S B_n dS. \tag{3.4}$$

The elementary surfaces ΔS_i can be given a vectorial representation by multiplying them by a unit vector along their normals. Then (3.3) can be expressed as the scalar product of the vector $\mathbf{B}$ and the vectorial area element $\Delta \mathbf{S}_i$,

Fig. 3.11 The magnetic flux through a small surface is equal to the product of the component of the field perpendicular to the surface times its area.

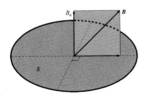

Fig. 3.12 For a surface in which the magnetic field $\mathbf{B}$ is not constant, its area can be divided into small regions ΔS_i and the flux calculated through each of them. The total flux is the sum of all the partial fluxes when all ΔS_i tend to zero.

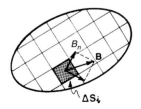

$$\Phi = \sum_{i=1}^{N} \mathbf{B}_i \cdot \Delta \mathbf{S}_i. \qquad (3.5)$$

Consequently, one can write (3.4) in the form

$$\Phi = \int_S \mathbf{B} \cdot d\mathbf{S}. \qquad (3.6)$$

The integral extends over the whole surface S.

3.6 Maxwell's Equations

We learned so far about the intimate relationship between electricity and magnetism. The rigorous analysis performed by Maxwell, based on the works of Faraday, Ampère, Gauss, etc., led him to the formulation of four basic equations which govern all electromagnetic phenomena at the classical level.

It is convenient to distinguish these laws in two cases: when the charges and fields are in vacuum (here we are speaking about the classical vacuum, which we may define as the absence of macroscopic matter; the quantum vacuum is a more complex entity); or when they are immersed in a medium. In what follows we shall refer to the Maxwell equations when the charges, currents, and fields are in vacuum. Later, we shall analyze the effects of a medium on the electric and magnetic fields.

We denote the electric and magnetic fields by $\mathbf{E}$ and $\mathbf{B}$, respectively. It is convenient to mention at this point that, similarly to the definition of magnetic flux through a surface, we can consider the electric flux produced by the electric field $\mathbf{E}$ through a surface $\mathbf{S}$.

3.6.1 *Gauss's Law for Electric Fields*

Let us consider the first Maxwell equation, which is the well-known Gauss law, establishing that the flux of the electric field vector through a closed surface around a charge is proportional to the value of that charge:

$$\oint_S \mathbf{E} \cdot d\mathbf{S} = 4\pi q. \qquad (3.7)$$

Here the integral extends over a closed surface.

If the surface is a sphere, it is very easy to calculate the integral in (3.7) in the case where the charge is concentrated at the centre: the flux is equal to the area of the sphere multiplied by the magnitude of the field E, and (3.7) gives

$$4\pi r^2 E = 4\pi q, \qquad (3.8)$$

whence

$$E = \frac{q}{r^2},\tag{3.9}$$

which is the field created by a point-like charge q at a distance r. We could obtain the same expression from (3.1) and (3.2) with $C = 1$, since Gauss's law is the integral version of Coulomb's law.

Similarly for a sphere of radius R with charge q, if we calculated the flux of the vector $\mathbf{E}$ through a sphere of radius $r > R$, we would get a similar expression to (3.9). This is equivalent to the case in which the charge of the sphere is concentrated at its centre, for $r > R$. If the sphere has an inner charge q' distributed uniformly in its volume and we calculate the electric field at a point $r_1 < R$, we get similar expressions to (3.8) and (3.9), but with q replaced by the charge q' inside the sphere of radius r_1.

The same situation occurs with a static gravitational field, for which Gauss's law is also valid. At the end of the first chapter, we assumed that we could make a hole right through the Earth, passing through its centre, and discussed the motion of an object through that hole. The gravitational attraction on the object diminishes with the distance to the Earth's centre, because each point of the object is attracted by the mass of the Earth included in a sphere concentric with the Earth and with radius determined by the position of the object at each instant. The gravitational effect on the body due to the mass of the Earth outside such a sphere is zero. In the centre of the Earth the gravitational field is zero, but the object continues its motion because of its kinetic energy.

If the interaction law were not inversely proportional to the square of the distance, Gauss's law would not be valid. As pointed out previously, Gauss's law is entirely equivalent to Coulomb's law (3.1). However, it should be pointed out that for a space with a different number of dimensions than three, Coulomb's law would have a different r-dependence, due to the different expressions for the surface.

In our analysis, we have assumed the charged sphere to be placed in vacuum (although the sphere is actually a medium) and we have not said anything about its composition (metallic or non-metallic). If the charged sphere contained free charges within it, it could not be metallic, since in this case the charge is distributed at its surface, and the electric field inside would be zero.

3.6.2 Gauss's Law for Magnetism

Maxwell's second equation establishes that the magnetic flux through a closed surface is zero:

$$\oint_S \mathbf{B} \cdot d\mathbf{S} = 0.\tag{3.10}$$

This is equivalent to the statement that there are no free magnetic charges. Free magnetic charges, so-called *magnetic monopoles*, have not been found in Nature. Theoretically, Paul Dirac considered a model of electrodynamics containing such objects. The definition of the magnetic field as the curl of a vector potential $\mathbf{A}$, i.e.

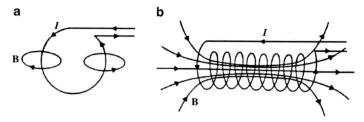

Fig. 3.13 A wire loop creates a magnetic field as shown in **a**. A set of loops as in **b** behave as a magnet. This kind of cylindrical coil of wire is called a solenoid.

$\mathbf{B} = \nabla \times \mathbf{A}$, and the new equation $\nabla \cdot \mathbf{B} = m$, where m is the magnetic charge of the monopole, are not completely compatible.[1] The vector potential $\mathbf{A}$ of the monopole has a singular (divergent) line or string, from infinity up to the point where the monopole is located. This corresponds to the idea of the monopole as a solenoid which is infinitely long and infinitesimally thin. When the quantization condition is imposed, it leads to the equation

$$\frac{em}{\hbar c} = \frac{n}{2},$$

where e is the electric charge and n is an integer. Thus, the very existence of the Dirac monopole implies the quantization of electric charge. In modern gauge field theories the existence of monopoles is theoretically admissible. But so far, there is no experimental evidence for them.

In 2009, was reported the existence of magnetic monopoles as quasi-particles in materials called spin-ices (see the discussion of ferromagnetism below; quasi-particles exist only in condensed matter). Usually, in other media, since there are no free magnetic charges or monopoles, there are no pure sources of magnetic lines. A current-carrying loop creates a magnetic field and the lines of force are closed as shown in Fig. 3.13.

At this point we must observe that the existence of an elementary unit charge also implies that the electric flux through a closed surface around a charge q must be quantized (it must be an integer multiple of e). The magnetic flux through an open surface also appears to be quantized. In quantum mechanics we find that, for charges in a magnetic field $\mathbf{B}$, there is a characteristic length $r_0 = \sqrt{\hbar c / eB}$ and a characteristic area $S_0 = r_0^2 = \hbar c / eB$ orthogonal to the field $\mathbf{B}$. The flux across such an area is quantized,

$$\Phi = B S_0 = \hbar c / e.$$

This suggests that flux quanta may be found, but for these to be observable, manifest quantum conditions are required, as in superconductivity. However, the observed flux quanta correspond to paired electrons, $\Phi = \hbar c / 2e$.

---

[1] For readers not familiar with vector calculus, the divergence and curl are defined in Sect. 5.8.

3.6.3 Faraday's Law

Faraday's law represents the third Maxwell equation. However, before discussing it, we must recall the concept of electromotive force. If we consider a charge moving arbitrarily in a closed circuit *abca* under the action of an electric field **E**, the work done by the unit of charge when moving around *abca* in a given direction, is called the electromotive force. Remark that the electromotive force is not really a force, but rather a *voltage*. Since the force per unit charge is the electric field, the electromotive force can be obtained if the electric field is known at each point of the curve *abca* (Fig. 3.14).

Assume that the charge moves round the curve *abca* in the direction indicated by the arrow, i.e., counterclockwise, which we call the positive direction. Let us divide the curve into small arcs and take the corresponding chords as elementary vectors $\Delta \mathbf{l}_i$ in the direction of rotation. If the electric field is $\mathbf{E}_i$ on the element of arc $\Delta \mathbf{l}_i$, the electromotive force is approximately given by the sum of the scalar products $\sum_i \mathbf{E}_i \cdot \Delta \mathbf{l}_i$, with the sum extending over all the segments of the closed curve *abca*. Now, the electromotive force is the limit of such a sum as $\Delta \mathbf{l}_i$ approaches zero:

$$\mathcal{E} = \lim_{\Delta \mathbf{l}_i \to 0} \sum \mathbf{E}_i \cdot \Delta \mathbf{l}_i. \tag{3.11}$$

This limit is the line integral of **E** along the curve *abca* and it is written as

$$\mathcal{E} = \int_{abca} \mathbf{E} \cdot d\mathbf{l}. \tag{3.12}$$

Faraday's law establishes that, if in the region of space where we have considered the curve *abca* there is a magnetic field variable in time, and if $\Phi_{abca} = \int_S \mathbf{B} \cdot d\mathbf{S}$ is the flux of this magnetic field through the area bounded by *abca*, then an electromotive force is produced in the circuit, proportional to the rate of change of the magnetic flux:

$$\mathcal{E} = -\frac{1}{c} \frac{\partial \Phi_{abca}}{\partial t}, \tag{3.13}$$

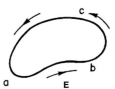

Fig. 3.14 An electric charge moving round the circuit *abca* under the action of an electric field **E** performs work. If this work is divided by the charge, the result is called the electromotive force in the circuit *abca*.

where c is the speed of light. Thus, the change in flux per unit time produces an electromotive force. So, if the number of magnetic lines of force which cross the area S bounded by $abca$ varies in time, an electromotive force is created in the curve $abca$, and the same would occur along any other curve bounding an area containing a time-varying magnetic field. It should be emphasized that, if we have a change in time of the magnetic flux in a region of space where there is no electric circuit, an electric field is still induced in that region of space. This electric field has closed lines, since there are no electric charges producing it.

The negative sign in (3.13) accounts for an interesting phenomenon. If the flux through the area S is growing, an electromotive force will be induced in the negative direction, leading to the appearance of an electric current in the same direction. But this current will in turn create a magnetic field in such a direction that its flux through S will tend to oppose the change of flux produced by the external field through this surface. This is Lenz's law.

3.6.4 Ampère–Maxwell Law

Maxwell's fourth equation establishes a relation between the magnetic field and the current producing it. According to Ampère's law, a current I creates a magnetic field **B** around it (Fig. 3.15).

For an infinitely long conducting wire, the magnetic lines of force are concentric circles centred around the wire. The relation between the current I and the magnetic field **B** is given by the proportionality between the line integral of **B** around an arbitrary contour $abca$ and the current crossing the surface S bounded by $abca$:

$$\int_{abca} \mathbf{B} \cdot d\mathbf{l} = \frac{4\pi}{c} I. \tag{3.14}$$

If the conducting wire can be considered as an infinitely long straight line and $abca$ is a circle of radius R, the field **B** is tangent to this circle at each of its points. Then from (3.14), we obtain

$$2\pi RB = \frac{4\pi}{c} I,$$

leading to

Fig. 3.15 Ampère's law: the magnetic field **B** created by an infinitely long linear current I is tangent to the magnetic lines of force, which are circles centred on the conducting wire.

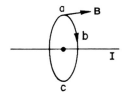

$$B = \frac{2I}{Rc}. \tag{3.15}$$

The expression (3.15) gives the intensity of the magnetic field at a distance R from a conducting wire carrying a current I. But the expression (3.14) is valid only if the electric field flux through the surface S bounded by $abca$ does not vary in time. If there is an electric field flux variable in time through S, it also contributes to the magnetic field $\mathbf{B}$, and (3.14) must be rewritten as

$$\int_{abca} \mathbf{B} \cdot d\mathbf{l} = \frac{4\pi}{c} I + \frac{1}{c} \frac{\partial}{\partial t} \int_S \mathbf{E} \cdot d\mathbf{S}, \tag{3.16}$$

where the term

$$\frac{1}{c} \frac{\partial}{\partial t} \int_S \mathbf{E} \cdot d\mathbf{S} \tag{3.17}$$

expresses the rate of change of the electric field flux through the surface S bounded by the curve $abca$.

The term (3.17) was introduced by Maxwell, who called it the *displacement current*. It establishes a symmetric relationship between (3.13) and (3.16) under the interchange of $\mathbf{E}$ and $\mathbf{B}$. This symmetry is not perfect, since in (3.13) the magnetic field flux appears with a minus sign. Even more importantly, in (3.13) there is no magnetic current, and this is valid for most media, where magnetic monopoles do not exist. (In the new media in which monopole quasi-particles have been reported, (3.13) must be suitably corrected.)

Maxwell's equations must all be considered simultaneously to obtain the electric and magnetic fields created by given charge and current distributions. If charges and currents are zero, Maxwell's equations lead to the electromagnetic wave equations, which we shall see in Chap. 4.

3.7   Lorentz Force

We now consider the force exerted on a moving charged particle by external electric and magnetic fields. If the velocity of the particle is $\mathbf{v}$ and the electric and magnetic fields are $\mathbf{E}$ and $\mathbf{B}$, respectively, this force has the form

$$\mathbf{F} = q \left(\mathbf{E} + \frac{1}{c} \mathbf{v} \times \mathbf{B} \right). \tag{3.18}$$

The expression (3.18) is called the Lorentz force: its electric component is proportional to the electric field, whereas the magnetic component is proportional to the vector product of the velocity of the particle and the field $\mathbf{B}$.

In particular, if there is no magnetic field, the Lorentz force is purely electric, $\mathbf{F} = q\mathbf{E}$. For instance, an attractive Coulomb force is exerted between an electron and a proton, and has the expression:

$$F = -\frac{q^2}{r^2}. \tag{3.19}$$

where $-q$ is the charge of the electron (equal in modulus to that of the proton) and r is the distance between them. If we multiply (3.19) by a unit vector along the line joining the two particles, we have the vector expression for the Coulomb force as in (3.1). If we consider the problem of the motion of an electron around the nucleus due to this force, we have a problem equivalent to Kepler's, and we obtain the planetary model of the atom due to Rutherford.

But this model has a difficulty, as a consequence of the laws of electrodynamics described by Maxwell's equations. The electron would radiate electromagnetic energy when moving around the nucleus and would finally fall onto it. In other words, such an atom would be unstable, and for that reason alternative mechanisms had to be found in order to describe it. This led to the invention of quantum mechanics, in which new ideas and concepts are introduced to solve the problem of the atom's stability, drastically changing our classical conceptions about the dynamics of atomic and subatomic particles.

If the electric field is zero, the force is purely magnetic. A charge q in a magnetic field $\mathbf{B}$ would be subject to a force

$$\mathbf{F} = q\frac{\mathbf{v}}{c} \times \mathbf{B}. \tag{3.20}$$

The most general motion produced by such a force is a helix (Fig. 3.16). If $\mathbf{B}$ is constant, the helix is the result of a circular motion perpendicular to $\mathbf{B}$ and a linear uniform motion parallel to it. Because of this motion, the charged particle emits radiation. The motion of charges in magnetic fields and the consequent production of radiation plays an important role in the particle accelerators of high energy physics laboratories, and also at cosmic scale in several cases, for instance, in objects like *pulsars*.

Let us consider the last two Maxwell equations, referring to charges in motion. A positive charge moving with velocity $\mathbf{v}$ as indicated by the arrow (Fig. 3.17) can be considered as a current element that flows in the direction of $\mathbf{v}$ and creates a magnetic field shown in the figure (right hand rule: if the thumb gives the direction of the current, the other fingers indicate the direction of the created field). If the charge is negative, its motion is equivalent to a current in the opposite direction, and the magnetic field created will also be in the opposite direction to the previous case. But in both cases, the moving charge produces an electric field around it.

On the other hand, a positive charge moving in a magnetic field $\mathbf{B}$, as we have already seen, would describe a circle (or a helix), with the direction of rotation as indicated in Fig. 3.18 (left hand rule: if the thumb gives the direction of the field,

Fig. 3.16 An electric charge in a magnetic field follows a circular path. If the charge initially has a velocity component parallel to the field, the resulting trajectory is a helix.

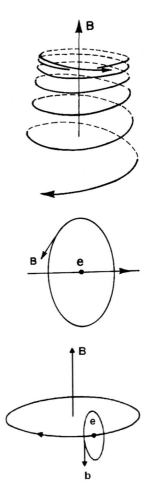

Fig. 3.17 An electric charge in motion is an elementary current that creates a magnetic field around it, as indicated in the figure.

Fig. 3.18 A charge moving in an applied magnetic field **B** creates a magnetic field **b** whose flux through the surface bounded by the orbit is opposed to the flux of **B**.

the other fingers indicate the direction of the current). Due to its motion, the charge creates a magnetic field **b** whose flux through the area bounded by the orbit is opposed to the flux of **B**. If the charge is negative, it will rotate in the opposite direction, but the magnetic field created by the effective current will still give a flux through the orbit that will oppose that of the applied field **B**.

Before ending our brief discussion of Maxwell's equations, we should point out that it is possible to give a mechanical description to the basic equations of electromagnetism. One could consider two interacting systems: the charged particles and fields. In these systems the Lagrangian formulation described at the end of the Chap. 1 contains a term for the particles, another term for the fields, and a third term for the interaction between particles and fields. The particles would have coordinates x, y, z and the fields are described by *functional coordinates* which are the so-called *vector potential* $\mathbf{A}(\mathbf{r}, t)$ and the *scalar potential* $\phi(\mathbf{r}, t)$. The fields $\mathbf{E}$ and $\mathbf{B}$ are related to the potentials by the expressions

$$\mathbf{E} = -\nabla\phi + \frac{1}{c}\frac{\partial \mathbf{A}}{\partial t}, \quad \mathbf{B} = \nabla \times \mathbf{A}.$$

If the value of the electromagnetic field is known, one set of Euler–Lagrange equations will give the expression of the Lorentz force as equations of motion for the charges. If the distribution of the charges is known, the other system of Euler–Lagrange equations will give the expressions for the fields $\mathbf{E}$ and $\mathbf{B}$ in terms of $\mathbf{A}$ and ϕ. Maxwell's equations are interpreted in this way as the equations of motion of the electromagnetic field. This Lagrangian approach must be formulated in the framework of the special theory of relativity (see Chap. 5), but its detailed presentation is outside the scope of the present book.

Motion of a charged particle in a constant magnetic field. We start from the Lorentz force in the case of zero electric field (3.20) and study in more detail the motion of a particle of charge e in a constant magnetic field $\mathbf{B}$ along the z axis, when the initial velocity of the charge is not parallel to $\mathbf{B}$. This can be written

$$\mathbf{F} = \dot{\mathbf{p}} = \frac{e}{c}\mathbf{v} \times \mathbf{B}. \tag{3.21}$$

By writing $\mathbf{p} = m\mathbf{v} = m(\mathbf{i}v_x + \mathbf{j}v_y + \mathbf{k}v_z)$, calculating the vector product $e\mathbf{v} \times \mathbf{B} = eB(\mathbf{i}v_y - \mathbf{j}v_x)$, and calling $\omega = eB/mc$, then equating terms, we get

$$\begin{aligned} \dot{v}_x &= \omega v_y, \\ \dot{v}_y &= -\omega v_x, \\ \dot{v}_z &= 0, \end{aligned} \tag{3.22}$$

from which it is clear that $v_x = v_\perp \cos(\omega t)$, $v_y = v_\perp \sin(\omega t)$, and $v_z = const$, where $v_\perp = \sqrt{v_x^2 + v_y^2}$. This implies motion along a cylindrical spiral, and for $v_z = 0$, it reduces to a circle in the plane orthogonal to $\mathbf{B}$. The rotation is counterclockwise if the charge is positive, and clockwise if it is negative. The components of the velocities v_x, v_y depend on the magnetic field through the frequency

$$\omega = eB/mc. \tag{3.23}$$

As a result, the total energy of the oscillating particle can be written as $E = m(v_\perp^2 + v_z^2)/2$, which does not depend on B. However, its frequency ω is by definition B dependent. We shall see in Chap. 6 that, in the quantum case, the energy is proportional to ω, this being the quantum oscillator. The radius of the circle described by the charged particle in coordinate space depends on B^{-1}, since $x = (v_\perp/\omega) \sin(\omega t)$ and $y = (v_\perp/\omega) \cos(\omega t)$, leading to

$$r = v_\perp (mc/eB). \tag{3.24}$$

One obtains the same result for r by equating the centripetal and Lorentz forces $mv_\perp^2/r = ev_\perp B/c$. The present problem is interesting in connection with several applications. For instance, in the case of motion in an inhomogeneous magnetic field, one may deduce the trajectory approximately by considering the field to be sectionally constant. The rotating charged particles, being accelerated, emit radiation (synchrotron radiation), and this is present in the Van Allen radiation belts in which particles spiral around magnetic lines of force.

In many applications, charged particles move in the presence of both magnetic and electric fields. This is used in the mass spectrometer, which separates ions according to their mass-to-charge ratio, and in the cyclotron, which is a device that can accelerate charged particles to very high speeds. The high energy particles produced are then used to collide with other particles or among themselves, depending on the aim of the experiment.

3.8   Fields in a Medium

The electric dipole moment is a very important concept needed to understand the behaviour of electric and magnetic fields in a medium. If in a region of the field there are charges $q_1, q_2, \ldots, q_n$ of positive or negative sign, whose position vectors with respect to the origin O of some system of reference $Oxyz$ are $\mathbf{r}_1, \mathbf{r}_2, \ldots, \mathbf{r}_n$, respectively (Fig. 3.19), the electric dipole moment is defined as the vector

$$\mathbf{p} = q_1\mathbf{r}_1 + q_2\mathbf{r}_2 + \ldots + q_n\mathbf{r}_n. \tag{3.25}$$

The case in which there are two equal and opposite charges is very important. Here, the electric dipole moment is independent of the position of the coordinate origin O. Actually, this is true for any number of charges if their total sum is zero. In atoms, if the distribution of charges is symmetric, they do not have a dipole moment. When atoms are placed in a strong enough electric field, the charge distribution inside them becomes significantly distorted, creating an atomic dipole moment. The same would occur with molecules in an electric field. However, there are some molecules, like those of water, which in normal conditions have a permanent dipole moment, giving rise to special properties under the action of external electric fields.

If we multiply $\mathbf{p}$ by the number of dipoles per unit volume N, we obtain a vector called the polarization density,

$$\mathbf{P} = N\mathbf{p}.$$

The dipole moment $\mathbf{p}$ and the polarization density $\mathbf{P}$ change if an electric field $\mathbf{E}$ is applied. For not very intense fields, in a linear, homogeneous and isotropic medium such as a dielectric (insulating substance of polarizable molecules), there is a linear relation between $\mathbf{P}$ and $\mathbf{E}$. The constant of proportionality is called the electric susceptibility χ_e:

$$\mathbf{P} = \chi_e \mathbf{E}. \tag{3.26}$$

Consider a positive charge Q embedded in a dielectric, for instance, water. The charge Q attracts the polarized water molecules and they distribute as shown in Fig. 3.20. At some distance from Q, the net charge inside a surface S is not Q, but is modified by an additional charge, because of the dielectric polarization. A test charge q placed on the surface S would detect a smaller charge than Q.

The effective field at a distance R from Q would be the vector

$$\mathbf{D} = \mathbf{E} + 4\pi \mathbf{P}, \tag{3.27}$$

where $\mathbf{E}$ is a vector with the magnitude $E = Q/R^2$, and $4\pi\mathbf{P}$ is an additional field due to the net charge included in the sphere of radius R, because of the polarization around Q. The vector $\mathbf{D}$ is the effective electric field, called the electric displacement. The expression (3.27) is still valid when the medium is not a dielectric.

Similarly, magnetic dipole moments $\boldsymbol{\mu}$ exist, associated with either spin or a closed current. In the latter case, the magnetic dipole moment is defined as the product of the current and the vector area bounded by it, all divided by the speed of light c (Fig. 3.21):

$$\boldsymbol{\mu} = \frac{I\mathbf{a}}{c}. \tag{3.28}$$

Fig. 3.19 A system of charges whose positions are referred to the coordinate system $Oxyz$ by means of their position vectors relative to the origin.

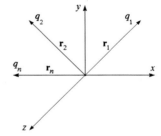

Fig. 3.20 A charge Q in a polarized medium attracts the molecules as indicated in the figure. In a spherical surface S centred at the charge Q, the net charge included differs from the value of Q.

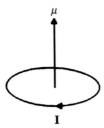

Fig. 3.21 A magnetic moment vector $\mu = I\mathbf{a}/c$ is associated with a circular current I of vector area $\mathbf{a}$.

The electric and magnetic dipole moments play similar roles in the study of the properties of matter: a closed current behaves like a magnet, and the magnetic moment is a physical quantity which characterizes the intensity of this magnet. Under the action of a magnetic field $\mathbf{B}$, the dipole orients itself parallel to the field.

Consider an electron on its orbit around the nucleus. Associated with this motion is a dipole orbital moment

$$\mu_L = \frac{e}{2mc}\mathbf{L}, \tag{3.29}$$

where e is the electron charge, m is its mass, c is the speed of light, and $\mathbf{L}$ is the electron's orbital angular momentum. Moreover, the electron has also an intrinsic angular momentum or spin $\mathbf{S}$ which leads to a magnetic moment. This is a quantum effect:

$$\mu_S = \frac{eg_s}{mc}\mathbf{S}, \tag{3.30}$$

where $S = \pm\hbar/2$ and $g_s \sim 2.002$ is the gyromagnetic ratio of the electron. The customary unit for elementary magnetic momenta is the *Bohr magneton*,

$$\mu_B = \frac{e\hbar}{2mc} = 9.274 \times 10^{-21} \text{erg G}^{-1}.$$

The total magnetic moment of the electron is

$$\mu = \frac{e}{2mc}(\mathbf{L} + 2\mathbf{S}).$$

A similar expression to (3.30) would give the proton magnetic moment, but as its mass M_p is much greater than m ($M_p \sim 1840\,m$), it has a magnetic moment three orders of magnitude smaller than the electron. In the low energy (non-relativistic) case of electron motion inside an atom placed in a magnetic field, two independent magnetic moments appear, orbital and spin, interacting with each other and contributing to the total energy of the electron in the atom. These two contributions, play an important role in determining the magnetic properties of matter.

Similarly to the electric polarization, the magnetization $\mathbf{M}$ is defined by

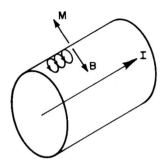

Fig. 3.22 A current I in a medium creates a magnetic field **H** which results from the sum of the magnetic field **B** and the field created by the induced elementary currents (or aligned magnets) in the medium. The net effect of these elementary magnets is to create a field $4\pi\mathbf{M}$ opposed (or parallel) to **B**. The effective field is $\mathbf{H} = \mathbf{B} - 4\pi\mathbf{M}$.

$$\mathbf{M} = N\boldsymbol{\mu}, \tag{3.31}$$

where N is the number of elementary currents (or particles bearing a magnetic moment) per unit volume.

The magnetization (in diamagnetic and paramagnetic media) depends linearly on the field **B**, i.e.,

$$\mathbf{M} = \chi_m \mathbf{B}. \tag{3.32}$$

The constant χ_m is called the magnetic susceptibility, being positive for paramagnetic materials and negative for diamagnetic ones. Usually χ_m is small, giving rise to a correspondingly weak magnetization.

Let us consider a current I creating a magnetic field **B** at a point of space. If we now introduce a magnetic body, there is an additional contribution from the magnetization **M** in the form of an 'effective current'. This leads to an effective field which is defined as

$$\mathbf{H} = \mathbf{B} - 4\pi\mathbf{M}. \tag{3.33}$$

The reason for this is the following: the field **B** induces elementary currents in the medium that determine the onset of some magnetization **M**. The net result is that the field would not have anymore its initial value **B** at the given point. The effective field is **H** given by (3.33), where $4\pi\mathbf{M}$ is the field created by the elementary currents induced in the medium by the external field **B**, originally created by the macroscopic current I (Fig. 3.22).

3.9 Magnetic Properties

We have seen in the previous section that the action of an external magnetic field on a medium depends on the existence or non-existence of permanent magnetic moments in the medium, as happens in ferromagnetic or paramagnetic substances. In the latter case we have what is called a diamagnetic medium.

The alignment of the electron magnetic moment parallel to the applied external field determines the paramagnetic and ferromagnetic properties. But the electron, due to its electric charge, tends to move in a circular orbit leading to a current that also creates an elementary magnetic moment opposed to the applied field.

Depending on the characteristics of the electronic states of the substance, one of the above-mentioned effects will dominate over the other, giving rise to the macroscopic magnetic properties of the materials.

3.9.1 Diamagnetism

As a result of the action of the external field, a magnetic moment **M** of opposite direction to **B** is created, and the magnetic susceptibility is therefore negative. This arises because the external field induces elementary currents in the substance. The external field also acts independently on the spin magnetic moments, but the net effect of this action on the system of electrons of a diamagnetic medium is negligible, as compared to the effect produced by the orbital motion of the charges.

From Lenz's law, the field created by the moving charges is opposed to the applied external field **B**. Inside the substance, the magnetic field decreases and the medium expels the magnetic lines of force. The magnetic lines of force within a diamagnetic substance have opposite direction to the external field **B**, and the net result of embedding a diamagnetic substance in a magnetic field **B** is the expulsion of the lines of force (Fig. 3.23a).

The phenomenon of diamagnetism is particularly important in superconductors, described later on.

3.9.2 Paramagnetism

In some substances there exist permanent magnetic dipoles associated with the electron spin. Under the action of an external magnetic field **B**, the magnetic dipoles line up parallel to the field. This effect is referred to as paramagnetism. Not all the dipoles are aligned parallel to the field, since the ordering action of the magnetic field is opposed by the disordering action of thermal motions, which increases with temperature.

In the presence of a paramagnetic medium, the lines of force of the external field **B** tend to concentrate inside the substance (Fig. 3.23b). If the magnetic field is switched off, the dipoles become disordered again and the substance does not retain magnetic properties.

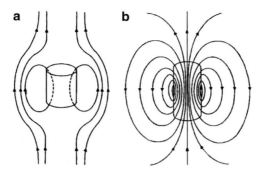

Fig. 3.23 **a** A diamagnetic body placed in a magnetic field expels the lines of force. **b** A paramagnetic body placed in a magnetic field attracts the lines of force and tends to concentrate them inside it.

3.9.3 Ferromagnetism

In some substances there is a spontaneous tendency for parallel neighbouring spins to couple. This is a purely quantum effect (see Chap. 6). The effect can be represented as the product of a quantity J, the so-called exchange integral, times the scalar product of the spins $s_i \cdot s_j$, leading to the creation of elementary dipoles, similar to what happens in the case of an externally applied aligning field. This effect leads to the phenomenon known as ferromagnetism. When a magnetizing field $\mathbf{H}$ is applied to a ferromagnetic substance, the substance acquires a macroscopic magnetization $\mathbf{M}$ parallel to the field. When the external field $\mathbf{H}$ is removed, the substance preserves some magnetization $\mathbf{M}$ and behaves as a permanent dipole, like a common magnet.

In a ferromagnetic substance there are elementary regions or domains with spontaneous magnetization (Fig. 3.24a). Ferromagnetic materials exhibit the phenomenon of *hysteresis*, which is depicted in Fig. 3.24b. If an external magnetic field $\mathbf{H}$ is applied, the domains align themselves with $\mathbf{H}$ up to some maximum value called saturation magnetization $\mathbf{M}_S$. If $\mathbf{H}$ decreases, the ferromagnet maintains some magnetization, and even when the magnetizing field $\mathbf{H}$ becomes zero, part of the alignment is retained, as a memory, and would stay magnetized indefinitely. To demagnetize the ferromagnet, it would be necessary to apply a magnetic field in the opposite direction. For a large enough negative field $-\mathbf{H}$, we can reach a negative saturation magnetization $-\mathbf{M}_S$. The change in magnetization from $-\mathbf{M}_S$ to $\mathbf{M}_S$ follows a similar path to the previous one, from negative to positive magnetization, closing the cycle of hysteresis. (The element of memory in a hard disk drive is based on this effect). If the temperature is increased, the ferromagnetic property disappears at some temperature T_c characteristic of each ferromagnetic substance. For $T > T_c$ the behaviour is paramagnetic. This critical temperature T_c is called the Curie temperature, in honour of Pierre Curie (1859–1906).

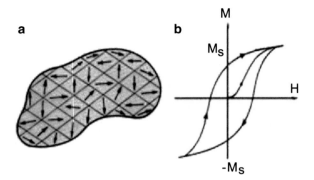

Fig. 3.24 **a** In a ferromagnetic material there are domains with spontaneous magnetization. Under an external field, the magnetic moments of the domains line up parallel to the applied field. If the external field is turned off, the ferromagnetic substance retains the acquired magnetization and behaves as a permanent dipole. **b** Hysteresis loop for a ferromagnet.

As the temperature is decreased, the magnetic susceptibility of a ferromagnetic substance varies, becoming infinite at the Curie temperature T_c. Furthermore, a spontaneous magnetization **M** appears in the substance even in the absence of an external field. Here occurs a phenomenon called a *second order phase transition*. The ferromagnetic substance reaches its minimal energy, or ground state (see Chap. 6), when all dipoles are oriented in one direction, at a nonzero value of its magnetization (see the sections on phase transitions and spontaneous symmetry breaking below).

3.9.4   Ferrimagnetism, Antiferromagnetism, and Magnetic Frustration

In some substances, a fraction of the magnetic moments become aligned in the opposite direction to the field, giving rise to ferrimagnetism (antiparallel magnetic moments not equal to those parallel to the field), with a net magnetization in the absence of an external field, and also antiferromagnetism (equal antiparallel and parallel moments), with vanishing total magnetization if no external field is applied (Fig. 3.25). The antiparallel ordering may be explained from the negative sign of the exchange integral J (see Chap. 6). Antiferromagnetic substances under the action of an external magnetic field may display ferrimagnetic behaviour due to a net difference between the magnetizations of the parallel and antiparallel lattices. In contrast to the usual ferromagnetic case, in which the ground state is non-degenerate, antiferromagnetic materials may have a degenerate (non-unique) ground state. For instance, in a rectangular lattice one can order the spins pairwise, but in a triangular one, this is not possible and magnetic frustration occurs. Frustration is also understood as the inability of the system to find a single ground state (Fig. 3.26).

Fig. 3.25 In
antiferromagnetic
substances, equal parallel
and antiparallel magnets are
disposed at alternate points
in the lattice. The four
neighbours of a dipole have
opposite directions, leading
to a zero net magnetization
in the absence of an external
field.

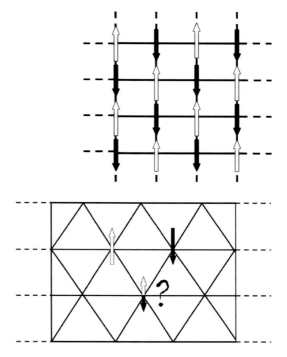

Fig. 3.26 In a frustrated
antiferromagnet lattice,
neighbouring dipoles have
several possible orientations,
leading to a degenerate
ground state.

3.9.5 Spin Ices and Monopoles

There are materials called spin ices consisting of tetrahedra of ions. Each ion has
a non-zero spin, and the interactions between neighbouring ions are such that two
of them point inside and two outside, which is a similar structure to ordinary ice,
where the oxygen atom has two neighbouring hydrogens nearby, and two further
away, since they belong to another molecule. This leads to some residual entropy in
ice, as pointed out by Linus Pauling in 1935.

Spin ices are frustrated ferromagnets. It is the strong local crystal field that produces the frustration, and not the nearest-neighbour interaction, as it is in antiferro-
magnets.

In condensed matter at low temperature, elementary quantum excitations above
the ground state are called quasi-particles, since they behave like particles carrying
quantized amounts of energy, momentum, electric charge, and spin. For instance,
the local electric charge defect creates a 'hole' which behaves like a positive particle
in a semiconductor. In the case of spin ices in magnetic fields, monopoles were
proposed in 2008. It allowed to account for a mysterious phase transition observed
experimentally and understood afterwards as a liquid–gas transition of the magnetic
monopoles. Later experiments were reported in 2009 in which the magnetic moments
lined up in tube-like structures resembling Dirac strings. At the end of each tube, a

defect formed, with a magnetic field that looked like that of a monopole. Monopoles were found to interact through a Coulomb-type potential and to carry a 'magnetic charge'.

3.10 Second Order Phase Transitions

As seen in Sect. 8.8, first order phase transitions occur, for instance, when ice melts and transforms into water, or when water transforms into vapour. There is a discontinuous change in the ordering of the atoms of the substance, and also a change in its volume or in the specific volume v, which is the total volume V divided by the number of molecules N. In second order phase transitions there is no discontinuous change in the state of the substance or in its specific volume. However, there is a sudden internal reordering. In general, in a second order phase transition, there is an increase in order and a change in the internal symmetry of the substance with the arising of an order parameter.

This is the case for ferromagnetic substances. At temperatures equal to or lower than the Curie temperature, they exhibit a reordering of their internal symmetry, since all the magnetic moments within the domains become parallel to a given direction and an order parameter arises: the magnetization $\mathbf{M}$. The ferromagnetic substance, when cooled below the critical temperature of the phase transition T_c, passes discontinuously from the disordered (paramagnetic) phase in which $\mathbf{M} = 0$ to the ordered (ferromagnetic) phase in which $\mathbf{M} \neq 0$, and it remains in this phase for all temperatures $T < T_c$.

The phenomenon called *ferroelectricity* and *ferroelectric* materials are analogous to ferromagnetism and ferromagnetic materials. In the case of ferroelectric materials, the elementary dipoles are not paired spins, but elementary cells (domains) in the crystal structure which have a spontaneous electric polarization. When ferroelectric materials are cooled below a certain temperature T_c, they manifest a macroscopic permanent (remanent) electric polarization, with the order parameter $\mathbf{P} \neq 0$, similar to ferromagnetic materials.

3.11 Spontaneous Symmetry Breaking

Spontaneous magnetization or electric polarization (in magnetic or electric domains) leads to the idea of *spontaneous symmetry breaking*, a very important concept in modern physics, closely connected to phase transitions. We shall consider the Gibbs free energy G per unit volume of a ferromagnetic substance. One can write it as a function of temperature T and magnetization M, at constant pressure, and expand it in powers of $M = \sqrt{\mathbf{M}^2}$ as

$$G(T, M) = G(T, 0) + \alpha_0 M + \alpha_1 M^2 + \beta_0 M^3 + \beta_1 M^4 + \cdots . \tag{3.34}$$

We neglect the terms of order higher than M^4. It can also be proved that, if the states with $M = 0$ and $M \neq 0$ have different symmetry, then $\alpha_0 = 0$. For a second order phase transition (like ferromagnetism), we must also have $\beta_0 = 0$. We define the quantity $G_1 = G(T, M) - G(T, 0)$. The condition for having an extremum for $\alpha_0 = \beta_0 = 0$ is

$$\frac{\partial G_1}{\partial M} = 2\alpha_1 M + 4\beta_1 M^3 = 0. \tag{3.35}$$

Then we get two solutions:

$$M = 0 \quad \text{and} \quad M^2 = -\frac{\alpha_1}{2\beta_1}. \tag{3.36}$$

If $\alpha_1 > 0$, the solution $M = 0$ corresponds to a minimum, since $\frac{\partial^2 G_1}{\partial M^2} = 2\alpha_1 > 0$. But if $\alpha_1 < 0$ and $\beta_1 > 0$, the solution $M = 0$ corresponds to a maximum, and there is a minimum in M_0, where

$$M_0^2 = -\frac{\alpha_1}{2\beta_1}. \tag{3.37}$$

The ground state exhibits a spontaneous magnetization M_0 which corresponds to the minimum of the Gibbs free energy: there is spontaneous symmetry breaking, since the equilibrium does not correspond to the symmetrical case $M = 0$, but to $M \neq 0$. Figure 3.27 depicts the approximate form of $G_1(T, M)$.

As G_1 depends on temperature, the value of M corresponding to the minimum decreases upon heating. In general, $\alpha_1 = f(T)(T - T_c)$, where $f(T_c) > 0$, i.e., there is some critical temperature at which the symmetry is restored (the spontaneous magnetization vanishes), and that is the Curie temperature T_c.

The case of ferroelectricity is quite similar to ferromagnetism, with the magnetization M replaced by the electric polarization P, but for some materials with $\beta_0 \neq 0$. We shall see that this case may lead to a first order phase transition, if $\beta_0 < 0$. (This happens, e.g., in BaTiO$_3$.) We have

$$G_1(T, P) = \alpha_1 P^2 + \beta_0 P^3 + \beta_1 P^4 + \cdots . \tag{3.38}$$

Fig. 3.27 The Gibbs free energy of a ferromagnet has its minimum at a nonzero value of the magnetization M_0.

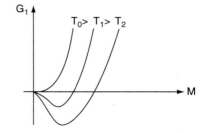

Fig. 3.28 The Gibbs free
energy G_1 versus electric
polarization in a first order
transition.

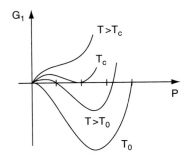

The quantity α_1 usually vanishes at some temperature T_0 and has thus the form
$\alpha_1 = \frac{1}{2}\gamma(T - T_0)$, but we have the same physical behaviour for $\alpha_1 = \frac{1}{2}\gamma'(T^2 - T_0^2)$.
In this case, the condition for extremum is

$$\frac{\partial G_1}{\partial P} = \gamma(T - T_0)P + 3\beta_0 P^2 + 4\beta_1 P^3 = 0. \tag{3.39}$$

Starting from (3.39), which is a third degree equation, we can discuss the dependence
of the symmetry breaking parameter P on temperature. At $T = T_0$, the symmetric
phase is metastable: we have a maximum for $P = 0$ and a minimum at some point
$P = P_0 \neq 0$. As the temperature is increased, the value of the symmetry-breaking
parameter decreases and the minimum of G_1 increases in such a way that there are two
points of intersection of G_1 with the P axis. These two points approach each other and
coincide at the critical temperature $T_c > T_0$. For $T > T_c$, the nonsymmetric phase
becomes metastable. The system becomes paraelectric (Fig. 3.28). Similarly, as both
phases (symmetric and nonsymmetric) have the same free energy, when decreasing
the temperature to T_c, there is a sudden symmetry breaking in some regions of the
material, where a nonzero polarization arises spontaneously.

3.12   Superconductivity

At the beginning of this chapter, we referred briefly to the properties of some substances with regard to electrical conductivity. In particular, we mentioned *electrical
resistance* as a consequence of opposition to the motion of charge carriers through
the metal.

A notable property of many metals and alloys is that the electrical resistance falls
abruptly at low temperatures. This phenomenon was observed for the first time by
Heike Kamerlingh-Onnes (1853–1926) in Leyden in 1911, when studying the resistance of mercury at liquid helium temperature. Near 4 K, the resistance of mercury
diminished abruptly and became practically zero: the mercury had become a *superconductor*. Actually, all superconductors have exactly zero resistivity to low applied

currents. We must remark that superconductivity is not a classical phenomenon. Like ferromagnetism and atomic spectral lines, superconductivity is a quantum phenomenon.

In a superconductor the current could flow for almost indefinite time. In some cases, this time has been calculated to be of the order of 100,000 years, and even more.

The superconductive property appears at different temperatures in different substances. For instance, for certain alloys of niobium, aluminum, and germanium it appears at 21 K, and for some semiconductors, at 0.01 K.

In a superconductor, an ordered phase appears when going down to temperatures below a critical value T_c. This is characterized by the formation of a very large number of opposite spin electron pairs (called Cooper pairs), which move almost freely through the substance. This model of superconductivity is called BCS theory after a famous paper by John Bardeen (1908–1991), Leon N. Cooper (b. 1930), and John Robert Schrieffer (1931–2019) in 1957. Bardeen, Cooper, and Schrieffer were awarded the Nobel Prize in 1972 for this theory.[2] Superconductivity, like ferromagnetism and ferroelectricity, is another case of spontaneous symmetry breaking, and one can consider a model of free energy showing the features pointed out above – the Ginzburg–Landau model, due to Lev Landau (1908–1968) and Vitaly Ginzburg (1916–2009). In this case the order parameter would be the 'condensate of Cooper pairs' rather than the magnetization.

3.13   Meissner Effect: Type I and II Superconductors

In 1933, Walther Meissner (1882–1974) and Robert Ochsenfeld (1901–1993) discovered that, if a superconductor material cools down in a magnetic field, the magnetic lines of force are expelled outside the superconductor for temperatures equal to or lower than T_c. The material behaves as if the magnetic field were zero inside (Fig. 3.29). However, the magnetic field is not strictly zero at the surface of the superconductor, but decreases quickly with the depth until it vanishes inside.

The Meissner–Ochsenfeld effect is different in superconductors of types I and II. For type I superconductors there is only one critical field H_c at which the material ceases abruptly to be a superconductor and becomes resistive. Elementary superconductors, such as aluminium and lead, are typical type I superconductors. In type II superconductors, if the field is increased to $H > H_{c1}$, there is a gradual transition from the superconducting to the normal state in an increasing magnetic field, since normal and superconducting regions are mixed, up to a second critical field H_{c2}, where superconductivity is destroyed.

---

[2]Incidentally, John Bardeen is the only person who received twice the Nobel Prize in Physics. First time, he shared the prize with William Shockley and Walter Brattain in 1956, for the invention of the transistor.

Fig. 3.29 The Meissner–Ochsenfeld effect. If a superconducting material is cooled in a magnetic field, the lines of force are expelled from the superconductor.

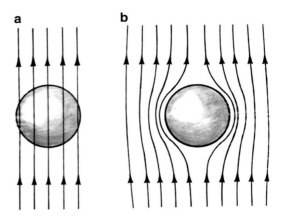

In 1986, K.A. Muller and J.G. Bednorz from IBM reported a superconducting effect at relatively high temperatures (40 K) in a compound of the form $La_{2-x} Sr_x CuO_4$, $x = 0.15$. Later, C.W. Chu found the effect at even higher temperatures ($T_c = 92$ K) in the compound $YBa_2Cu_3O_{7-\delta}$. These discoveries and the prospect of finding the superconducting effect at even higher temperatures promise revolutionary consequences for future technology.

All high-temperature superconductors are type II superconductors. They usually manifest themselves as superconductors at higher temperatures and magnetic fields than type I superconductors, and this allows them to conduct higher currents. Type II superconductors are usually made of metal alloys or complex oxide ceramics, and are mostly complex copper oxide ceramics. However, in 2008, Hideo Hosono and colleagues from the Tokyo Institute of Technology found lanthanum oxygen fluorine iron arsenide ($LaO_{1-x}F_xFeAs$) to be a superconductor with a high critical temperature of 26 K. The discovery inspired international research on iron-based superconductors and the critical temperature was raised to 55 K. The superconductive phase transition is usually of first order for type I, and of second order for type II.

3.14 Appendix of Formulas

In this chapter we use only CGS units. Let us recall some properties of vectors. We define the operators nabla $\nabla = \mathbf{i}\frac{\partial}{\partial x} + \mathbf{j}\frac{\partial}{\partial y} + \mathbf{k}\frac{\partial}{\partial z}$ and Laplacian $\Delta = \nabla \cdot \nabla = \frac{\partial^2}{\partial x^2} + \frac{\partial^2}{\partial y^2} + \frac{\partial^2}{\partial z^2}$. Since $\mathbf{r} = \mathbf{i}x + \mathbf{j}y + \mathbf{k}z$ and $r = \sqrt{x^2 + y^2 + z^2}$, the action of ∇ on $\mathbf{r}$ as a scalar product gives its divergence as a scalar, viz., $\nabla \cdot \mathbf{r} = 3$, whereas its action on the scalar r gives its gradient $\nabla r = (\mathbf{i}x + \mathbf{j}y + \mathbf{k}z)/r = \mathbf{r}/r = \mathbf{r}_0$. From this, whenever $r \neq 0$,

$$\Delta\left(\frac{1}{r}\right) = \nabla \cdot \nabla\left(\frac{1}{r}\right) = \nabla \cdot \left(\frac{-\mathbf{r}}{r^3}\right) \tag{3.40}$$

$$= -\frac{\nabla \cdot \mathbf{r}}{r^3} + 3\mathbf{r} \cdot \frac{\nabla r}{r^4} = -\frac{3}{r^3} + \frac{3}{r^3} = 0.$$

Usually, the Gauss law is written as $\nabla \cdot \mathbf{E} = 4\pi\rho$, where ρ is the charge density. Recall also the so-called *divergence theorem*, which states that the integral of the divergence of a vector $\mathbf{A}$ taken within a volume V bounded by a closed surface S is equal to its flux through the boundary surface:

$$\int_V \nabla \cdot \mathbf{A} \, d^3x = \oint_S \mathbf{A} \cdot d\mathbf{S}. \tag{3.41}$$

Thus, if $V(r) = e/r$ is the potential from which the electric field $\mathbf{E} = -\nabla V(r) = e\mathbf{r}/r^3$ is obtained, we have

$$\int \nabla \cdot \mathbf{E} \, dV = \oint \mathbf{E} \cdot d\mathbf{S}. \tag{3.42}$$

If we assume the charge e to be located at the coordinate origin, the flux of the electric field $\mathbf{E}$ through a spherical surface of radius R is

$$\oint \mathbf{E} \cdot d\mathbf{S} = 4\pi e.$$

However, due to (3.40), the integrand on the left-hand side of (3.42) vanishes at all points, except at the coordinate origin. We conclude that

$$\nabla \cdot \mathbf{E} = e\nabla \cdot (\mathbf{r}/r^3) = 4\pi e\delta(\mathbf{r}),$$

where $\delta(\mathbf{r})$ is called the Dirac delta function, which is not a function in the usual sense, but a functional. It is characterized by the properties

$$\delta(\mathbf{0}) = \infty$$

and for $\mathbf{r} \neq 0$, $\delta(\mathbf{r}) = 0$, while for an infinite volume V, if $g(\mathbf{r})$ is a continuous function, we have

$$\int_V g(\mathbf{r})\delta(\mathbf{r})d\mathbf{r} = g(0),$$

and in particular, for $g(\mathbf{r}) \equiv 1$, we have

$$\int_V \delta(\mathbf{r})d\mathbf{r} = 1.$$

These properties are not satisfied by any ordinary function. The Dirac delta function will be defined in another way in Chap. 4.

Problems

Problem 3.1 A sphere of radius R carries a charge Q distributed uniformly through-out its volume. Calculate (a) the charge density; (b) the electric field inside and outside the sphere. Use Gauss' theorem.

Problem 3.2 In a polarizable medium (as is shown in Fig. 3.20) a static charge q produces a screened electric field of the form

$$\mathbf{E} = \frac{q e^{-\lambda r}}{r^2} \mathbf{r}_0, \tag{3.43}$$

where $\mathbf{r}_0$ is the unit vector along r, and λ is a constant, namely the inverse of the Debye length. Find the charge density at every point of the medium.

Problem 3.3 A cosmic ray electron moves at a speed 2.5×10^{-2} times the speed of light ($c = 3 \times 10^{10}$ cm s^{-1}) in a direction perpendicular to the Earth's magnetic field and interacts with it at a height where $B = 0.1$ G. Calculate the radius of the circle described and the frequency of rotation.

Problem 3.4 A cylindrical wire of permeability μ carries a steady current I. If the radius of the wire is R, find the observable magnetic field inside and outside the wire.

Problem 3.5 What is the difference, if any, between the geometry of the electric field lines generated by an electric charge as in (3.2) and those generated by a magnetic field through Faraday's law?

Problem 3.6 A rectangular coil with N loops, length a, and width b rotates with frequency f in a magnetic field B. Show that a current arises in the coil that is driven by an electromotive force

$$\mathcal{E} = 2\pi N f b a B \sin(2\pi f t).$$

Literature

1. E. Purcell, *Electricity and Magnetism*, 2nd edn. (McGraw-Hill, New York, 1985). A clear and rigorous introductory course on electricity and magnetism
2. R.P. Feynman, *The Feynman Lectures on Physics*, vol. II. (Addison Wesley, Reading, Massachusetts, 1969). A classic textbook written in Feynman's enthusiastic and illuminating style
3. L.D. Landau, E.M. Lifshitz, *The Classical Theory of Fields* (Pergamon, Oxford, 1980). An excellent text on the classical theory of the electromagnetic field
4. R.K. Pathria, *Statistical Mechanics* (Butterworth-Heinemann, Oxford, 1996). The topics of first and second order phase transitions are clearly introduced in this book
5. C. Kittel, *Solid State Physics*, 8th edn. (Wiley, New York, 2005). Contains a very good introduction to ferromagnetism, diamagnetism, ferroelectricity, and superconductivity

6. J.D. Jackson, *Classical Electrodynamics*, 3rd edn. (Wiley, New York, 1998). An updated first-class treatise on classical electrodynamics
7. M. Chaichian, I. Merches, A. Tureanu, *Electrodynamics* (Springer, Berlin Heidelberg, 2014). This book is recommended to complement several topics dealt with in the present chapter
8. Y.-k. Lim (ed.), *Problems and Solutions on Electromagnetism* (World Scientific, 2005). Some problems from this excellent book were adapted and included

Chapter 4
Electromagnetic Waves

In the previous chapter we saw that an electric charge at rest creates a surrounding static field. A charge moving with constant velocity creates a constant electromagnetic field displacing with the charge. However, an accelerated charge produces a field of a different nature: a radiation field, which propagates far from its source and becomes independent of it. A charge moving at constant velocity is displaced with its field, but does not emit radiation. Charge acceleration is a necessary and sufficient condition for the generation of electromagnetic radiation. We shall see later that radiation behaves in dual form, as waves and as particles (photons), but this chapter will deal mainly with the wave behaviour of radiation.

Around an accelerated charge, one can identify a set of surfaces moving at the velocity of light, called wave fronts. In fact, each point of the wave front behaves as a generator of new waves. If we assume that the wave front is stopped by a screen with two perforated holes, the radiation passing through these holes behaves like two independent point sources which generate new waves.

It is interesting to mention the important case of the so-called dipole radiation. This is the case, e.g., of an oscillating charge whose velocity is small compared with the speed of light. If $\mathbf{p}$ is the dipole moment, the emitted power (energy per unit time) is given by the expression

$$\frac{dE}{dt} = \frac{2}{3c^3}\ddot{\mathbf{p}}^2, \tag{4.1}$$

where c is the speed of light and the dots over symbols denote time derivatives, so that $\ddot{\mathbf{p}}$ is the second derivative of $\mathbf{p}$ with respect to time.

Light is the most common form of electromagnetic radiation. Newton made profound investigations of the behaviour of light, and assumed that it was composed of corpuscles emitted by the bodies. On the other hand, Christiaan Huygens (1629–1695), Augustin-Jean Fresnel (1788–1827), Thomas Young (1773–1829), and Gustav Kirchhoff (1824–1887) were the creators of the wave hypothesis, because the phenomena of interference and diffraction that they studied are manifestations of wave motion.

© The Author(s), under exclusive license to Springer-Verlag GmbH, DE, part of Springer Nature 2021
M. Chaichian et al., *Basic Concepts in Physics*, Undergraduate Lecture Notes in Physics, https://doi.org/10.1007/978-3-662-62313-8_4

For a long time, however, because of Newton's prestige, the corpuscular hypothesis had wide acceptance among physicists, as compared with the wave hypothesis, in spite of the evidence of the phenomena which demonstrated the wave nature of light.

In the second half of the nineteenth century, the electromagnetic theory elaborated by Maxwell, and the experiments performed by Hertz, demonstrated the electromagnetic nature of light. However, the existence of the æther was accepted as the propagating medium for such waves. At the beginning of the twentieth century, as a consequence of the special theory of relativity, Einstein proved that the *luminiferous æther* hypothesis was not needed. In addition, his successful quantum explanation of the photoelectric effect meant a partial return to the corpuscular model, but keeping the wave model alive, in a dual way. We shall return to this point in Chap. 6, which is connected to the origin of quantum mechanics.

4.1   Waves in a Medium and in Æther

The idea of the æther to support electromagnetic waves propagating in vacuum originated from the analogy with the elastic wave propagation in a medium (solid, liquid, or gas).

The propagation of longitudinal waves is typical of gases, like, for example, sound propagating in air. Although the molecules move continuously in all directions, one could imagine wave propagation as a succession of compressions and dilations of the gas along the direction of wave motion.

Sound propagation is a regular and systematic motion in which the compressions and dilations are transmitted successively from one part of the gaseous medium to its immediate neighbour, without the individual molecules moving along the wave. This is different from the random motion of individual molecules with different velocities in all directions of space.

In solid bodies, in addition to longitudinal waves, transverse waves also propagate: the material points in the medium oscillate perpendicularly to the direction of propagation. These transverse waves propagate with a different velocity compared to the longitudinal ones (Fig. 4.1).

Transverse waves on the surface of a liquid can be observed when an object is thrown into it. The wave propagates, but each particle of the medium oscillates

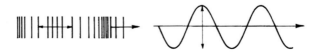

Fig. 4.1   Longitudinal and transverse waves. In the first case, the oscillatory motion is produced in the direction of propagation of the wave, and in the second, perpendicular to it.

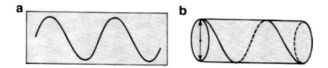

Fig. 4.2 Polarized transverse waves. The oscillations are produced either all in a plane that contains the direction of polarization, as represented in (**a**), or in different planes (the points oscillate in various directions in the plane perpendicular to the direction of propagation of the wave), as shown in (**b**).

perpendicularly to the direction of propagation of the wave, without any motion of the particles in the direction of the wave (Fig. 4.2).

In transverse waves, all the points of the medium can oscillate in the same plane, and the wave is said to be linearly polarized. However, it may also happen that the points of the medium oscillate, describing circles or ellipses in planes perpendicular to the direction of propagation. These waves are said to be *circularly* or *elliptically polarized*, respectively.

The analogy with the propagation of elastic waves motivated many nineteenth century physicists to postulate the existence of the æther, which was assumed to be an immaterial fluid filling the empty space between the bodies and serving as a medium to support the propagation of light. However, because light was composed of transverse waves, to eliminate the possibility of longitudinal waves, they had to assume that the æther had greater rigidity than steel (among other necessary properties).

4.2 Electromagnetic Waves and Maxwell's Equations

As mentioned above, the nature of light as an electromagnetic wave became evident after the theoretical work performed by Maxwell, and confirmed experimentally by Hertz.

A linearly polarized wave, for instance, is described by two oscillating electric and magnetic fields perpendicular to each other (Fig. 4.3). At each point where the wave propagates we can assume an electric vector that oscillates from positive to negative values, and another, magnetic vector, perpendicular to it, performing similar oscillations. In the case of a circularly or elliptically polarized wave, these vectors would describe circles or ellipses, respectively, in planes perpendicular to the plane of wave propagation (Fig. 4.4).

The frequency ν of the electromagnetic wave is the number of oscillations per second made by these vectors at each point. The angular frequency $\omega = 2\pi\nu$ is also widely used. There is a relation between the wavelength λ, the frequency ν, and the speed of light c, viz.,

$$\lambda = \frac{c}{\nu}. \tag{4.2}$$

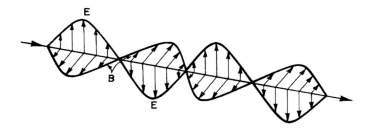

Fig. 4.3 In an electromagnetic wave, the electric and magnetic vectors oscillate perpendicularly to each other, and to the direction of propagation. The figure illustrates a linearly polarized wave.

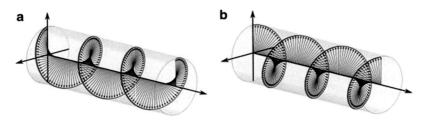

Fig. 4.4 The electric field vector of a circularly polarized electromagnetic wave: **a** right-handed polarization; **b** left-handed polarization.

The speed of light (and of all other electromagnetic waves) in vacuum is a universal constant, approximately equal to 300,000 km/s.

There is, then, an inverse relationship between wavelength and frequency. The unit of frequency (one cycle or oscillation per second) is the hertz, in honour of the German physicist Heinrich Hertz (1857–1894).

A radio wave of frequency 600 kHz (or 600,000 Hz) has the wavelength

$$\lambda = 500 \,\text{m}.$$

For a frequency of 20 MHz (20×10^6 Hz) the wavelength will be

$$\lambda = 15 \,\text{m}.$$

Visible light frequencies are considerably higher, of the order of hundreds of terahertz (1 terahertz (THz) = 10^{12} hertz). Red corresponds to about 400 THz and violet to 600 THz. Visible light wavelengths are around 4,000 Å for violet and around 7,500 Å for the extreme red, where one angstrom (Å) is 10^{-10} m or 10^{-8} cm.

The angstrom characterizes the dimensions of the atom very well. For instance, the diameter of a hydrogen atom (assumed spherical) is about one Å. Thus, the wavelengths of visible light are several thousand times greater than the atomic dimensions.

A typical wavelength of very soft X rays or the extreme ultraviolet (EUV) is about of 100 Å, corresponding to a frequency of 10^{16} Hz. Hard X rays have wavelengths of the order of 0.1 Å, corresponding to a frequency of 10^{19} Hz. Gamma rays with even higher frequencies and smaller wavelengths are also observed, for example, in electron–positron annihilation: $e^+ + e^- \rightarrow 2\gamma$.

4.2.1 Wave Propagation

In Chap. 3 we introduced Maxwell's equation in integral form. They can be written as well in differential form, using some theorems of calculus. Gauss's laws for the electric and magnetic fields read:

$$\nabla \cdot \mathbf{E} = 4\pi\rho, \tag{4.3}$$
$$\nabla \cdot \mathbf{B} = 0,$$

while the Maxwell–Faraday and Maxwell–Ampère equations are, respectively:

$$\nabla \times \mathbf{E} = -\frac{1}{c}\frac{\partial \mathbf{B}}{\partial t}, \tag{4.4}$$
$$\nabla \times \mathbf{B} = \frac{1}{c}\frac{\partial \mathbf{E}}{\partial t} + \frac{4\pi}{c}\mathbf{J},$$

with the notations of the Chap. 3. In the absence of currents and charges, the last two Maxwell equations, which contain time derivatives, can be re-written as wave equations for the fields $\mathbf{E}$ and $\mathbf{B}$. Let us assume that the fields $\mathbf{E}$ and $\mathbf{B}$ vary in space only on the x-axis. In this case, the wave equations deduced from (4.4) are

$$\frac{\partial^2 \mathbf{E}}{\partial x^2} = \frac{1}{c^2}\frac{\partial^2 \mathbf{E}}{\partial t^2}, \qquad \frac{\partial^2 \mathbf{B}}{\partial x^2} = \frac{1}{c^2}\frac{\partial^2 \mathbf{B}}{\partial t^2}. \tag{4.5}$$

These two equations can be obtained from a more general wave equation, involving the vector potential $\mathbf{A}$, and it is easily proved from its solutions that $\mathbf{E}$ and $\mathbf{B}$ are orthogonal, $\mathbf{E} = \mathbf{B} \times \mathbf{u_k}$, where $\mathbf{u_k}$ is the unit vector of the direction of propagation. The more general solution for $\mathbf{E}$ (for $\mathbf{B}$ we have a similar solution), is

$$\mathbf{E} = \mathbf{e}[f(x - ct) + g(x + ct)], \tag{4.6}$$

where $\mathbf{e}$ is a constant vector. (Notice that the equations (4.5) and the general solution (4.6) do not change in form if the sign of the coordinates x, y, z, or time t, is changed. The equations are said to be covariant under the parity transformation P and the time reversal transformation T, which will be detailed in Chap. 7.) If we consider $g = 0$ and $x > 0$, the resulting equation (4.6) describes a wave propagating from present (or past) to future. This is called *progressing* solution. Let us consider two points, x and $x + \Delta x$, and the times t and $t + \Delta t$ corresponding to two positions of the wavefront, and determine the relation between Δx and Δt, that is, how much should the wave advance so that the value of $\mathbf{e}f(x - ct)$ does not change. We have

$$(x + \Delta x) - c(t + \Delta t) = x - ct, \tag{4.7}$$

i.e., $\Delta x = c\Delta t$. If $f = 0$ and $g \neq 0$, and $x > 0$, we obtain a *regressing* solution, going backward in time, since the condition would now be

$$(x + \Delta x) + c(t + \Delta t) = x + ct, \tag{4.8}$$

or $\Delta x = -c\Delta t$. Obviously, as there is no evidence of such a wave, we must have $g = 0$. Therefore, we neglect the regressing wave by using an argument from everyday experience, which is very close to the notion of irreversibility discussed in Chap. 2. This means that we assume an *arrow of time* in choosing the physically appropriate solutions of the electromagnetic wave equation. In the next chapter, we shall see that, if we have a signal propagating with a speed faster than c, such a signal would be seen by some observers to travel to the past, violating causality.

As a simple example of a progressing wave, let us consider a sinusoidal wave

$$\mathbf{E}(x, t) = \mathbf{e} \sin(kx - \omega t). \tag{4.9}$$

This expression describes a wave propagating at the speed of light whenever the following relation is satisfied:

$$k = \frac{\omega}{c}. \tag{4.10}$$

The quantity $k = 2\pi/\lambda$ is equal to the number of wavelengths contained in 2π units of distance. This is called the *wave number*. To each value of λ there corresponds one value of k, and vice versa. For a wave propagating in three dimensions, one can define a wave vector, $\mathbf{k} = (2\pi/\lambda)\mathbf{u_k}$. The argument of the oscillatory function would then be $\mathbf{k} \cdot \mathbf{r} - \omega t$. However, for simplicity, in what follows we shall consider propagation in one dimension only.

4.2.2   Coherence

We now consider a source of light emitting monochromatic radiation, that is, radiation of a precise frequency. If at the same instant, two different points of the source with coordinates x_1 and x_2 emit radiation, the quantity $kx - \omega t$ will not be the same for the emitted waves. We define the phase difference to be the quantity

$$(kx_1 - \omega t) - (kx_2 - \omega t) = k(x_1 - x_2). \tag{4.11}$$

If $k(x_1 - x_2) = \Delta\varphi$, where $\Delta\varphi = $ const., there is stationary interference. If $\Delta\varphi = 0$ (or $2\pi n$, where $n = 1, 2, \ldots$), the two rays interfere constructively, leading to a stronger amplitude. If $\Delta\varphi = \pi(2n + 1)$, the two rays interfere destructively and the amplitudes cancel each other. Between these values, there are several possibilities for the stationary interference. If $k(x_1 - x_2) \neq$ const., the waves are said to be incoherent.

A phase difference could also occur when the rays are emitted at different instants, t_1 and t_2. In general, we have

$$(kx_1 - \omega t_1) - (kx_2 - \omega t_2) = \Delta\varphi, \tag{4.12}$$

and the condition of coherence is again $\Delta\varphi =$ const.

The coherence when $\Delta\varphi = 0$ is especially important. Most light sources produce incoherent radiation. However, at the beginning of the 1960s, the laser (Light Amplification by Stimulated Emission of Radiation) was invented as a result of the work of Charles Hard Townes (1915–2015) in the United States, and Nikolai Basov (1922–2001) and Alexandr Prokhorov (1916–2002) in Russia. They were awarded the 1964 Nobel Prize in Physics for this work.

The laser is a source of monochromatic coherent (constructive, $\Delta\varphi = 2\pi$) light. The basic principles of lasers can be understood from the laws of emission and absorption of radiation discovered by Albert Einstein in 1917. We shall return to the basic principles of lasers in Chap. 6.

4.3   Generation of Electromagnetic Waves

4.3.1   Retarded Potentials

If $\rho(\mathbf{r}')$ is the electric charge density at the point $\mathbf{r}'$ of coordinates (x', y', z') of a body at rest, it creates an electric potential at a point $\mathbf{r}$ of coordinates (x, y, z) of the form

$$\phi(\mathbf{r}) = \int_V \frac{\rho(\mathbf{r}')}{R} d\mathbf{r}',$$

where $R = |\mathbf{r} - \mathbf{r}'|$ and V is the volume of the charged body. The electric field is obtained as $\mathbf{E} = -\nabla\phi(\mathbf{r})$. By defining the Laplacian operator as $\Delta = \partial^2/\partial x^2 + \partial^2/\partial y^2 + \partial^2/\partial z^2$, we obtain

$$\nabla \cdot \mathbf{E} = -\Delta\phi = 4\pi\rho(\mathbf{r}),$$

which is the first Maxwell equation in differential form.

If the electric potential is generated by a time-dependent charge density $\rho(\mathbf{r}, t)$, the previous Maxwell equation becomes

$$\left(\Delta - \frac{1}{c^2}\frac{\partial^2}{\partial t^2}\right)\phi = -4\pi\rho(\mathbf{r}, t).$$

Its solution is obtained by integrating on the volume V the potential created at $\mathbf{r}$ by the density of charge $\rho = \rho(\mathbf{r}', t - \frac{R}{c})$. The result is the *retarded* potential $\phi(\mathbf{r}, t)$ propagating from past to the future, and a similar equation may be written for the vector potential $\mathbf{A}(\mathbf{r}, t)$ in terms of the current density $\mathbf{j}(\mathbf{r}', t - R/c)$:

$$\phi(\mathbf{r}, t) = \int_V \frac{\rho(\mathbf{r}', t - R/c)}{R} d\mathbf{r}', \tag{4.13}$$

$$\mathbf{A}(\mathbf{r}, t) = \int_V \frac{\mathbf{j}(\mathbf{r}', t - R/c)}{R} d\mathbf{r}'. \tag{4.14}$$

Here, $t' < t$ and $t - t' = \frac{R}{c}$ is precisely the time during which the electromagnetic field spreads from point $\mathbf{r}'$ to point $\mathbf{r}$. Besides retarded solutions, the Maxwell equations possess unphysical *advanced* potential solutions evolving backward in time with $t' > t$, or $t' - t = \frac{R}{c}$, determined by taking $\rho = \rho(\mathbf{r}', t + \frac{R}{c})$. The advanced solutions are discarded, as we explained previously.

Equations (4.13) and (4.14) can be used to find the potentials and the corresponding electric and magnetic fields created at a point $(\mathbf{r}, t)$ by a single charge moving arbitrarily. The expressions for $\mathbf{E}$ and $\mathbf{B}$ consist of two terms: at large distances the first one is proportional to the charge velocity and varies as $1/R^2$, while the second, proportional to the charge acceleration, varies as $1/R$. This last term is related to the radiation of electromagnetic waves by the charge. In the *wave zone*, far from external time-dependent charges and currents, and restricted to some finite volume, the electromagnetic field behaves as plane electromagnetic waves.

4.3.2 Mechanisms Generating Electromagnetic Waves

Radio waves are generated by electronic means using a special amplifier circuit and an antenna. The conducting electron gas of the metallic antenna oscillates at the frequency of an oscillating field generated by an electronic device. As a consequence, it emits radiation waves which propagate through the atmosphere.

The atmosphere is not necessary for the propagation of these waves because, like light, they can propagate in vacuum. Radio waves can be reflected by the ionosphere containing charged ions, and this property is sometimes utilized for radio transmissions between very distant points. A vacuum tube device called a magnetron is used to generate microwaves. Magnetron tubes produce electron oscillations by means of a combination of electric and magnetic fields.

Everybody knows that visible light can be generated in several ways: by means of chemical reactions as in combustion, by heating a body as in a filament lamp, etc.

One can also produce light by accelerating electrons in very intense magnetic fields, and in nuclear fission and fusion, etc.

X rays can be generated in collisions between a high energy electron beam and a metallic target. Gamma rays are produced in quantum transitions within atomic nuclei, or by accelerating electrons to very high energies, and also in collisions and disintegrations of elementary particles, etc.

A continuum background of 'breaking radiation', or *bremsstrahlung*, that can reach frequencies of 10^{24} Hz is produced in particle accelerators. This corresponds to wavelengths of the order of 10^{-14} cm. So, the electromagnetic spectrum extends from wavelengths of thousands of kilometers to the short X or gamma rays that are 10^{-23} times smaller; or considered from the opposite point of view, from frequencies 30 Hz (VLFW, Very Low Frequency Waves) in certain radio waves, to over 10^{24} Hz for X and gamma radiation produced in some particle accelerators.

However, very long waves may exist, say of wavelength 1,000 km, or frequency lower 30 Hz, and the low frequency limit gives constant fields. For example, let us assume a wave of frequency equal to $1/3,600$ Hz (which makes one oscillation per hour). To a good approximation, such a field can be considered constant over an interval of a few seconds. And similarly, any constant field can be considered as a low frequency wave, in the limit as the frequency tends to zero.

There are also gamma rays of frequency higher than 10^{24} Hz. On this scale, visible light covers a narrow bandwidth in the region from 10^{14} to 10^{15} Hz.

4.4 Wave Properties

4.4.1 Interference

We have already mentioned the phenomenon of interference when we discussed coherence. We return to this phenomenon by considering two waves propagating in the same direction in a medium. The effects of the two oscillatory motions add up, and the two extreme situations are depicted in Fig. 4.5. In the case **a** the resulting wave oscillates with greater amplitude since its motion is produced by the sum of the two oscillations, which are said to be *in phase* (there is no phase difference between them). In such a situation we say that there is *constructive interference*. In the case **b**, on the other hand, the points do not oscillate at all, being under the action of opposite effects (the waves are *out of phase*). The whole oscillatory motion vanishes, as there is *destructive interference*.

Let us examine the case of two waves, 1 and 2, oscillating in the same plane, but forming an angle θ between their directions of propagation, as depicted in Fig. 4.6. If the two waves are in phase at some point P, at the point Q along the direction of wave 2, the phase difference between the two waves will be $\Delta\varphi = 2\pi x/\lambda$, where d is the distance between the points P and Q, $x = d \sin\theta$, and λ is the wavelength. The waves are in phase (interfere constructively) whenever

Fig. 4.5 Interference
between two waves,
constructive in (**a**) and
destructive in (**b**).

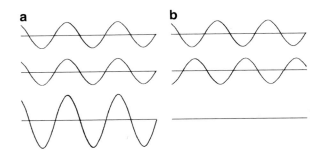

Fig. 4.6 For two parallel
coherent beams, forming an
angle θ, if they are in phase at
point P the phase difference
at point Q, at a distance d
from P, is $\Delta\phi = \frac{2\pi}{\lambda}d\sin\theta$.

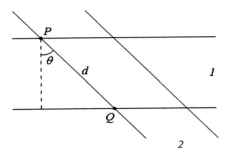

$$d\sin\theta = \pm n\lambda, \quad n = 0, 1, 2...,$$

and are out of phase (interfere destructively) when

$$d\sin\theta = \pm(2n+1)\lambda/2.$$

In this case, the phase difference is due to a difference in the path length.

By using the Euler formula $e^{ix} = \cos x + i\sin x$, it is useful to express a wave as the real part of a complex exponential, i.e $A\cos(kx - \omega t) = A\,\mathrm{Re}\,e^{i(kx-\omega t)}$. Thus, we may work with complex exponential expressions, with the understanding that at the end, we take the real part of the resulting complex number. This will be useful also when we speak about the intensity of the wave.

Let us consider the case of interference of two light waves in space, described by their electric field components:

$$\mathbf{E}_1(\mathbf{r}, t) = \mathbf{e}_1(\mathbf{r})e^{i[\varphi_1(\mathbf{r})-\omega t]}$$

and

$$\mathbf{E}_2(\mathbf{r}, t) = \mathbf{e}_2(\mathbf{r})e^{i[\varphi_2(\mathbf{r})-\omega t]}.$$

Both waves, $\mathbf{E}_1(\mathbf{r}, t)$ and $\mathbf{E}_2(\mathbf{r}, t)$, are assumed to have the same polarization. This means that $\mathbf{e}_1$ and $\mathbf{e}_2$ are parallel vectors. Their common direction is defined by a unit vector $\mathbf{u}$, which gives the direction of the *linear polarization* (see Sect. 4.4.3).

Fig. 4.7 Young's double-slit experiment.

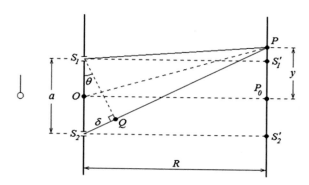

The field resulting from the superposition of the two waves is

$$\mathbf{E}(\mathbf{r}, t) = \mathbf{E}_1(\mathbf{r}, t) + \mathbf{E}_2(\mathbf{r}, t). \tag{4.15}$$

The intensity I is defined as the energy flux density, which is proportional to the square of the field, E^2. Thus, the intensity of light at a point $\mathbf{r}$ is given by the square of the modulus of the complex field $\mathbf{E}(\mathbf{r}, t)$:

$$I(\mathbf{r}) = \frac{c}{4\pi} \mathbf{E}(\mathbf{r}, t) \mathbf{E}^*(\mathbf{r}, t) = I_1(\mathbf{r}) + I_2(\mathbf{r}) + 2\sqrt{I_1(\mathbf{r}) I_2(\mathbf{r})} \cos(\varphi_1(\mathbf{r}) - \varphi_2(\mathbf{r})), \tag{4.16}$$

where $I_1(\mathbf{r}) = \frac{c}{4\pi} \mathbf{e}_1^2(\mathbf{r})$ and $I_2(\mathbf{r}) = \frac{c}{4\pi} \mathbf{e}_2^2(\mathbf{r})$. We have maximal interference when

$$\Delta\varphi = \varphi_1(\mathbf{r}) - \varphi_2(\mathbf{r}) = \pm 2\pi n, \tag{4.17}$$

where n is a natural number. For $I_1 = I_2$, the maximum intensity is $I = 4I_1$. The minima result when

$$\Delta\varphi = \pm(2n + 1)\pi, \tag{4.18}$$

and for $I_1 = I_2$ the minima have zero intensity.

A famous experiment by Thomas Young (1773–1829) was crucial for the acceptance of light as a wave, in the beginning of the nineteenth century. To this day, *Young's double-slit experiment* is the prototype for interference observation of any waves. The setup is as in Fig. 4.7. Let us consider, along with Young, a source of monochromatic light, such that the beam of light falls on two slits, S_1 and S_2, separated by a distance a. The resulting beams on the other side of the pierced screen, being separated from a common beam, are thus coherent, which is essential for interference to take place. Another screen is placed at a distance R from the first one, such that $R \gg a$, and is used for the visualization of the interference pattern. Let us consider an arbitrary point P on the latter screen and calculate the intensity of light at that point. According to (4.16), where we take the two beams to have the same amplitude $\mathbf{e}_1 = \mathbf{e}_2 = \mathbf{e}$, the result will be:

$$I(P) = 4I_0 \cos^2 \frac{\Delta\varphi(P)}{2},$$ (4.19)

where I_0 is the intensity at the centre of the visualization screen C and $\Delta\varphi(P)$ is the difference of phase at the point P between the two beams coming from the slits S_1 and S_2. Since these beams are split from a single one, the only source for the difference of phase is the path difference, which can be easily calculated. We denote by y the distance between the point P and the centre of the screen P_0. The path difference in the notation of Fig. 4.7 is S_2Q. Remark that for $R \gg a$, the angle $\widehat{S_1QS_2}$ approaches a right angle, and $\widehat{S_2S_1Q} = \widehat{P_0OP} = \theta$. Then the path difference $\delta = S_2Q$ turns out to be, in our approximation:

$$\delta = a \sin\theta \approx a \tan\theta \approx a\frac{y}{R}.$$ (4.20)

Recall the formulas (4.17) and (4.18) for the formation of a maximum or a minimum of intensity, respectively. The phase difference between the two beams in the double-slit setup is

$$\Delta\varphi = \frac{2\pi}{\lambda}\delta,$$ (4.21)

which, plugged into (4.17) and (4.18), tells us that if the path difference is an integer multiple of the wavelength, $\delta = \pm n\lambda$, the interference is constructive, while if $\delta = \pm(n + 1/2)\lambda$, the interference is destructive. This conditions, together with (4.20), give the location y on the screen of the interference maxima and minima.

The interference pattern which appears on the visualization screen, not far from the optical axis, will be formed of bright fringes alternating with dark fringes. At the point P_0, where the difference of path is zero, appears of course a bright fringe.

Remark that the interference pattern depends on the wavelength of the light. If one uses white light, the maxima for different colours will be formed at different points, and thus the pattern will show all the colours of the spectrum. At the centre there will be a bright fringe for all colours, i.e. a white spot, but the next red bright fringe will be further from the centre than the corresponding fringe for violet, with all the other colours in between them (Fig. 4.8).

4.4.2 Diffraction

Interference and diffraction are basically the same phenomenon, but it is commonly understood that interference involves few sources of light, while diffraction involves a large number of them. Diffraction is produced when light passes near the edges of some object, for instance, slits in an opaque screen. Due to diffraction, the region of geometrical shadow is not dark, but covered by diffraction fringes.

There are two main approximations to this complex phenomenon: the Fraunhofer approximation, which applies to the field far from the source (border or slit) and the

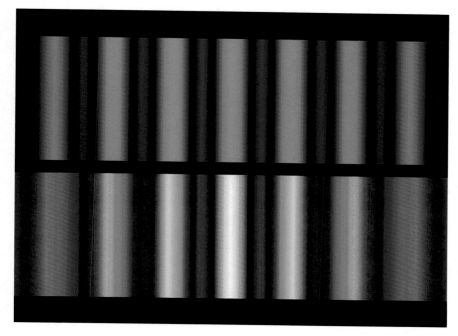

Fig. 4.8 Interference pattern in the double-slit experiment with green light and with white light.

Fresnel approximation, which applies to the field near the source. We shall discuss here the Fraunhofer case, and start with the problem of the diffraction through a rectangular slit.

Let us consider a parallel beam of monochromatic waves, passing perpendicularly to a rectangular slit on a screen, whose centre O is taken as the origin of a system of coordinates $Oxyz$. The beam propagates along x. The slit has width D along the y axis and length G along the z axis. A section through the slit is presented in Fig. 4.9. According to *Huygens' principle*, every point of a light front becomes secondary source, producing spherical waves. Thus, the field E_P created at the point P of a second screen parallel to the first one, located at a distance $R \gg D, G$ from it, is given by the superposition of all the secondary waves produced at each point of the slit. The problem becomes an interference one, in which we consider each point of the extended rectangular slit as a secondary source—like a Young experiment with an infinity of slits!

Let us start with the vector $\mathbf{E} = \mathbf{e}_0 \, e^{i(\omega t - \mathbf{k} \cdot \mathbf{r})}$, where $\mathbf{e}_0$ is the source amplitude, which is considered to be the same for all the secondary sources (the incident beam is parallel). For clarity, let us consider for the moment that we observe the diffracted light in the vertical plane, at an angle θ, i.e. at the points $P(R, y, 0)$, such that $\sin \theta = y/R$. Taking as reference the path which passes through the centre of the slit O, all the other paths will be shorter or longer than it by a distance $\delta(y') = y' \sin \theta = y' \frac{y}{R}$,

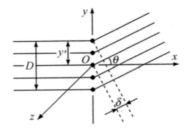

Fig. 4.9 Path differences in the single-slit diffraction. The point $P(R, y, 0)$ would be far at the right, measured at the second screen located at a distance R from the first (not represented on this figure). The beam forming the angle θ with the x axis looks as made of parallel rays at short distances from the slit, but it converges at P.

where y' is the coordinate of the secondary source (Fig. 4.9). The diffracted field at the point P is obtained by the summation of all these contributions:

$$E_P(R, x, y) = e_0 e^{i(\omega t - \varphi)} \int_{-\frac{D}{2}}^{\frac{D}{2}} \int_{-\frac{G}{2}}^{\frac{G}{2}} e^{ik(yy' + zz')/R} dy' dz'. \tag{4.22}$$

By defining the angles $\alpha(y) = kyD/2R$ and $\beta(z) = kzG/2R$, we easily obtain

$$E_P(R, x, y) = ADG \frac{\sin \alpha}{\alpha} \frac{\sin \beta}{\beta} e^{i(\omega t - \varphi)}, \tag{4.23}$$

where A is a constant whose expression can be straightforwardly found.

For the points on the x-axis, $\alpha = \beta = 0$. In the limit $\alpha \to 0$ we have $\sin \alpha/\alpha \to 1$ and similarly for β. Thus, at the point P_0 of the screen the magnitude of the field will be $E(P_0) = ADG e^{i(\omega t - \varphi)}$ and the corresponding intensity of light I_0 is proportional to $(ADG)^2$.

The light intensity at an arbitrary point P can be written in terms of I_0 as:

$$I_P = I_0 \frac{\sin^2 \alpha}{\alpha^2} \frac{\sin^2 \beta}{\beta^2}, \tag{4.24}$$

and the diffraction pattern will be given by the maxima and minima of the function

$$F(\alpha) = \frac{\sin^2 \alpha}{\alpha^2}.$$

As $F(0) = 1$, while for any other values of α holds the relation $\sin \alpha < \alpha$, we conclude that for $\alpha = 0$ there is an absolute maximum, and this occurs when $y = 0$. As $F(\alpha) \geq 0$, the minima are expected to be located at points where $\sin \alpha = 0$, that is, whenever $\alpha = \pm n\pi$, for $n = 1, 2....$ Other extrema are obtained by taking the derivative:

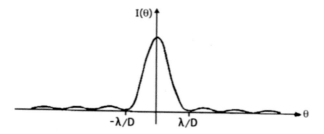

Fig. 4.10 The intensity of light plotted against the angle θ in the diffraction pattern. The characteristic width is approximately $2\lambda/D$.

$$\frac{dF}{d\alpha} = 2\frac{\sin \alpha}{\alpha}\frac{\alpha \cos \alpha - \sin \alpha}{\alpha^2} = 0. \tag{4.25}$$

Thus, there are other extrema for $\tan \alpha = \alpha$. We conclude that the second maxima are located at points with $|\alpha| < 3\pi/2$. Notice that

$$F\left(\frac{3\pi}{2}\right) \simeq \frac{4}{9\pi^2} \sim 0.045.$$

The second maxima of intensity are near five per cent of the principal one. Other maxima are even smaller.

In the case $D \ll G$, the slit behaves as infinitely long along the z axis. Thus $\beta = 0$ and $\sin \beta/\beta = 1$. The problem becomes two-dimensional, since only the x and y axes are involved. The field reduces to

$$E_P(R, x, y) = ADG\frac{\sin \alpha}{\alpha}e^{i(\omega t - \varphi)}, \tag{4.26}$$

and the intensity is $I = I_0 \sin^2 \alpha/\alpha^2$. The problem is reduced to the xy-plane and we may write $\alpha = \pi D \sin \theta/\lambda$, the angle θ having the same significance as above. For small θ we have $\alpha \sim 2\pi D\theta/\lambda$. The diffraction pattern intensity is depicted in Fig. 4.10.

By combining the previous results on interference and diffraction, for monochromatic light, it can be shown that in the two-slit interference the intensity on the second screen can be written in a more complete form as

$$I(\theta) = 4I_0^2\frac{\sin^2 \alpha}{\alpha^2}\cos^2\left(\frac{\pi d \sin \theta}{\lambda}\right). \tag{4.27}$$

The resulting curve is depicted at the centre of Fig. 4.11.

If one looks at a light source through the slit formed by joining two adjacent fingers, fringes can be observed. These result from the diffraction phenomenon. The intensity of the diffraction pattern has the form indicated in Fig. 4.10.

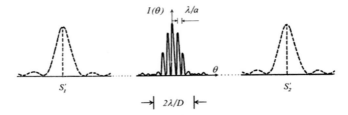

Fig. 4.11 The intensities of the diffraction patterns that would appear on the second screen when S_1 or S_2 are opened alternatively is represented by dotted lines. When both slits are opened at the same time, only the pattern represented at the centre (not drawn to scale) is produced. The variation of the intensity is indicated with respect to the angle θ. The angular separation between the interference fringes is of the order of λ/a, and the characteristic width of the main pattern is $2\lambda/D$.

Fig. 4.12 Diffraction of sound waves from the voice of a person A approaching a corner. The voice is heard by another person R that cannot see him.

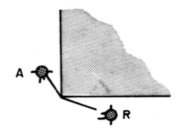

The interference and diffraction phenomena occur not only for electromagnetic waves, but for any type of wave. For instance, for sound waves, whose wavelength is of the order of one meter, the diffraction occurs at the boundary of any common object. Thus, if a person A approaches a street corner and speaks (Fig. 4.12), his voice can be heard by another person R, even though R is coming from an orthogonal direction and is not visible to A. The sound waves from A's voice have been diffracted through a significant angle and arrive at R with enough intensity to be perceived. This does not happen for light waves, whose diffraction angle would be extremely small.

4.4.3   Polarization

We have mentioned already the polarization in earlier paragraphs, as a constant vector characterizing the oscillation of the waves in a certain direction. In this section we shall explain in more detail this property of electromagnetic waves.

Let us define the electric vector of a plane wave as

$$\mathbf{E} = \mathrm{Re}(\mathbf{e}\,e^{i\varphi}). \tag{4.28}$$

If $\mathbf{e} = \mathbf{e}_1 + i\mathbf{e}_2$ is a complex vector and we demand that its square $\mathbf{e}^2$ be real, we must have $\mathbf{e}^2 = \mathbf{e}_1^2 - \mathbf{e}_2^2 + 2i\mathbf{e}_1 \cdot \mathbf{e}_2$ such that $\mathbf{e}_1 \cdot \mathbf{e}_2 = 0$, that is $\mathbf{e}_1$ and $\mathbf{e}_2$ are perpendicular.

Fig. 4.13 Two circular polarizations taken in opposite directions can generate a linear polarization as illustrated in the figure. The horizontal components cancel each other and the vertical ones are added. By a similar procedure, an elliptic polarization can be generated.

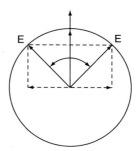

We assume that $\mathbf{e}_1$ is along the x axis and $\mathbf{e}_2$ is along the y axis. Obviously,

$$E_x = e_1 \cos \varphi, \quad E_y = \pm e_2 \sin \varphi. \tag{4.29}$$

The sign in front of e_2 depends either on the direction chosen for e_2, or on the phase φ. We may assume the phase $\varphi = \omega t - k_z z$ as corresponding to a wave propagating along the z-axis. Thus, (4.29) indicates that in the xy-plane the polarization vector describes an ellipse of semiaxes e_1 and e_2, orthogonal to the propagation axis, since we have

$$\frac{E_x^2}{e_1^2} + \frac{E_y^2}{e_2^2} = 1. \tag{4.30}$$

The wave is said to be *elliptically polarized*. For $e_1 = e_2$ we get a circle, and it is said that the wave is *circularly polarized*. The ratio of the components of the original complex amplitude of the wave is $\frac{E_x}{E_y} = \pm i$ for, respectively, positive (counterclockwise) and negative (clockwise) rotations. If either $\mathbf{e}_1$ or $\mathbf{e}_2$ is zero, we have a *linearly polarized*, or plane polarized wave.

We may define the circular polarization unit vectors as

$$\mathbf{e}^{\pm} = \frac{1}{\sqrt{2}}(\mathbf{e}_1 \mp i\mathbf{e}_2). \tag{4.31}$$

Then $\mathrm{Re}(\mathbf{e}^+ e^{i\varphi}) = \frac{1}{\sqrt{2}}(\mathbf{e}_1 \cos \phi + \mathbf{e}_2 \sin \varphi)$. We observe that it represents a wave with positive circular polarization. The sum of two opposite circular polarizations $\mathbf{e}^+ + \mathbf{e}^-$ gives a linear polarization $\sqrt{2}\mathbf{e}_1$. They can be also linearly combined to give elliptical polarizations (see Fig. 4.13).

There exist so-called optically active media, in which the left- and right-hand circularly polarized light have different speeds of propagation, consequently different indices of refraction, n_L and n_R. This phenomenon is known as circular birefringence. Such media are said to be chiral, which means that the molecules cannot be superposed on their mirror images by a combination of rotations and translations. In other words, the parity symmetry is broken in the medium. Suppose that a beam of monochromatic linearly polarized light is directed on such a medium.

The linear polarization is the superposition of the two circular polarizations, which propagate with different speeds, but have the same frequency ω. As a result, the left- and right-hand polarizations will have different wave lengths, namely $\lambda_{L,R} = \frac{\lambda}{n_{L,R}}$, where $\lambda = c/\omega$ is the wavelength in vacuum. Due to the difference in speed, there appears a phase difference between the two circular polarizations $\Delta\phi = 2\pi L \frac{\Delta n}{\lambda}$, where $\Delta n = n_L - n_R$ and L is the width of optically active material traversed by the light. As a result, the superposition of the two circular polarization remains linearly polarized, but the plane of polarization is rotated by the amount $\Delta\phi/2$.

Optical activity can be also induced in certain media by the application of a static magnetic field **B** and then the phenomenon is called *Faraday effect*. It occurs in a variety of media, like condensed matter, plasma, and in interstellar space. The angle of rotation of the polarization plane β is proportional to the projection of the magnetic field on the direction of propagation of the electromagnetic wave in that medium, namely $\beta = \nu B L$, where L is the width of the region where the magnetic field and the wave interact and ν is the *Verdet constant*. If the Verdet constant is positive, it corresponds to an R-rotation (counterclockwise) if the wave propagation is parallel to the magnetic field, and to an L-rotation (clockwise) when the direction of propagation is anti-parallel. In a plasma, β is proportional to the charge density and changes its sign if the sign of the charge is changed.

In addition to the transverse (linear, circular, and elliptical) polarizations, in a medium composed of electrically charged particles, as a plasma, there can be also longitudinally polarized waves (their polarization vector is parallel to the wave vector **k**). Longitudinal waves are purely electrostatic. In vacuum, only transverse waves propagate.

4.4.4 Spectral Composition

To each colour there corresponds a frequency, and in the visible region, to each frequency there corresponds a colour. Both the visible and the invisible radiation received from the Sun are mixtures of many frequencies.

From the mathematical point of view, a fixed frequency (monochromatic) wave extends from the past infinity to the future infinity (Fig. 4.14). If we plot the amplitude versus frequency, the spectrum of a monochromatic wave is a point (Fig. 4.15). A wave with a constant frequency ω_0 between the instants t_1 and t_2, and zero before and after, can be written as a superposition of many waves with different frequencies (Fig. 4.16). Between $-\infty$ and t_1 and between t_2 and $+\infty$, the waves interfere destructively, but between t_1 and t_2 the set of waves of different frequencies is arranged in such a way that only the frequency ω_0 survives (Fig. 4.17).

A pulse like the one depicted in Fig. 4.18 can be described as a superposition of an infinite number of waves of different frequencies. The spectrum is shown in Fig. 4.19, and the shorter the pulse, the broader the spectrum curve.

Fig. 4.14 A wave of fixed frequency (monochromatic) extends in time from $-\infty$ to $+\infty$.

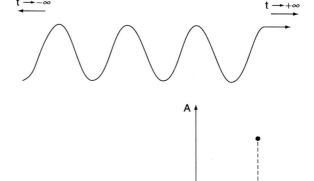

Fig. 4.15 Spectrum of a monochromatic wave.

Fig. 4.16 Discrete spectrum corresponding to a wave of finite duration of the type illustrated in the Fig. 4.17.

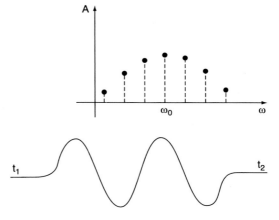

Fig. 4.17 A wave of constant frequency and finite duration between the instants t_1 and t_2.

One can establish that, if $\Delta t = t_2 - t_1$ is the duration of the pulse and $\Delta \omega$ is the spectral width or the frequency interval required to represent the pulse as a superposition of waves, the following relation is satisfied:

$$\Delta \omega \Delta t \approx 1. \tag{4.32}$$

For a fixed frequency, the bandwidth is $\Delta \omega = 0$. Then, since $\Delta \omega \approx 1/\Delta t$, in order to have $\Delta \omega = 0$, we must have $\Delta t = \infty$. Analogously, if the pulse is extremely short $\Delta t \approx 0$, its representation requires an infinite bandwidth.

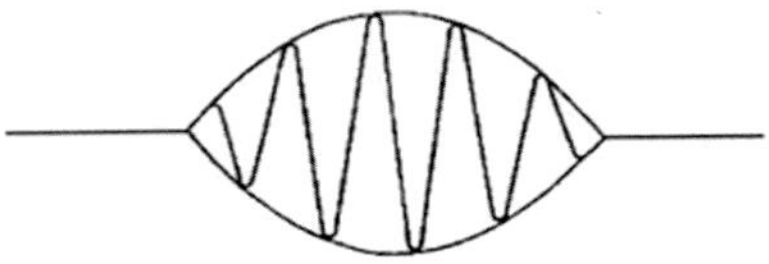

Fig. 4.18 Pulse of radiation.

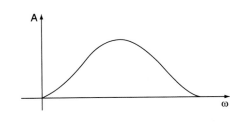

Fig. 4.19 Continuous
spectrum of frequencies
corresponding to the pulse of
Fig. 4.18.

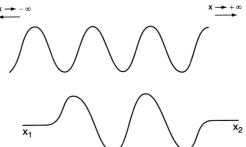

Fig. 4.20 A pure wave of
wavelength λ propagating in
space extends from $-\infty$ to
$+\infty$.

Fig. 4.21 A wave in space
extending from x_2 to x_1.

Similar reasoning can be made for a wave in space. Mathematically, an oscillation
of wavelength λ is assumed to extend from $-\infty$ to $+\infty$ (Fig. 4.20). If the wave
extends only between x_1 and x_2, one can represent it as a superposition of a large
number of waves of different wavelengths (Fig. 4.21). Then, in full analogy with the
relation (4.32), one could write

$$\Delta k \, \Delta x \approx 1. \tag{4.33}$$

That is, a wave of fixed wavelength λ has $\Delta k = 0$, and it is obtained by taking
$\Delta x \to \infty$ in the relation $\Delta k \approx 1/\Delta x$. Conversely, a very short pulse of extension
$\Delta x \approx 0$ requires a very large interval Δk.

In Chap. 5, we shall see that the set (x, y, z, ict) are spacetime coordinates in
the Minkowski four-dimensional space. In addition, the set $(p_x, p_y, p_z, iE/c)$ are the
components of a four-momentum vector, as are $\hbar(k_x, k_y, k_z, i\omega/c)$. But momenta
and coordinates are conjugate dynamical variables. Thus, (4.32) and (4.33) suggest
that analogous relations can be expected to hold in a form of wave mechanics, such
as quantum mechanics. The analogous relations to (4.32) and (4.33) lead to the
Heisenberg uncertainty principle, as we shall see in Chap. 6.

4.5 Fourier Series and Integrals

Most of the previous results can be understood more clearly mathematically by means of the theory of Fourier series and Fourier integrals.

Given an arbitrary function $f(x)$ defined on some interval $(-l \leq x \leq l)$, one can write it as a series of sine and cosine functions:

$$f(x) = \frac{b_0}{2} + \sum_{n=1}^{\infty} a_n \sin \frac{n\pi x}{l} + \sum_{n=1}^{\infty} b_n \cos \frac{n\pi x}{l}, \tag{4.34}$$

where

$$b_0 = \frac{1}{l} \int_{-l}^{l} f(\xi) d\xi,$$

$$a_n = \frac{1}{l} \int_{-l}^{l} f(\xi) \sin \frac{n\pi \xi}{l} d\xi, \quad b_n = \frac{1}{l} \int_{-l}^{l} f(\xi) \cos \frac{n\pi \xi}{l} d\xi.$$

In (4.34), we have an expansion of $f(x)$ in Fourier series on the given interval. In the interval $(l \leq x \leq 3l)$, the series (4.34) reproduces the value of the function $f(x)$ in $(-l \leq x \leq l)$, and so on. That is, (4.34) represents a periodic function. If we use the exponential representation of sine and cosine by means of the Euler formula

$$e^{i\theta} = \cos\theta + i\sin\theta,$$

we can write the Fourier series as

$$f(x) = \sum_{n=-\infty}^{\infty} c_n e^{in\pi x/l}, \quad c_n = \frac{1}{2l} \int_{-l}^{l} f(\xi) e^{-in\pi \xi/l} d\xi. \tag{4.35}$$

If $l \to \infty$, the sum (4.35) converges to an integral. It is customary to write the function $f(x)$ and its *Fourier transform* $\tilde{f}(k)$ as

$$f(x) = \frac{1}{\sqrt{2\pi}} \int_{-\infty}^{\infty} \tilde{f}(k) e^{ikx} dk, \tag{4.36}$$

$$\tilde{f}(k) = \frac{1}{\sqrt{2\pi}} \int_{-\infty}^{\infty} f(\xi) e^{-ik\xi} d\xi.$$

From here it is easy to understand that, if $f(t)$ represents a radiation pulse at a given instant, its spectral composition will be given by its Fourier transform:

$$\tilde{f}(\omega) = \frac{1}{\sqrt{2\pi}} \int_{-\infty}^{\infty} f(t) e^{i\omega t} dt. \tag{4.37}$$

Similarly, if we simultaneously consider the position and time, the wave can be written in terms of its Fourier transform $\tilde{f}(k, \omega)$ as

$$f(x, t) = \frac{1}{2\pi} \int_{-\infty}^{\infty} \int_{-\infty}^{\infty} \tilde{f}(k, \omega) e^{i(kx-\omega t)} dk d\omega. \tag{4.38}$$

Suppose now that $\tilde{f}(k, \omega) = 1$, i.e., the spectrum is constant: all values of wave numbers and frequencies contribute the same weight. It follows that

$$f(x, t) = \frac{1}{2\pi} \int_{-\infty}^{\infty} \int_{-\infty}^{\infty} e^{i(kx-\omega t)} dk d\omega = 2\pi \delta(x) \delta(t), \tag{4.39}$$

where the function

$$\delta(z) = \frac{1}{2\pi} \int_{-\infty}^{\infty} e^{ikz} dk$$

is the so-called Dirac delta function, which satisfies

$$\int_{-\infty}^{\infty} \delta(z) dz = 1, \quad \int_{-\infty}^{\infty} \delta(z - z_0) g(z) dz = g(z_0).$$

Thus, (4.39) represents a pulse of infinite amplitude in space and time, at the point x and at the instant t.

In contrast, a spectral density $\tilde{f}(k, \omega) = \delta(k - k_0)\delta(\omega - \omega_0)$ would give a plane wave with wave number k_0 and frequency $\omega_0, f(x, t) = \frac{1}{2\pi} e^{i(k_0 x - \omega_0 t)}$.

In three dimensions, the delta function is easily defined if we replace kx by $\mathbf{k} \cdot \mathbf{r} = k_x x + k_y y + k_z z$, and integrate over k_x, k_y, and k_z:

$$\delta(\mathbf{r}) = \frac{1}{(2\pi)^3} \int_{-\infty}^{\infty} e^{i\mathbf{k}\cdot\mathbf{r}} dk_x dk_y dk_z.$$

Let us consider the (spatial) Fourier transform of the Coulomb field created by a point charge e located at the centre of coordinates. This is a three-dimensional transform and the procedure is similar to the one-dimensional case. The charge density is $\rho = e\delta(\mathbf{r})$, since the charge is precisely localized, and the scalar potential ϕ satisfies the Poisson equation

$$\Delta\phi(\mathbf{r}) = -4\pi\rho(\mathbf{r}), \tag{4.40}$$

which can be deduced from Gauss's law, $\nabla \cdot \mathbf{E}(\mathbf{r}) = 4\pi\rho(\mathbf{r})$.

Let us write ϕ in terms of its Fourier transform $\tilde{\phi}(k)$. We have:

$$\phi(\mathbf{r}) = \frac{1}{(2\pi)^{3/2}} \int_{-\infty}^{\infty} e^{i\mathbf{k}\cdot\mathbf{r}} \tilde{\phi}(k) d\mathbf{k}, \tag{4.41}$$

where $d\mathbf{k} = dk_x dk_y dk_z$. If we apply the Laplacian operator (divergence of the gradient, $\Delta = \nabla \cdot \nabla$) to both sides of (4.41), we obtain

$$\Delta\phi(\mathbf{r}) = -\frac{1}{(2\pi)^{3/2}} \int_{-\infty}^{\infty} k^2 e^{i\mathbf{k}\cdot\mathbf{r}} \tilde{\phi}(k) d\mathbf{k}. \tag{4.42}$$

From here we deduce that $\widetilde{\Delta\phi}(k) = -k^2 \tilde{\phi}(k)$.

Now, by taking the Fourier transform of both sides of (4.40) we get

$$\widetilde{\Delta\phi}(k) = -\frac{1}{(2\pi)^{3/2}} \int_{-\infty}^{\infty} 4\pi e\delta(\mathbf{r}) e^{-i\mathbf{k}\cdot\mathbf{r}} d\mathbf{r} = -\sqrt{\frac{2}{\pi}} e. \tag{4.43}$$

Finally,

$$\tilde{\phi}(k) = \sqrt{\frac{2}{\pi}} \frac{e}{k^2}. \tag{4.44}$$

By applying the gradient to the result, we obtain the Fourier transform of the electric field:

$$\tilde{\mathbf{E}}(\mathbf{k}) = -ie\sqrt{\frac{1}{\pi}} \frac{\mathbf{k}}{k^2}. \tag{4.45}$$

Thus, the Fourier transform of the electric field is a longitudinal vector, as it is also in coordinate space where $\mathbf{E}(\mathbf{r}) = -e\mathbf{r}/r^3$. One interprets this as a *longitudinal polarization* of the electromagnetic field. Notice that the above Fourier expressions (4.44) and (4.45) of the Coulomb field do not refer to propagating waves – it is assumed that $\omega = 0$ (the field is static).

4.6 Reflection and Refraction

The well-known laws of light reflection and refraction can be deduced by starting from Fermat's principle, which establishes that in propagating between two points A and B, light takes such a trajectory that the elapsed time is a minimum (or a maximum). If the light ray trajectory touches the surface of another medium from which it is reflected, it is very easy to prove that the minimum time corresponds to the case when the angles of incidence and reflection as measured with respect to the normal to the surface are equal:

$$i = r. \tag{4.46}$$

Let us consider the plane AOB that contains the points A and B (Fig. 4.22). By taking B' to be the symmetric point of B with respect to the plane of reflection, it is easy to see that $AOB' = AOB$ is the minimum trajectory. Any other point C would give a longer trajectory ACB'. Then, since the velocity of light is constant, the smaller the trajectory, the shorter the travel time, and the actual trajectory is AOB.

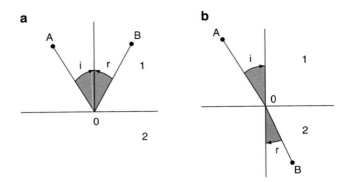

Fig. 4.22 Trajectory of a ray of light between two points A and B when the light is reflected (**a**) and when it is transmitted by refraction (**b**). In the first case the light moves with the same velocity along its trajectory, since it is in the same medium 1, although it touches the surface or interface separating the medium 1 from the medium 2. In the second case, the light travels the part AO of its trajectory in the first medium and the part OB in the second, and its velocity is different in the two media.

Something similar occurs with refraction. We consider the simple case of two isotropic media (in anisotropic media the speed of light is not the same in all directions). If the speed of light is v_1 in medium 1 and v_2 in medium 2, we find that the relation

$$\frac{\sin i}{\sin r} = \frac{v_1}{v_2} \tag{4.47}$$

is the one that minimizes the travel time between points A and B (Fig. 4.22). This can easily be demonstrated by the reader.

The quantity $n_{21} = v_1/v_2$ is called the index of refraction of the medium 2 with respect to the medium 1. When we speak about the index of refraction of a medium, it is assumed that the first medium is the vacuum, that is, the ratio of the velocity of light in vacuum c to the velocity of light in the medium, i.e., $n = c/v$. In some materials, charged particles can travel faster than $v = c/n$, but still slower than c. They produce a shock wave called Cherenkov radiation.

Total internal reflection. Let us assume a light beam that propagates in the medium 1 and is incident on the surface of the medium 2, where it is partially reflected and partially refracted. If the index of refraction of the first medium is larger than that of the second, $n_1 > n_2$, this means that, as $\sin i / \sin r = n_2/n_1$, there exists an incident angle i_0 such that $r = \pi/2$. In other words, $\sin i_0 = n_2/n_1$. For $i > i_0$, $\sin r$ is purely imaginary. The refracted wave propagates only parallel to the surface. The ratio of the electric vectors of the incident (i) and reflected (r') waves is $E_i/E_{r'} = 1$. Thus, the incident energy is equal to the reflected one. As there is no energy flow across the surface, the wave energy is totally reflected. The total internal reflection is exploited in several technical applications, for instance, in optical fibers.

4.7 Dispersion of Light

Finally, we mention briefly the phenomenon of light dispersion. For radiation propagating in vacuum, we have $c = \omega/k$ for all frequencies. But in a medium like water, air, glass, etc., the speed of wave propagation depends on the frequency. This is equivalent to saying that the angular frequency ω is a more complicated function of k than the above linear relation satisfied in vacuum. In this case, the medium is said to be dispersive, and each frequency has a different speed of propagation. Radiation of different frequencies passing from one medium to another is deflected by some angle in such a way that, the greater the speed in the medium, the larger the deflection angle, as we saw before. This is why we observe the seven colours of the spectrum when light passes through a prism or through drops of water in a rainbow.

The *group velocity* is a number characterizing the velocity of propagation of a wave train in a dispersive medium:

$$v_g = \partial\omega/\partial k. \tag{4.48}$$

The quantity $v_p = \omega/k$ is called the *phase velocity*.

Let us see the difference between group velocity and phase velocity in a qualitative way. We shall refer to Fig. 4.18. The phase velocity corresponds to the inner wave and the group velocity to the pulse propagation. It is not difficult to imagine a pulse moving at some velocity and carrying within it a wave moving at a different velocity which may be much greater than that of the pulse. This wave appears at the beginning of the pulse and disappears at the end.

For a gas of electrons of density N, of the kind occurring in the ionosphere in an oversimplified model, the index of refraction squared ($n^2 = \epsilon_r$, where ϵ_r is the relative electric permittivity of the medium) is a complex number:

$$n^2 = 1 + \frac{Ne^2}{m} \frac{1}{(\omega_0^2 - \omega^2) - i\omega\gamma}, \tag{4.49}$$

where m is the mass of the electron, i is the imaginary unit and ω_0 and γ are characteristic parameters of the medium. It can be shown that, for values of the frequency $\omega < \omega_0$, the real part of n^2 is greater than unity and grows up to a maximum at a point $\omega \lesssim \omega_0$, and it is zero at $\omega = \omega_0$, while for $\omega_0 < \omega$, it is smaller than unity. An anomalous dispersion occurs, in which the electromagnetic radiation is absorbed. The absorption is given by the imaginary part of n^2, which is proportional to $\gamma\omega$. It increases from $\omega \lesssim \omega_0$, having a maximum at $\omega \gtrsim \omega_0$.

In that interval of values of ω around ω_0, the group velocity does not represent the velocity of propagation of the signal transporting the energy (in that region, v_g may become greater than c). This was demonstrated by Arnold Sommerfeld and Léon Brillouin, invalidating the claims that a group velocity greater than c, when ω is close to ω_0, means the transport of energy at a speed greater than that of light, which would go against Einstein's special relativity theory. Sommerfeld and Brillouin demonstrated precisely the opposite: the velocity of propagation of the signal (information transfer) never exceeds the value c. This also applies to more recent experiments with laser beams propagating through some media, for which $v_g > c$.

4.8 Black Body Radiation

When a piece of iron is heated by fire, it begins to take on a reddish colour, then becomes incandescent red, and finally white. We observe that as the temperature increases, its colour varies. This corresponds to the fact that the frequency of the radiation emitted also increases.

However, it is not correct to say the *frequency*, since it actually emits radiation over a broad band of frequencies. A proportionality exists between the frequency of maximum intensity and the absolute temperature. In other words, the incandescent red iron looks red because it emits more red light than any other colour.

Several nineteenth century physicists, among whom we should mention Gustav Kirchhoff (1824–1887), Ludwig Boltzmann (1844–1906), John William Strutt, known as Lord Rayleigh (1842–1919), Wilhelm Wien (1864–1928), Walther Nernst (1864–1941), and James Jeans (1877–1946), studied the emission of radiation by a black body.

A black body absorbs all incident radiation, i.e., it does not reflect any radiation. For instance, a good model of a black body is a closed cavity within an object, such as a gasoline can. Any radiation that enters through its opening suffers internal reflections, and is finally absorbed by the walls (Fig. 4.23).

The study of radiation emitted by a black body when it is heated up is of great theoretical interest, since the black body is also a perfect emitter.

The black body behaves similarly to a piece of iron (which can to a certain extent be considered as a black body), in the sense that the frequency of the emitted radiation of maximum intensity is proportional to the absolute temperature.

Figure 4.24 depicts the energy density emitted by a black body as a function of frequency for the temperatures T_1, T_2, and T_3. It is observed that the maximum energy increases with temperature, in such a manner that the maximum occurs at increasing frequencies. The frequency is actually proportional to the absolute temperature, i.e., $\nu_{max}/T = $ const., according to Wien's law.

The shape of the curves in Fig. 4.24 implies a serious contradiction with the theories accepted at the end of the nineteenth century, although on the other hand, it leads to entirely logical conclusions.

The established theory assumed that the black body radiation (in today's language, the *photon gas*) was a thermodynamic system, similar to an ideal gas composed of oscillators. Furthermore, by applying the established principles of thermodynamics (in particular, the so-called principle of equipartition of energy, which attributes an equal energy $E = kT/2$ to each degree of freedom), it followed that the total energy of the black body radiation would be infinite.

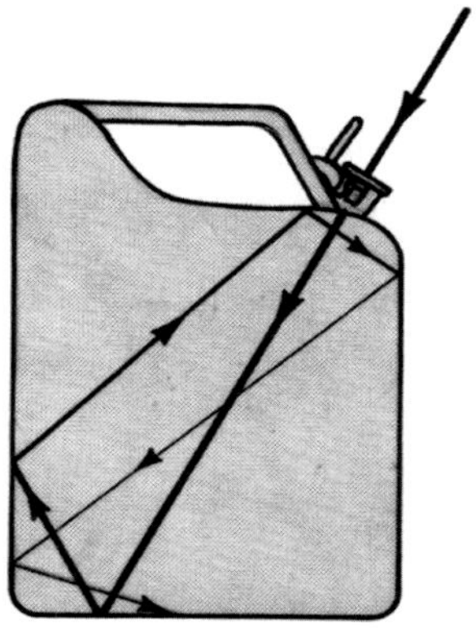

Fig. 4.23 An object containing a cavity, like a gasoline can, gives a good idea of a black body. The incident radiation suffers multiple reflections and internal absorptions, and is finally absorbed completely.

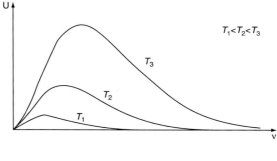

Fig. 4.24 Energy density of a black body at different absolute temperatures T_1, T_2, and T_3. The maximum is reached at some frequency which is proportional to the temperature, according to Wien's law.

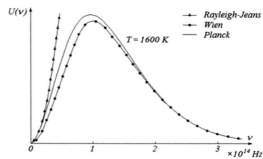

Fig. 4.25 The dotted lines represent the empirical Rayleigh–Jeans and Wien laws. The dark line represents the Planck law.

This was absurd, since it would mean that, by heating a body, it could be made to emit an infinite amount of energy, a fact that was demonstrated experimentally to be false. On the other hand, the conclusions of the established theory were valid for low frequencies. In other words, according to experimental results, there was a good correspondence with the predictions of classical thermodynamics for low frequency radiation: the energy density of black body radiation increased with frequency (the Rayleigh–Jeans law). However, as the frequency increased a maximum occurred, and then the energy density decreased (Fig. 4.25).

Wien identified an empirical law that described the behaviour at high frequencies very well. To reconcile the predictions of Rayleigh–Jeans with those of Wien, Max Planck (1858–1947) introduced a revolutionary hypothesis: that radiation is not continuously emitted and absorbed by the bodies. For radiation of frequency ν, the energy can be emitted in amounts proportional to $h\nu$, i.e.

$$E = nh\nu, \tag{4.50}$$

where $n = 1, 2, \ldots$, and h is Planck's constant with value 6.628×10^{-27} erg $\cdot$ s.

Starting from this hypothesis, Planck deduced a law that interpolated between the Rayleigh–Jeans and Wien laws, and was entirely compatible with the experimental results:

$$U(\nu, T) = \frac{8\pi}{c^3} \frac{h\nu^3}{e^{h\nu/kT} - 1}. \tag{4.51}$$

However, Planck's hypothesis contained a new feature, the quantization of radiation energy, that Planck himself did not accept, and later led to a partial return to the corpuscular hypothesis regarding the nature of light. This partial return was triggered by Einstein in his explanation of the *photoelectric effect*. This effect concerns electron emission by metals illuminated by radiation of frequency greater than or equal to some threshold frequency. For radiation below the threshold frequency, there is no electron current, although the radiation intensity may be high. On the other hand, even with very weak intensity of the incident radiation, if the frequency is greater than or equal to the threshold frequency, there is an electron current due to electrons emitted by the metal.

To explain this effect, Einstein assumed that the radiation energy was concentrated in the form of granules or quanta of energy according to the law $E = h\nu$, following Planck's hypothesis in a more advanced form. Therefore, the electron could only absorb energy in multiples of $h\nu$.

If the work needed to extract an electron is W then, since E is greater than or equal to W, the ejected electron moves with a kinetic energy

$$T = h\nu - W. \tag{4.52}$$

When the radiation intensity increases, the number of emitted electrons also increases as long as the frequency is greater than the threshold frequency $\nu_0 = W/h$. For lower frequencies than ν_0, even for very intense radiation, electrons are not emitted, since T would be negative.

Einstein's idea was a partial return to the corpuscular model of light, because the granules of light or photons have an energy proportional to their frequency. The dual wave–particle property is inherent in the new theory of the microscopic world which, after the work by Planck and Einstein, brought quantum theory into existence.

Problems

Problem 4.1 Consider the double slit experiment with green light $\lambda = 5460$ Å. Assume that the slits are separated by $d = 0.1$ mm, and the screen is at $D = 20$ cm. What is the angular position of i) the first minimum? ii) the fifth maximum?

Problem 4.2 Integrate (4.51) over ν and obtain the so-called Stefan–Boltzmann law, which gives the total energy density of black body radiation at temperature T.

Problem 4.3 Black body radiation may be considered as a gas of free massless particles, viz., photons, assumed to be in equilibrium at temperature T (see Chap. 8 for more detail), with chemical potential $\mu = 0$. It is easily seen from (4.50) and (4.51) that the average density of photons in the frequency interval $(\nu, \nu + d\nu)$ can be written as $dN(\nu) = dU(\nu, T)/h\nu$. Hence,

$$N = \frac{8\pi V}{\pi^2 c^3} \int_0^\infty \frac{\nu^2 d\nu}{e^{h\nu/kT} - 1} = 0.244 \left(\frac{kT}{\hbar c}\right)^3. \tag{4.53}$$

(i) Assuming the Sun's surface to be a black body at a temperature of about 6000 K, calculate the photon density near it.
(ii) Using the results in Chap. 2, find an expression for the free energy density $F = U - TS$ of black body radiation, where the entropy density is $S = -\partial F/\partial T$. Calculate also the heat capacity at constant volume $C_V = \partial U/\partial T$.

Literature

1. F.S. Crawford, *Waves*, Berkeley Physics Course (McGraw-Hill, New York, 1965). An excellent book devoted to wave motion
2. J.D. Jackson, *Classical Electrodynamics*, 3rd edn. (Wiley, New York, 1998). An updated and first-rate treatise on classical electrodynamics. The problem of classical electromagnetic radiation is discussed at an advanced level
3. M. Chaichian, I. Merches, D. Radu, A. Tureanu, *Electrodynamics* (Springer, Berlin Heidelberg, 2014). This book is recommended to complement several topics dealt with in the present chapter
4. B. Rossi, *Optics* (Addison Wesley, Reading, Massachusetts, 1956). An excellent treatise on optics
5. L.D. Landau, E.M. Lifshitz, *The Classical Theory of Fields* (Pergamon Press, Oxford, 1965). The important topics of optics are discussed in an original way
6. R.P. Feynman, *The Feynman Lectures on Physics, Definitive Edition* (Pearson, Addison Wesley, 2006). In the first volume, the discussion about wave interference and diffraction is a masterpiece

Chapter 5
Special Theory of Relativity

Albert Einstein was born in Ulm, Bavaria, on 14 March 1879. The outstanding relevance of Einstein's work for modern science is recognized in the words of the Russian physicist Igor E. Tamm:

> To consider Einstein as the most outstanding physicist of the twentieth century would be, perhaps, to underestimate him. To be fair, he can only be compared with Newton. Newton and Einstein represent the summits of human progress in the understanding of Nature, the summits that dominate a period of 300 years of development of the exact sciences. And in spite of the distance in time, there is a close affinity between them.

The name of Einstein is mainly associated with the theory of relativity. However, we owe him much other work that also significantly influenced the evolution of physics throughout the twentieth century. His theoretical investigation of Brownian motion served as the basis for the experimental work done by Jean Perrin (1870–1942) to determine precisely Avogadro's number. This in turn provided definite evidence in favour of the Maxwell–Boltzmann kinetic theory. Einstein's theory of the photoelectric effect introduced the revolutionary idea that electromagnetic radiation had a dual behaviour as wave and corpuscle, taking the first steps toward quantum theory. The papers dealing with these three topics were published by him in 1905. In 2005, the scientific community celebrated the centennial anniversary of this 'marvellous year'.

In 1917, one year after the publication of his most celebrated work, the general theory of relativity, he investigated the emission and absorption of radiation by matter, making an important contribution to the quantum theory of radiation.

Einstein's scientific work in his later years (he passed away on 18 April 1955) was dedicated to elaborating a unified theory of the electromagnetic and gravitational fields. Although his initial goal was not achieved, the unification idea was successfully re-born within the modern quantum theory of fields (Fig. 5.1).

M. Chaichian et al., *Basic Concepts in Physics*, Undergraduate Lecture Notes in Physics, https://doi.org/10.1007/978-3-662-62313-8_5

Fig. 5.1   Albert Einstein.

5.1   Postulates of Special Relativity

When Einstein was 15 years old, he often wondered what would happen if a person tried to catch a ray of light. The young Einstein probably established a parallel with another question that was easier to answer: What would happen if one tried to catch a bullet shot by a gun?

The answer to this question would be: in order to do it, he would have to travel in a vehicle moving at the same velocity as the projectile. In this way, he would move beside it, being at rest with respect to the projectile. It would then be sufficient to stretc.h out his arm and take the projectile in his hand, whence his motion would continue without change. However, for the ray of light, Einstein got the answer 10 years later, in 1905, and it was negative: a person could never reach such a speed, and even moving at very high speed, say 150,000 km/s, the light would always move at the same speed, $c = 300,000$ km/s, with respect to him.

The reader may wonder whether this means that the fact of moving with respect to a light source does not produce any physical effect on the light observed. The answer is that there is indeed some influence. If the source moves away from the observer, the *colour* of the light changes: the light shifts toward red or, put another way, the observed frequency of the wave diminishes. If the source approaches him, the light shifts toward violet, i.e., its frequency increases.

But the answer given by Einstein to the question he asked himself in his adolescence had far-reaching implications, since it produced deep modifications to the concepts of space and time that everyone had accepted since Newton, and put an end to a set of internal contradictions in the physics of the nineteenth century. These were mainly due to the hypothesized existence of the so-called *luminiferous æther*, which was hypothesized to be an imponderable fluid filling the empty space between bodies, and considered to be the medium *supporting* the propagation of light waves.

Let us recall the importance of inertial frames of reference in classical mechanics and reconsider the validity of Galileo's principle of relativity, which establishes that the laws of mechanics are the same in all inertial frames. This means that the oscillations of a pendulum, for instance, in an inertial frame, are similar in any other frame moving at constant velocity with regard to it. A consequence of Galileo's principle of relativity and of the notions of absolute time and space were the Galileo transformations (1.24). The equations of mechanics, like those governing the motion of a planet around the Sun under the action of Newton's gravitational force, do not change in form, i.e., they are said to be *covariant* under such a transformation. From this it follows that, if an object moves with velocity $\mathbf{V}$ with respect to an inertial frame, and if this frame in turn moves with velocity $\mathbf{V}'$ with respect to another inertial frame, the velocity of the object with respect to this second frame satisfies the velocity sum law $\mathbf{V}'' = \mathbf{V} + \mathbf{V}'$, that is, the principle of relativity of Galileo implies the additivity of the velocities when the motions are considered as referred to several inertial frames.

On the other hand, in the case of electromagnetic phenomena, Galileo's principle of relativity is not valid. In particular, it is not satisfied by light propagation. Maxwell's equations and the electromagnetic wave equation are not covariant under Galileo's transformations. For this reason, the existence of an absolute frame of reference was hypothesized, and it was assumed that electromagnetic waves would move in the *luminiferous æther* at the velocity of 300,000 km/s. It was expected that light would have different velocities if measured in a frame at rest or moving with respect to the æther, and also that the result would be different if the velocity of light were measured by an observer moving in the same sense as the Earth's rotation or along a perpendicular direction.

Experiments of this kind were performed at the end of the nineteenth century, the most famous being the Michelson–Morley experiment, devised by Albert A. Michelson (1852–1931) and Edward W. Morley (1838–1923), but leading to negative results: the effect due to the expected difference in the velocities of light along and perpendicular to the Earth's rotation was not found. The physics community thus had to come to terms with the following facts:

1. Newtonian mechanics and Galileo's principle of relativity were valid (verified in mechanical experiments and astronomical observations);
2. The laws that govern electromagnetic phenomena (described by the Maxwell equations) were also valid, and verified experimentally. But these equations did not satisfy the Galilean relativity principle and hence it was expected that the speed of light would be different for an observer at rest and an observer in motion;
3. The experiments carried out in order to measure such a difference of velocities gave negative answers, as if the speed of light were the same for both observers.

Apparently the statements 1, 2, and 3 could not all be valid simultaneously in the theoretical framework of the epoch, and Einstein proposed to solve this contradiction in 1905 by formulating two basic principles or postulates:

1. The speed of the light emitted by a source is the same for all observers, whatever their state of motion;
2. The laws of physics (including electromagnetic phenomena) are valid in all inertial frames.

In his considerations Einstein implicitly admitted another postulate, of no less importance: the transformations of spacetime coordinates between inertial frames should be *linear transformations*, as the Galilean transformations.

Hence, Einstein generalized the Galilean principle of relativity to all physical phenomena, including electromagnetic phenomena, and in this way it became clear that, by assuming the validity of his two postulates, all the previously mentioned contradictions would disappear.

Einstein arrived at these conclusions on purely theoretical arguments. At that time he was employed in the Office of Patents at Bern, Switzerland. He once declared that he did not know at that time about the results of the Michelson–Morley experiment.

The differences between Einstein's and Galileo's principles of relativity had remarkable consequences. Not only was the controversial luminiferous æther no longer necessary, but there was no reason to assume its existence. Further, the validity of his postulates also implied the disappearance of the concepts of *absolute space* and *absolute time* as independent entities in Newtonian mechanics: space and time now formed a combined entity, called *spacetime*, in which they were intimately related, and the fundamental laws of physics could be written as mathematical expressions in a four-dimensional space.

Einstein created a new mechanics such that, when the speeds of the particles are small compared with the speed of light c, it coincides with Newtonian mechanics, while it differs greatly for velocities close to c. These ideas were initially rejected by many physicists, but they were finally accepted when they faced the evidence of their experimental confirmation.

One crucial consequence of Einstein's postulates is the relativity of simultaneity. In order to illustrate it, let us consider a hypothetical train travelling at a very high velocity (in order for the effect to be noticeable, the train must move at a velocity comparable to c). This is illustrated in Fig. 5.2, where two observers are considered, for whom the speed of light is the same, inside the train and outside it.

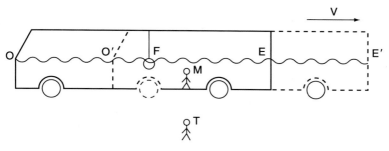

Fig. 5.2 For the observer T, the light signal arrives first at the end O', then at E'. For the observer M it arrives at both points, E and O, simultaneously. From this, we conclude that time elapses differently for the two observers.

For the observer M who is sitting in the train, a light signal emitted by the lamp F located at the middle of the train would reach both ends of the train simultaneously. That is, provided with adequate detectors, he could check that the light arrives at the same instant at the points O and E.

But for the observer T located in the railway station, things occur in a different manner. When the ray travelling toward the right reaches the end of the train, this point will be in the position E'; the light travelling toward the left will arrive at the other end at the point O'. Bearing in mind that the distance FE' is greater than FO', for the observer T the light will arrive first at O', then at E'. That is, for him the two events are not simultaneous.

Michelson interferometer. The Michelson–Morley experiments sought to detect the hypothesized motion of light relative to the luminiferous æther, but such a motion was never found. The idea was that a difference in the speeds of monochromatic light along and perpendicular to the horizontally moving interferometer arm would be detected from the interference pattern. This negative result gave experimental support to the basic postulates of special relativity, as formulated by Einstein. Interestingly, Michelson interferometry (see Fig. (5.3)) also played a fundamental role in the experimental observation of a much more elusive prediction, of general relativity: gravitational waves. Gravitational waves were finally detected using laser interferometry as an essential experimental tool in LIGO (see Chap. 10).

5.2 Lorentz Transformations

We have already seen in the first chapter that a frame of reference is determined by a system of three coordinate axes, to fix the position of the objects with respect to them, and a clock in order to measure the time at which the events occur. In classical mechanics, a single clock serves for all frames of reference. In relativistic,

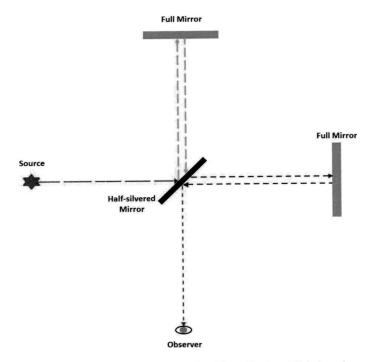

Fig. 5.3 The Michelson interferometer. Invented by Albert Abraham Michelson, it uses a beam-splitter to split the light from a source into two beams, which are directed in the two arms of the interferometer. Each of these is reflected back toward the beamsplitter, which then combines their amplitudes. The detector registers the interference pattern.

mechanics, each frame requires an appropriate clock. The clocks in different frames of reference run differently.

Let us assume a frame of reference K, and consider two events: the departure of a light signal from a point A and the arrival of that signal at another point B. The coordinates of the first event in such a reference frame would be (including time as a fourth coordinate):

$$x_1, y_1, z_1, t_1,$$

and those of the second event:

$$x_2, y_2, z_2, t_2.$$

Then these numbers must satisfy

$$(x_2 - x_1)^2 + (y_2 - y_1)^2 + (z_2 - z_1)^2 - c^2(t_2 - t_1)^2 = 0. \tag{5.1}$$

We can see this easily, since the distance between A and B is

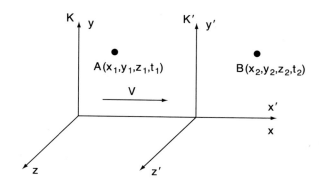

Fig. 5.4 Two inertial reference frames K and K' oriented so that their x and x' axes coincide. The frame K' moves with constant velocity V with respect to K. Two events A and B have coordinates (x_1, y_1, z_1, t_1) and (x_2, y_2, z_2, t_2) in the frame K, and (x'_1, y'_1, z'_1, t'_1) and (x'_2, y'_2, z'_2, t'_2) in K'.

$$d = \sqrt{(x_2 - x_1)^2 + (y_2 - y_1)^2 + (z_2 - z_1)^2}, \tag{5.2}$$

and the interval of time between the two events is

$$\Delta t = t_2 - t_1. \tag{5.3}$$

As the signal propagates at the speed of light, we have

$$d = c\Delta t, \tag{5.4}$$

which can be re-written in the form (5.1).

If the events are now measured in another frame of reference K' moving with velocity V with regard to K, in this new frame the two events will have the coordinates:

$$x'_1, y'_1, z'_1, t'_1 \text{ and } x'_2, y'_2, z'_2, t'_2,$$

and they should satisfy the equation

$$(x'_2 - x'_1)^2 + (y'_2 - y'_1)^2 + (z'_2 - z'_1)^2 - c^2(t'_2 - t'_1)^2 = 0. \tag{5.5}$$

Let us assume that K' moves parallel to the x axis with the constant speed V with respect to the frame K (Fig. 5.4). In order that the relations (5.1) and (5.5) be satisfied by the coordinates of the events in the K and K' frames, they should be related by a linear transformation, the so-called Lorentz transformation (or FitzGerald–Lorentz transformation), whose initial formulation was proposed by George F. FitzGerald (1851–1901) and Hendrik A. Lorentz (1853–1928), in an attempt to interpret the Michelson–Morley experiment as a contraction of all bodies parallel to their direction of motion:

$$x' = \frac{x - Vt}{\sqrt{1 - V^2/c^2}},$$
$$y' = y,$$
$$z' = z,$$
$$t' = \frac{t - (V/c^2)x}{\sqrt{1 - V^2/c^2}}.$$

(5.6)

This is the transformation which replaces the Galilean transformation (1.24) in Einstein's theory of relativity.

If x_1', y_1', z_1', t_1', and x_2', y_2', z_2', t_2' are replaced by their expressions in terms of x_1, y_1, z_1, t_1, and x_2, y_2, z_2, t_2 according to (5.6), then (5.5) is converted to (5.1). This means that the expression (5.1) is *invariant* under the transformations (5.6), which depend on the velocity V.

For small velocities compared with the speed of light, the Lorentz transformation (5.6) can be approximated by the Galilean transformation

$$x' = x - Vt,$$
$$y' = y,$$
$$z' = z,$$
$$t' = t.$$

The Lorentz transformations are a consequence of the constancy of the speed of light in all the inertial frames, and of the linearity of the coordinate transformations.

If two events, which we shall call 1 and 2, are not related by the departure and arrival of a light signal, then their coordinates would not satisfy the equality (5.1), and we would have one of the two possibilities:

$$S_{12}^2 > 0, \quad \text{or} \quad S_{12}^2 < 0,$$

(5.7)

where

$$S_{12}^2 = c^2(t_2 - t_1)^2 - (x_2 - x_1)^2 - (y_2 - y_1)^2 - (z_2 - z_1)^2$$

is called the *spacetime interval*. If we calculate the interval between the two events observed from the frame K', its value is the same as the one calculated for the frame K.

By applying the Lorentz transformation to the coordinates of the events 1 and 2, one can check that S_{12} does not change. The interval between two events is the same for all inertial frames. It is a *relativistic invariant*.

Given two events 1 and 2, if $S_{12}^2 > 0$, it is called a *time-like interval*, and the two events can be causally connected, i.e., they can be related to each other by means of a signal travelling at lower speed than light. In particular, it is always possible to find a reference frame in which the two events occur at the same point of space.

As an example, let us suppose that a traveller throws some object through a window of a train, and five seconds later, throws another object through the same window.

For an external observer, the two events occurred at different points of space and at different times. For the traveller, the two events occurred at the same point of space, but at different times. The interval between the two events is time-like.

If $S_{12}^2 < 0$, the interval is said to be *space-like*. The two events cannot then be related causally, since the spatial distance between the two points at which they occur is greater than the product of the velocity of light by the difference of time between them:

$$(x_2 - x_1)^2 + (y_2 - y_1)^2 + (z_2 - z_1)^2 > c(t - t_1)^2. \tag{5.8}$$

When the interval between two events is space-like, it is always possible to find a frame of reference in which the two events occur at the same instant of time, although at different points.

There is still the possibility that $S_{12} = 0$, in which case the interval is called *light-like*, since the two spacetime points can be related by a light signal.

The interval between two events in spacetime is a generalization of the distance between two points in ordinary space. In Euclidean geometry, the distance $|PQ|$ of two points $P = (x_1, y_1)$ and $Q = (x_2, y_2)$ in a plane is expressed as a sum of squares by the theorem of Pythagoras:

$$(PQ)^2 = (x_1 - x_2)^2 + (y_1 - y_2)^2. \tag{5.9}$$

The generalization to higher dimensions is also a sum of squares. For example, for two points $P = (x_1, y_1, z_1, w_1)$ and $Q = (x_2, y_2, z_2, w_2)$ in four-dimensional Euclidean space:

$$(PQ)^2 = (x_1 - x_2)^2 + (y_1 - y_2)^2 + (z_1 - z_2)^2 + (w_1 - w_2)^2. \tag{5.10}$$

In spacetime, the formula for the distance or interval between two points $P = (x_1, y_1, z_1, ct_1)$ and $Q = (x_2, y_2, z_2, ct_2)$ is the following:

$$(PQ)^2 = S_{12} = c^2(t_1 - t_2)^2 - (x_1 - x_2)^2 - (y_1 - y_2)^2 - (z_1 - z_2)^2.$$

The space is said to be pseudo-Euclidean. For this reason the interval can be either positive, negative, or zero.

In the case of the Euclidean plane, if we fix the point P and vary the point Q, so that $(PQ)^2 = $ const., the resulting curve will be a circle. In the pseudo-Euclidean plane, for $(PQ)^2 = $ const. $\neq 0$ the resulting curve will be a *hyperbola* determined by the equation:

$$(PQ)^2 = c^2(t_1 - t_2)^2 - (x_1 - x_2)^2 = \text{const.} \neq 0.$$

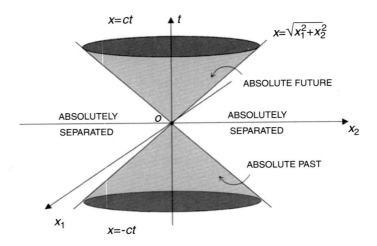

Fig. 5.5 Representation of the light cone with respect to the event O in a system with two space (x_1, x_2) and one time (t) coordinates. The events within the light cone are divided into two classes: absolute past and absolute future. The events located outside the cone are absolutely separated from O. Their intervals with respect to O are space-like.

5.3 Light Cone and Causality

If we choose some event as coordinate origin (we represented only two space coordinates in Fig. 5.5 for simplicity), the events that are separated from such an event by an interval $S^2 = 0$, would satisfy the equation

$$x^2 - c^2 t^2 = 0, \quad \text{where} \quad x^2 = x_1^2 + x_2^2, \tag{5.11}$$

that is,

$$\sqrt{x_1^2 + x_2^2} = \pm ct, \tag{5.12}$$

which gives a cone passing through the coordinate origin. In the four-dimensional spacetime, we have a higher-dimensional cone, called the *light cone*. All the events that one can connect to O by means of a light signal lie on the surface of this cone. The spacetime trajectories of all massless particles, such as photons, gravitons, etc., all lie on this cone.

The events inside the cone are separated from O by time-like intervals. For these, $x^2 - c^2 t^2 < 0$ and a causal relation between them and the origin is possible. Those corresponding to $t > 0$ form the absolute future, and there is no reference frame in which they could occur simultaneously with the event taking place at O $(t = 0)$. For $t < 0$, we have the events that correspond to the absolute past.

The concepts of absolute future and absolute past are not strictly derived from the basic postulates of special relativity, but are introduced from outside with the

Fig. 5.6 Although the events b, c are absolutely separated from O, their light cones intersect with that of O, so the consequences of b, c could be known by O at future times corresponding to the events O', O''.

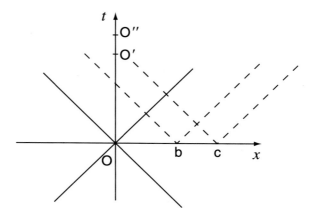

implicit notion of an *arrow of time*, which we already encountered in studying the solution of the electromagnetic wave equations, when we discarded the regressing solutions (see Chap. 4). The existence of this direction of time implies that time has a special feature, as compared to spatial coordinates.

The points outside the light cone are separated from O by means of space-like intervals. If x and t are the coordinates of such a point, then $c^2 t^2 - x^2 < 0$. They correspond to events absolutely separated from O. It is possible to find a reference frame in which space-like separated events occur simultaneously with O. In Fig. 5.6, for example, the events b and c are simultaneous with O. Suppose that the event O corresponds to "*I am here now*". It is not possible for me to know about the events b, c, etc., occurring simultaneously with me *now*. But O can be informed about their consequences, since the light cone originating, say, at b intersects my light cone, and all the events which are consequences of b and which correspond to this region of intersection can be related causally with my future positions in spacetime O', O'', etc., whence I can be informed about such events.

We have repeatedly mentioned the idea of a *causal connection* between two events: two events can be causally connected if they lie inside or on each other's light cone surface. Thus, the notion of causality is rigorously formulated in special relativity. Any *observable* process must obey causality. In the relativistic formulation of quantum theory, causality is one of the fundamental requirements.

5.4 Contraction of Lengths

Consider a rod of length ℓ' fixed relative to the frame K', which is moving with respect to K with the velocity V (Fig. 5.7). The rod lies parallel to the axes x, x'. We denote by 1 the event of observing the left end of the rod from the frame K. The coordinates are x_1, y_1, z_1, t_1. The event of observing the right end of the rod from K is denoted by 2. The coordinates would be x_2, y_1, z_1, t_1. The two observations are made simultaneously in K, since we wish to know the length of the rod in this frame.

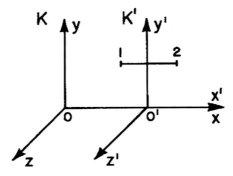

Fig. 5.7 The rod whose ends 1 and 2 are fixed to the frame K'. In order to measure its length from K, its ends 1 and 2 are observed at the same instant *according to the clocks of* K. The conclusion of the observers of K is that the length of the rod is shorter than when it was at rest. However, an observer in K' would justify this result with the following argument: for him the two measurements, according to the clocks of K', were not made simultaneously.

According to (5.6), we have:

$$x_1' = \frac{x_1 - Vt_1}{\sqrt{1 - V^2/c^2}},$$
$$x_2' = \frac{x_2 - Vt_1}{\sqrt{1 - V^2/c^2}}.$$

(5.13)

Subtracting the second relation from the first, we find

$$x_2' - x_1' = \frac{x_2 - x_1}{\sqrt{1 - V^2/c^2}}.$$

(5.14)

But $x_2' - x_1' = \ell'$ is the length of the rod in the frame K', so the length observed in the frame K is shorter:

$$x_2 - x_1 = \ell = \sqrt{1 - V^2/c^2}\,\ell',$$

(5.15)

since $\sqrt{1 - V^2/c^2} < 1$. We observe that the length of the moving rod seems to contract along the direction of motion.

5.5   Time Dilation: Proper Time

Consider now a clock fixed to the frame K', which is to be compared with various clocks in the frame K (Fig. 5.8). By an inverse Lorentz transformation, we find:

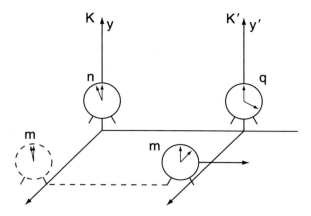

Fig. 5.8 The observer in K concludes that the moving clock m in K' is running slowly. This happens because, when the moving clock started, it was synchronized with a clock n fixed in K. After passing by another clock q fixed in K, the latter concludes that the moving clock lags. The observer in K' considers that the clocks n and q in K with which the time of the moving clock has been compared were wrongly synchronized.

$$t = \frac{t' + (V/c^2)x'}{\sqrt{1 - V^2/c^2}}. \tag{5.16}$$

Suppose the clock is fixed relative to K', having the space coordinates $x'_1, 0, 0$, and we observe it at time t'_1. In K, this time will correspond to t_1 given by

$$t_1 = \frac{t'_1 + (V/c^2)x'_1}{\sqrt{1 - V^2/c^2}}. \tag{5.17}$$

When the clock in K' indicates the time t'_2, in K we have

$$t_2 = \frac{t'_2 + (V/c^2)x'_1}{\sqrt{1 - V^2/c^2}}. \tag{5.18}$$

Subtracting one expression from the other, it follows that

$$t_2 - t_1 = \frac{t'_2 - t'_1}{\sqrt{1 - V^2/c^2}}. \tag{5.19}$$

The interval of time $t_2 - t_1$ is greater than $t'_2 - t'_1$. If t'_1 corresponds to the event *the clock in K' indicates 12:00 noon* and t'_2 to the event *the clock in K' indicates 12:05 p.m.*, then the time interval of five minutes in K' is estimated by the observers in K to correspond to a longer interval. If $\sqrt{1 - V^2/c^2} = 1/2$, i.e., $V \sim 0.87c$, it will be ten minutes in K. The conclusion of the observers in K is that moving clocks run slower than those at rest.

For the observers in K, the lengths of the objects parallel to the direction of motion in the moving frame are contracted and the clocks lag. But for the observers in the moving frame K', they are at rest and the moving frame is K, so they would be expected to arrive at similar conclusions, since both are inertial frames. How can this apparent paradox be explained?

Let us consider the case of length contraction. The events corresponding to the observation of the ends of the moving rod by observers in the frame K are (x_1, y_1, z_1, t_1) and (x_2, y_1, z_1, t_1). They determine a space-like interval

$$S_{12}^2 = -(x_2 - x_1)^2. \tag{5.20}$$

But the two events are not simultaneous for an observer located in the frame K'. They occur at different instants t_1', t_2' given by

$$\begin{aligned} t_1' &= \frac{t_1 - (V/c^2)x_1}{\sqrt{1 - V^2/c^2}}, \\ t_2' &= \frac{t_1 - (V/c^2)x_2}{\sqrt{1 - V^2/c^2}}. \end{aligned} \tag{5.21}$$

As $x_2 > x_1$, it follows that $t_1' > t_2'$, and for observers in the frame K', it seems that the position of the end 1 of the rod was measured after the position of the end 2. Viewed from the K' frame, the positions of the observers in the K frame would be

$$\begin{aligned} x_1' &= \frac{x_1 - Vt_1}{\sqrt{1 - V^2/c^2}}, \\ x_2' &= \frac{x_2 - Vt_1}{\sqrt{1 - V^2/c^2}}. \end{aligned} \tag{5.22}$$

The interval between the two events, as measured in K', turns out to be

$$S_{12}'^2 = c^2(t_2' - t_1')^2 - (x_2' - x_1')^2 = -(x_2 - x_1)^2 = S_{12}^2. \tag{5.23}$$

From this it is concluded that the observers in both frames obtain the same values for the measurements of the relativistic invariants (in this case, the interval). But the length and the time separately are not relativistic invariants, and they cannot give the same values when measured from the two frames.

Something similar occurs with time dilation. In this case it is important to note that the asymmetry of the measurement in the two frames lies in the fact that one moving clock in K' is compared with several clocks located in K. The observer travelling with the moving clock K' concludes that the discrepancy in the measurement of time with respect to the clocks in the frame K occurs because the clocks are not properly synchronized.

If the moving frame K' where the moving clock is located were to stop, it would indeed appear to be retarded with respect to the clocks of the frame K. This is not in contradiction with the equivalence of the two inertial frames mentioned earlier. When the frame K' decelerates in order to stop, it ceases to be an inertial frame and is no longer equivalent to K. In order to compare the two clocks at the same point and at the same time, it is necessary to accelerate K positively or K' negatively, and thus the clocks lose their synchronization.

The time measured in the frame of reference in which the clock is at rest is called *proper time*. We observed in the last example that the interval of proper time is shorter than the interval measured by a fixed observer. This leads to the twin paradox: if one of two twins leaves the Earth at a velocity close to that of light, when he returns after many years, he will appear much younger than his brother who remained on Earth.

The phenomenon of time dilation has not been verified for human beings, but for elementary particles. For instance, for μ mesons (or muons), the mean lifetime is about 2×10^{-6} s. If they move at velocities close to c, their average lifetime is prolonged by several orders of magnitude. If they are produced by cosmic rays in the upper atmosphere and if their energy is high enough, they can be observed at sea level as a result of time dilation. This allowed their discovery in 1936, in cosmic rays experiments.

5.6   Addition of Velocities

The impossibility of exceeding the speed of light is a consequence of Einstein's postulates. This is easily derived from the law of addition of velocities in relativistic mechanics. Taking the relations

$$x = \frac{x' + Vt'}{\sqrt{1 - V^2/c^2}}, \qquad t = \frac{t' + (V/c^2)x'}{\sqrt{1 - V^2/c^2}}, \tag{5.24}$$

after differentiating them with respect to (x', t'), let us divide the first equation by the second. It follows that

$$V_x = \frac{V'_x + V}{1 + V V'_x/c^2}, \tag{5.25}$$

where $V_x = \frac{dx}{dt}$ represents the velocity of a particle with respect to the rest frame K, while $V'_x = \frac{dx'}{dt'}$ represents the velocity of the same particle, but measured from the moving frame K'.

For $V \ll c$, the denominator approaches unity and we get approximately

$$V_x \approx V'_x + V. \tag{5.26}$$

This is the law of composition of velocities of classical mechanics. But from (5.25), if the particle moves at high speed with respect to K', for instance, $V'_x = c/2$, and the frame K' in turn also moves at the same velocity $V = c/2$ with respect to K, one has

$$V = \frac{c/2 + c/2}{1 + 1/4} = \frac{4}{5}c, \tag{5.27}$$

which is smaller than c.

But even by taking $V'_x = c$ and $V = c$, it would not be possible to exceed the speed of light. The result would be

$$V_x = \frac{c + c}{1 + c^2/c^2} = c. \tag{5.28}$$

That is, even if K' moved at the velocity of light with respect to K and the particle also moved at the velocity of light with respect to K', its velocity with respect to K would be precisely the velocity of the light. We see that it is not possible by means of the relativistic law of composition of velocities to exceed the speed of light c by summing velocities that are smaller than or equal to c.

5.7   Relativistic Four-Vectors

One of the most interesting geometrical consequences of the Lorentz transformations is that the transformations of the space and time coordinates are geometrically equivalent to a rotation in a four-dimensional space called spacetime or Minkowski space, in honour of Hermann Minkowski (1864–1909) who was the first to observe that Einstein's relativity theory requires us to consider time as a fourth dimension.

In the Euclidean geometry of the plane, if a vector has components (a, b) in a frame of coordinates (x, y), in another frame of coordinates (x', y') forming an angle α with the first, it will have coordinates (a', b') given by

$$\begin{aligned} a' &= a \cos \alpha + b \sin \alpha, \\ b' &= -a \sin \alpha + b \cos \alpha. \end{aligned} \tag{5.29}$$

whence $a'^2 + b'^2 = a^2 + b^2$.

In the theory of relativity, changing the description of an interval between two events from a frame of reference K to another K' means changing an interval with components $(x_2 - x_1, y_2 - y_1, z_2 - z_1, c(t_2 - t_1))$ to another one with components $(x'_2 - x'_1, y'_2 - y'_1, z'_2 - z'_1, c(t'_2 - t'_1))$ by means of a transformation similar to (5.29). To do this, let us make the change of variable $\tau = ict$ and consider the angle as

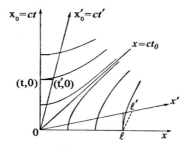

Fig. 5.9 Minkowski diagram. The Lorentz transformation from the system K to the system K' can be represented geometrically as a rotation of the axes x and ct through the angle $\alpha = \arctan V/c$, as illustrated in the figure. Here, V is the relative speed between the two systems, in the x directions. The grid of hyperbolas represent the curves with the property $c^2 t^2 - x^2 = \text{const.}$, for different values of the constant. The straight lines $x' = \text{const.}$ are parallel to the ct' axis, and the straight lines $ct' = \text{const.}$ are parallel to the x' axis in the oblique-angled system of coordinates.

$$\cos i\alpha = \frac{1}{\sqrt{1 - V^2/c^2}},$$
$$\sin i\alpha = \frac{iV/c}{\sqrt{1 - V^2/c^2}}. \tag{5.30}$$

Then

$$
\begin{aligned}
x_2 - x_1 &= (x_2' - x_1')\cos i\alpha + (\tau_2' - \tau_1')\sin i\alpha, \\
y_2 - y_1 &= y_2' - y_1', \\
z_2 - z_1 &= z_2' - z_1', \\
\tau_2 - \tau_1 &= -(x_2' - x_1')\sin i\alpha + (\tau_2' - \tau_1')\cos i\alpha \; .
\end{aligned} \tag{5.31}
$$

The transformations for $x_2 - x_1$ and $\tau_2 - \tau_1$ in (5.31) are similar to those in (5.6). The difference lies in the imaginary character of the variable $\tau = ict$ and in the fact that $\sin i\alpha$ and $\cos i\alpha$ are not actually trigonometric, but hyperbolic functions. Then we may write

$$\cos^2 i\alpha - \sin^2 i\alpha = 1, \tag{5.32}$$

but it is better to write this relation in terms of hyperbolic functions as

$$\cosh^2 \alpha - \sinh^2 \alpha = 1,$$

recalling the definitions $\cosh \alpha = (e^\alpha + e^{-\alpha})/2$, $\sinh \alpha = (e^\alpha - e^{-\alpha})/2$. The transformation (5.31), which is another way of writing the Lorentz transformation, leaves the interval S_{12} invariant, and we have a rotation through an imaginary angle in Minkowski space (Fig. 5.9).

All the physical quantities in the theory of relativity must have definite transformation properties, being scalars, vectors, tensors, etc., under Lorentz transformations. This means new relations between quantities that were apparently independent in non-relativistic physics, just like the new relations between space and time, which were not present in the mechanics of Galileo and Newton.

5.8   Electrodynamics in Relativistically Covariant Formalism

In the light of these concepts from special relativity, we now return to some points from Chap. 3. In the first place, we should note the importance of the electromagnetic field four-vector. Thus, for a system of electric charges in vacuum, we can define a vector, the vector potential **A**, from which the magnetic field **B** is obtained by applying to it a vector operator, called curl or rotor.

The curl of a vector can be defined by starting with a differential operator ∇, which is formed by taking the partial derivatives with respect to the coordinates x, y, z, multiplied respectively by the unit vectors $\mathbf{e}_1, \mathbf{e}_2, \mathbf{e}_3$ along these axes, and summed afterwards. Then the vector product of this operator with the vector **A** is the magnetic field **B**. That is, starting from

$$\nabla = \mathbf{e}_1 \frac{\partial}{\partial x} + \mathbf{e}_2 \frac{\partial}{\partial e} + \mathbf{e}_3 \frac{\partial}{\partial z}$$

and

$$\mathbf{A} = \mathbf{e}_1 A_x + \mathbf{e}_2 A_y + \mathbf{e}_3 A_z,$$

we define

$$\begin{aligned}\mathbf{B} = \operatorname{rot} \mathbf{A} &= \nabla \times \mathbf{A} \\ &= \mathbf{e}_1 \left(\frac{\partial A_z}{\partial y} - \frac{\partial A_y}{\partial z} \right) + \mathbf{e}_2 \left(\frac{\partial A_x}{\partial z} - \frac{\partial A_z}{\partial x} \right) + \mathbf{e}_3 \left(\frac{\partial A_y}{\partial x} - \frac{\partial A_x}{\partial y} \right).\end{aligned} \tag{5.33}$$

It is also important to recall the concept of gradient of a scalar function, since the electrostatic field is defined as the gradient of the scalar potential ϕ, viz.,

$$\operatorname{grad} \phi = \nabla \phi = \mathbf{e}_1 \frac{\partial \phi}{\partial x} + \mathbf{e}_2 \frac{\partial \phi}{\partial y} + \mathbf{e}_3 \frac{\partial \phi}{\partial z} . \tag{5.34}$$

Then the electric field is defined in general as

$$\mathbf{E} = -\frac{1}{c} \frac{\partial \mathbf{A}}{\partial t} - \nabla \phi . \tag{5.35}$$

In electrodynamics, the potentials **A** and ϕ form the four-vector of the electromagnetic field, with components $A_1 = A_x$, $A_2 = A_y$, $A_3 = A_z$, $A_4 = i\phi$. Then we can write **E** in the form

$$\mathbf{E} = \frac{1}{i} \left[\mathbf{e}_1 \left(\frac{\partial A_1}{\partial \tau} - \frac{\partial A_4}{\partial x} \right) + \mathbf{e}_2 \left(\frac{\partial A_2}{\partial \tau} - \frac{\partial A_4}{\partial y} \right) + \mathbf{e}_3 \left(\frac{\partial A_3}{\partial \tau} - \frac{\partial A_4}{\partial z} \right) \right]. \quad (5.36)$$

The expression (5.36) is a way of writing (5.35) so that it has a form similar to (5.33). In this way we illustrate the fact that the components of the vectors $\mathbf{E}$ and $\mathbf{B}$ are the elements of a more complex mathematical entity, called the electromagnetic field tensor, which we represent by means of a four-dimensional antisymmetric matrix:

$$F_{\mu\nu} = \begin{pmatrix} 0 & -B_z & B_y & iE_x \\ B_z & 0 & -B_x & iE_y \\ -B_y & B_x & 0 & iE_z \\ -iE_x & -iE_y & -iE_z & 0 \end{pmatrix}, \quad (5.37)$$

where $\mu, \nu = 1, 2, 3, 4$. This tensor is the fundamental physical quantity describing the electromagnetic field in vacuum. The electric field $\mathbf{E}$ is the spatial vector with components $E_j = i F_{4j}$, whereas the magnetic field $\mathbf{B}$ is the pseudovector with components $B_i = \frac{1}{2} \epsilon_{ijk} F_{jk}$, where $j, k = 1, 2, 3$. From the properties of $F_{\mu\nu}$ under Lorentz transformations, the interdependence of the electric and magnetic fields can be found. Actually, these are characterized in general by the relativistic invariants

$$\mathbf{E}^2 - \mathbf{B}^2 = \text{invariant}, \qquad \mathbf{E} \cdot \mathbf{B} = \text{invariant}. \quad (5.38)$$

For instance, if $\mathbf{E}$ and $\mathbf{B}$ are perpendicular in some frame of reference (so that $\mathbf{E} \cdot \mathbf{B} = 0$), they will be perpendicular in any other frame of reference.

On the other hand, if in one frame there is only an electric field ($\mathbf{E} \neq 0$ and $\mathbf{B} = 0$), in another frame of reference, a magnetic field $\mathbf{B}'$ may also appear, but in the new frame, the new vectors $\mathbf{E}'$ and $\mathbf{B}'$ must satisfy the condition

$$\mathbf{E}'^2 - \mathbf{B}'^2 = \mathbf{E}^2. \quad (5.39)$$

Similarly, if in the initial frame there is only a magnetic field $\mathbf{B}$, in another frame of reference we should observe electric and magnetic fields $\mathbf{E}'$ and $\mathbf{B}'$, but satisfying the condition

$$\mathbf{E}'^2 - \mathbf{B}'^2 = -\mathbf{B}^2. \quad (5.40)$$

The four-vector $A_\mu = (A_1, A_2, A_3, A_4)$ is usually not observable, but the components of the field tensor $F_{\mu\nu}$, i.e., $\mathbf{E}$ and $\mathbf{B}$, are observable. If one adds to the four-vector A_μ the four-gradient of a Lorentz-scalar function f of coordinates and time, the value of $F_{\mu\nu}$, that is, the values of $\mathbf{E}$ and $\mathbf{B}$, do not change. This means that if we carry out the transformation

$$A_1' \rightarrow A_1 + \frac{\partial f}{\partial x},$$

$$A_2' \rightarrow A_2 + \frac{\partial f}{\partial y},$$

$$A_3' \rightarrow A_3 + \frac{\partial f}{\partial z}, \tag{5.41}$$

$$A_4' \rightarrow A_4 + \frac{\partial f}{\partial x_4} = i \left(\phi - \frac{1}{c} \frac{\partial f}{\partial t} \right),$$

and substitute A_1', A_2', A_3', A_4' into (5.33) and (5.36), the same values for **E** and **B** are obtained as by using A_1, A_2, A_3, A_4.

This property is called *gauge invariance*, and it is very important in modern physics.

Let us remark at this point that there is another way for defining the four-vectors and tensors in special relativity. One may avoid the imaginary fourth-component by using the so-called covariant and contravariant quantities (we shall encounter them again in Chap. 10 in a more general case). A contravariant four-vector is written as $A^\mu = (A^0, A^1, A^2, A^3) = (A^0, \mathbf{A})$. Its covariant partner is A_μ, with the property $A_0 = A^0$ and $A_i = -A^i$, with $i = 1, 2, 3$. The scalar product of two four-vectors A^μ and B^μ is given by

$$A^\mu B_\mu = A^0 B_0 + A^1 B_1 + A^2 B_2 + A^3 B_3 = A^0 B^0 - A^1 B^1 - A^2 B^2 - A^3 B^3.$$

A covariant four-vector is obtained from its contravariant expression by multiplying it by the metric tensor $g_{\mu\nu}$, where $g_{00} = 1$, $g_{11} = g_{22} = g_{33} = -1$ and $g_{\mu\nu} = 0$, for $\mu \neq \nu$. Thus,

$$A_\mu = g_{\mu\nu} A^\nu,$$

where sum over repeated indices is assumed. Similarly, we may define how to get contravariant four-vectors from the covariant ones. The scalar product may be written as $A^\mu B_\mu = g_{\mu\nu} A^\mu B^\nu$. The correspondence with the imaginary components is simple: $A^0 = A_0 = A_4/i$, whereas the spatial components are the same as the contravariant ones $A^{1,2,3} = A_{1,2,3}$.

Four-vectors and tensors in the imaginary fourth-component notation are sometimes easier to handle. Also, they appear in Euclidean field theories (as temperature quantum field theory), and it is useful to be familiar with both notations.

5.9 Energy and Momentum

The momentum of a free particle of mass m moving at a velocity **V** is defined in the special theory of relativity as

$$\mathbf{p} = \frac{m\mathbf{V}}{\sqrt{1 - V^2/c^2}}, \tag{5.42}$$

and its energy as

$$E = \frac{mc^2}{\sqrt{1 - V^2/c^2}}. \tag{5.43}$$

The two quantities form a four-vector $P_\mu = \left(p_x, p_y, p_z, i\frac{E}{c}\right)$, with

$$P_\mu^2 = p^2 - \frac{E^2}{c^2} = -m^2c^2, \tag{5.44}$$

from which we obtain

$$E = c\sqrt{p^2 + m^2c^2}. \tag{5.45}$$

For low velocities $V \ll c$, so that $\sqrt{1 - V^2/c^2} \approx 1$, the expressions (5.42) and (5.43) yield the non-relativistic momentum

$$\mathbf{p} = m\mathbf{V},$$

and energy

$$E = mc^2 + \frac{mV^2}{2}, \tag{5.46}$$

that is, the *rest energy* mc^2 plus the expression for the kinetic energy of Newtonian mechanics. For $V = 0$, we have the expression

$$E = mc^2, \tag{5.47}$$

which relates the mass of a body at rest with its energy content. This expression is probably the most widely known consequence of the theory of relativity.

The largest amount of energy that a body is able to produce (for example, when it is transformed completely into radiation) is equal to the product of its mass and the square of the speed of light. This relation explains the production of enormous amounts of energy in nuclear fission processes (division of an atomic nucleus), in which a certain excess of the initial mass of the nucleus when compared to the sum of the masses of the final nuclei is totally converted into radiation energy.

But there are some other processes, such as particle–antiparticle pair creation, to which we will refer subsequently in more detail, in which a photon, which is massless, but has high enough energy, when passing near an atomic nucleus or through a magnetic field, disappears, while producing two new particles, an electron and a positron. For the process to occur, the energy of the photon should be greater than or equal to twice the rest energy of the electron (the mass of the positron is equal to the electron mass, and therefore, its rest energy is equal to that of the electron), that is

$$E_{photon} \geq 2\,mc^2, \tag{5.48}$$

where m is the electron mass.

5.10   Photons

From the expressions (5.42) and (5.43) for the energy and momentum of a particle in relativistic mechanics, one deduces the relationship

$$\mathbf{p} = \frac{E}{c^2}\mathbf{V}. \tag{5.49}$$

This expression is the analog of the non-relativistic expression

$$\mathbf{p} = m\mathbf{V}, \tag{5.50}$$

with the mass m replaced by the ratio E/c^2. If the particle moves at the speed of light, we have

$$\mathbf{p} = \frac{E}{c}\mathbf{n}, \tag{5.51}$$

where $\mathbf{n}$ is the unit vector in the direction of motion. Introducing (5.51) into (5.44), we observe that $m = 0$ for a particle moving at the speed of light. This is understandable, if we recall that the particle moves at the maximum possible velocity. If it had a nonzero mass, then the greater its velocity, the more difficult it would be to accelerate it, that is, to increase its velocity, and in fact it could not reach the speed of light.

We have already referred to the behaviour of electromagnetic radiation as waves and as particles, which we called photons. The particle character is more easily observable for high frequencies, as in the X and gamma rays.

Photons are typical relativistic particles, whose mass is zero. But their corpuscular character is relative, since the frequency of the radiation coming from a lamp, measured by an external observer, depends on the velocity of the lamp relative to the observer. For instance, if it emits red light, and approaches the observer at high velocity, the light could appear to be yellow, blue, etc., depending on its velocity. If the lamp approaches the observer at almost the speed of light, the observer will detect photons of X or gamma rays. This is a relativistic effect, and radiation of low frequency emitted by a lamp can be *seen* by an observer as having very high frequency (where the corpuscular character dominates) if the lamp approaches the observer at a velocity close to c. The opposite effect occurs if the lamp moves away from the observer.

Fig. 5.10 Right (R) and left (L) circular polarizations of the photon. In the first case, the spin direction is parallel to its momentum **k**. In the second, they are antiparallel.

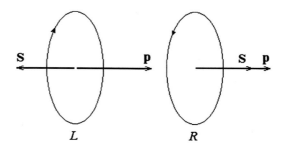

Let us analyze the characteristics of the photon in vacuum in more detail. The condition of gauge invariance, combined with the fact that it moves at the speed of light, implies that the photon has only two transverse degrees of freedom, which correspond to the two possible polarizations (for instance, left- and right-circular) of an electromagnetic wave, in the wave version of the electromagnetic radiation phenomenon (Fig. 5.10).

On the other hand, the photon has an intrinsic angular momentum or spin equal to 1 (in units of the Planck constant $\hbar$, as already mentioned). For any particle, one can define a quantity proportional to the projection of the spin on the direction of momentum, namely the *helicity*:

$$H = \frac{\mathbf{S} \cdot \mathbf{p}}{|\mathbf{p}|}.$$

For massless particles, the helicity is a relativistic invariant. When the relativistic particle has zero mass, its momentum and spin direction can only be parallel or antiparallel. This means that the helicity can have only two values, and we interpret them as characterizing two degrees of freedom. For a photon, these are identified with the two transverse circular polarizations.

A particle similar to the photon, but with mass different from zero, would move at a lower velocity than c, and in addition to the two transverse degrees of freedom, it would have a third one, corresponding to a longitudinal polarization.

If the radiation propagates in a medium, for instance in a plasma, these longitudinal oscillations may appear, since the radiation acquires an *effective mass*, and the velocity of propagation of the waves is less than the speed of light in vacuum.

5.11 Neutrinos

Neutrinos were considered for years as purely relativistic particles, that is, with zero rest mass. They differ from photons, which are bosons (have integer spin $\hbar$), because neutrinos are fermions (have half-integer spin $\hbar/2$). However, in recent

Fig. 5.11 For the neutrino, the spin is related to the momentum direction as in a left-hand screw (*L*), while for the antineutrino, the relation is as in a right-hand screw (*R*).

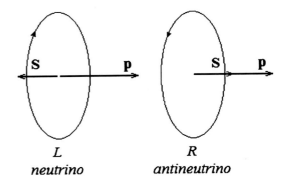

years observations of so-called neutrino oscillations suggest that they may have a very small mass.

Such oscillations have been measured experimentally, indicating a mass of around $\sim 10^{-5}$ of the electron mass (see Chap. 11). However, for the moment we shall consider the neutrinos to be massless, as is done in the Standard Model of particle physics. The neutrino spin is related to its momentum as shown in Fig. 5.11. The neutrino is said to have left helicity (*L*), while the antineutrino has right helicity (*R*). For spinors, or spin 1/2-fermions, the values of the helicity can be $H = \pm \hbar /2$. The plus sign corresponds to *R* and the minus sign to *L*.

It is easy to understand that such a massless neutrino cannot change its helicity, since that would be equivalent to reversing its direction of motion, and to do that, it would be necessary to be at rest at some moment, or in some reference frame. Neutrinos are found in Nature only with negative helicity, or left-handed. They are *chiral particles*. We shall return in more detail to chirality in Chap. 11.

Antineutrinos are produced in beta decay and their *R* helicity is related to parity non-conservation, which we shall discuss in Chap. 9.

5.12   Tachyons and Superluminal Signals

Is it actually forbidden by the special theory of relativity for particles to move faster than light? We have seen that particles moving at lower speeds than c will have a speed lower than c in any frame of reference. This is a consequence of the law of composition of velocities (5.25). If $v > c$, the expressions for the energy and momentum (5.43) and (5.42) can be real if the mass m is a pure imaginary number $m = iM$.

At the end of the 1950s, it was suggested that the existence of superluminal particles, called tachyons, would not contradict special relativity.

However, there is no evidence for tachyons as observable particles. We shall see in Chap. 7 that quantum theory assumes the existence of non-observable *virtual particles* which are not necessarily confined by the light cone, and their momentum/energy

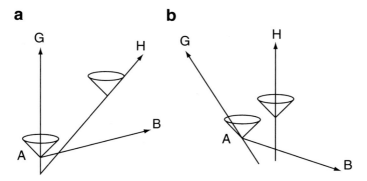

Fig. 5.12 (**a**) If G has a superluminal source, a signal sent in A is seen by H as if it traveled to the past when it is received at B. (**b**) The events as seen by H.

Fig. 5.13 If F also has a superluminal source, and when the signal sent by G from A is received at B, he sends a signal in the opposite direction, the latter is received by G before the signal is emitted from A. Thus, G could influence his past.

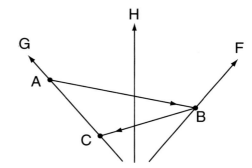

relation may correspond to a higher speed than c. But virtual particles do not carry energy at an *observable* speed higher than c.

However, if we assume the existence of a superluminal 'signal', this could transmit information to the past, violating causality. This was argued cogently by Roger Penrose (b. 1931). Consider an observer G emitting a superluminal signal at the point A of his world line. Another observer H, which coincided with G before sending the signal and moved after that with velocity $V \leq c$ (Fig. 5.12a) would observe that the signal arrived at B *before* having been emitted, and thus travelled backward in time. From the point of view of the observer H, the diagram is as shown in Fig. 5.12b.

If now a third observer F, moving in the opposite direction to G (Fig. 5.13), also emits a superluminal signal just at the moment when the signal sent by G is received at B, but in the opposite direction, this signal would arrive at a point C on the world line of G *before* being emitted at A. In this way G would have influenced his own past.

Let us justify some of these conclusions by using the expressions for the Lorentz transformations of spacetime coordinates. We assume that the observer H sees G as moving to the left with velocity $-\mathbf{V}$ and F as moving to the right with velocity $\mathbf{V}$. We assume also that, at the moment when the superluminal signal with velocity $V_g > c$

is sent by G from the point A, the spacetime coordinate origins of the three frames coincide. For G, the point A has coordinates $(0, 0)$. Let us denote the coordinates of the signal's arrival at B by $(x, t = x/V_g)$. The two events (separated by a space-like interval) have coordinates $(0, 0)$ and (x', t') in the frame of H, where, defining $\gamma = 1/\sqrt{1 - V^2/c^2}$,

$$x' = \gamma x \left(1 - \frac{V}{V_g}\right), \qquad t' = \gamma \frac{x}{V_g} \left(1 - \frac{V V_g}{c^2}\right). \tag{5.52}$$

If we assume $V V_g > c^2$, then $t' < 0$. Then H concludes that the superluminal signal arrived at B before it was emitted. If the observer F sends a superluminal signal from B in the opposite space direction, also with velocity V_g, it is easy to show by similar reasoning that it is received at C at some time $t < 0$, before being sent from A. In this way, G can influence his own past. This precludes observable superluminal signals, since they would violate Einstein causality. An equivalent problem with superluminal signals appears in general relativity, in connection with the hypothetical wormholes, but this topic is purely speculative.

5.13   The Lagrangian for a Particle in an Electromagnetic Field

Equations (5.42), (5.43) can be obtained from the Lagrangian in relativistic mechanics. For a free particle, we define the action as

$$S = -mc \int_1^2 ds, \tag{5.53}$$

where $ds = \sqrt{1 - V^2/c^2} dt$ is the infinitesimal space-time interval. For the Lagrangian of the free particle, we get $L = -mc^2\sqrt{1 - V^2/c^2}$, and from this, equations (5.42), (5.43) follow immediately since $\mathbf{p} = \partial L/\partial \mathbf{v}$, and $E = \mathbf{p} \cdot \mathbf{v} - L$.

In the presence of an electromagnetic field A_μ, the action of the particle contains the additional term $\frac{e}{c} \int A_\mu dx_\mu = \frac{e}{c} \int A_\mu (dx_\mu/dt) dt$, where $dx_\mu/dt = (\mathbf{v}, ic)$. The new Lagrangian is

$$L = -mc^2\sqrt{1 - V^2/c^2} + \frac{e}{c}\mathbf{A} \cdot \mathbf{v} - e\phi, \tag{5.54}$$

from which the generalized momentum is $\mathbf{P} = \frac{\partial L}{\partial \mathbf{v}}$, whence

$$\mathbf{P} = \frac{m\mathbf{v}}{\sqrt{1 - V^2/c^2}} + \frac{e}{c}\mathbf{A} = \mathbf{p} + \frac{e}{c}\mathbf{A}. \tag{5.55}$$

We can get the Hamiltonian from $H = \mathbf{v} \cdot \partial L / \partial \mathbf{v} - L$, and after substituting the appropriate terms we have

$$H = \frac{mc^2}{\sqrt{1 - V^2/c^2}} + e\phi. \tag{5.56}$$

Note that the Hamiltonian does not contain the term $\mathbf{A}$, so it does not depend on the magnetic field. We conclude that

$$\left(\frac{H - e\phi}{c}\right)^2 = m^2 c^2 + \left(\mathbf{P} - \frac{e}{c}\mathbf{A}\right)^2. \tag{5.57}$$

Replacing the Hamiltonian by the energy, we get an expression for the total energy of a charged particle of mass m in an electric potential $\phi = \frac{e'}{r}$. We shall assume that it is created by another particle of mass $M \gg m$. Then, denoting the product of their charges by $K = ee'$, the energy is

$$E = c\sqrt{p^2 + m^2 c^2} + \frac{K}{r}. \tag{5.58}$$

If we use polar coordinates in a plane, we have $p^2 = p_r^2 + L^2/r^2$, where L is the angular momentum, and we can find the equation of motion of the particle of mass m. It can be shown that, for the case $Lc > |K|$, the curve is given by the expression

$$r = \frac{c^2 L^2 - K^2}{EK + c\sqrt{L^2 E^2 - m^2 c^2 (L^2 c^2 - K^2)} \cos\left(\phi\sqrt{1 - K^2/c^2 L^2}\right)}. \tag{5.59}$$

If $E < mc^2$, the trajectory is confined to finite values of r, but it can never be closed. Defining $\theta = \phi\sqrt{1 - K^2/c^2 L^2}$, if ϕ takes values in $(0, 2\pi)$, the range of θ is smaller, in $(0, 2\pi\sqrt{1 - K^2/c^2 L^2})$. Thus, the radial frequency is smaller than the angular frequency, and the motion is in the form of a rosette, i.e., like a precessing elliptic curve. If $E > mc^2$, the trajectory is open, like a precessing hyperbolic curve.

Problems

Problem 5.1 Write down the inverse of the Lorentz transformation (5.6).

Problem 5.2 Doppler effect. Consider an electromagnetic wave propagating in a frame of reference K_0 in which the source is at rest. Its electric field is

$$\mathbf{E} = \mathbf{e}\, e^{i(k_1 x + k_2 y + k_3 z - \omega_0 t)}.$$

To describe the propagation of light, the phase term must be a relativistic scalar, which is the scalar product of two four-vectors, the coordinate four-vector x, y, z, ict and the wave four-vector with components k_i, $i = 1, 2, 3$, and $k_4 = i\omega/c$, where $\omega = (k_1^2 + k_2^2 + k_3^2)^{1/2}c$. The frequency of the light wave is measured by an observer moving with velocity V relative to the light source. We consider the observer to be moving parallel to the x axis, and we can write the component $k_1 = k \cos \alpha = \frac{\omega}{c} \cos \alpha$. Find the observed frequency ω as a function of the source frequency ω_0 in the rest frame, the velocities V and c, and the direction of the wave vector α.

Problem 5.3 Consider a particle of mass M at rest which decays spontaneously into two particles of masses m_1 and m_2. Since they are moving, their energies are bigger than their rest energies, $E_1 \geq m_1 c^2$, $E_2 \geq m_2 c^2$. Conservation of energy implies

$$Mc^2 = E_1 + E_2. \tag{5.60}$$

Momentum must be conserved, as well as energy, in the decay. Find the expressions for E_1 and E_2 in terms of the masses M, m_1, and m_2.

Problem 5.4 The Sun's core is estimated to have a radius $1/5$ of the solar radius. Its volume is expected to be roughly $(1/5)^3$ of the total volume of the Sun, which is $V_\odot = 1.422 \times 10^{18}$ km^3. Estimate the equivalent mass of the radiation enclosed if its temperature is $\sim 1.36 \times 10^7$ K.

Problem 5.5 Assume that (5.59) is valid for a relativistic Keplerian problem, that is, for the case of a gravitational field. This is an approximation, since the problem should actually be solved in the framework of general relativity (see Chap. 10). However, the present approximation is assumed for the moment to be special relativity plus Newtonian gravity. For the motion of a photon, we take $m = 0$ everywhere except in the quantity $K = -GMm$. Calculate the deviation from a straight line for a photon of energy $E > 0$ coming from a distant star and passing close by the limb of the Sun, of mass $M_\odot$.

Hint: Write $K = -GME/c^2$, that is, treat the photon as a particle interacting with the gravitational field via an "effective gravitational mass" E/c^2. (This is not strictly rigorous and we shall use it only in the present problem.) By calling $r_g = 2GM/c^2$ the Sun's gravitational radius, we obtain $K = -Er_g/2$, leading to $K^2/c^2L^2 = r_g^2/4R_\odot^2$. The motion is actually an open precessing hyperbolic motion, since the argument of the curve is such that, when the polar angle ϕ varies from 0 to 2π, it sweeps through an angle less than 2π, but since the difference is only a very small amount, namely, $\sim 2\pi r_g^2/R_\odot^2$, it can be approximated by a hyperbolic motion.

Literature

1. E.F. Taylor, J.A. Wheeler, *Spacetime Physics* (W.H. Freeman and Co., San Francisco, 1966). An excellent introductory book for special relativity

2. C. Kittel, C. Knight, M.A. Ruderman, *Mechanics, The Berkeley Physics Course* (McGraw-Hill, New York, 1973). Contains a very good introduction to special relativity
3. L.D. Landau, E.M. Lifschitz, *The Classical Theory of Fields* (Pergamon, Oxford, 1975). The part of this book devoted to the electromagnetic field is a masterpiece. An excellent introduction to special relativity at an advanced level
4. R. Penrose, *The Emperor's New Mind* (Penguin Books, New York, 1991). This interesting and deep essay discussed the compatibility between special relativity and quantum mechanics, as well as other topics

Chapter 6
Atoms and Quantum Theory

In the twentieth century began a new era for physics by modifying ideas established in previous centuries. The concepts of space and time were changed drastically by special relativity, while quantum behaviour and the dual nature of radiation were discovered. In Chap. 4 was described how the investigations on black body radiation and the photoelectric effect opened the way to understanding the quantum properties of the atomic world, which were first revealed by studying the emission and absorption of electromagnetic radiation.

While the wave nature of light had been demonstrated beyond doubt in a large number of experiments, in some new phenomena light appeared to have a corpuscular structure. The situation turned still more paradoxical when it became clear that, conversely, particles making up atomic structure, such as electrons, manifestly showed wave properties. Such wave–corpuscle duality was a characteristic of the atomic world.

It became clear later that it was not possible to determine simultaneously and with arbitrarily high precision the momentum and position of an atomic particle, e.g., the electron. A new form of mechanics had to be invented: quantum mechanics.

6.1 Motion of a Particle

Classically the motion of a particle is described by giving its position and its velocity (or its momentum) at any moment of time and, in principle, there is nothing to stop us from knowing where it is and how much momentum it has at each instant of time. But for a particle of the atomic world, this is not possible. To illustrate this point, it is preferable to start from two experiments.

First Experiment. Assume that we have fine sand and a screen with two small holes, in the form indicated in the Fig. 6.1. On a screen below it, we measure the probability

© The Author(s), under exclusive license to Springer-Verlag GmbH, DE, part of Springer Nature 2021
M. Chaichian et al., *Basic Concepts in Physics*, Undergraduate Lecture Notes in Physics, https://doi.org/10.1007/978-3-662-62313-8_6

Fig. 6.1 The probability of arrival of the grains of sand on the screen is P_1, if hole 2 is closed, and P_2 if hole 1 is closed. When both holes are open, it is $P_{12} = P_1 + P_2$.

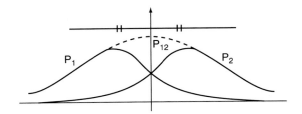

of arrival of the grains of sand when the two holes are open. We observe that they form a small hill of sand. If we are interested in the probability in a direction x (say, along a straight line on the screen, parallel to the line passing through both holes, as indicated in the figure), we will get the curve P_{12}. If we now close hole 2, a hill will be formed below hole 1, and the corresponding probability along the line joining the holes will be the curve P_1. The same analysis for the case in which 1 is closed and 2 is kept open would give a similar hill under hole 2 and the curve P_2.

The result of the experiment when the holes are open alternatively is the same as when both are open simultaneously. Furthermore, at each point,

$$P_{12} = P_1 + P_2. \tag{6.1}$$

Second Experiment. Suppose that a similar experiment is done with electrons, as in Fig. 6.2. Now, an "electron hill" would not form on the screen, but it would be possible in principle to measure the probability of arrival at each point, with adequate detectors.

If hole 2 is closed, we have the curve of probability P_1 along x (very similar to the one corresponding to the diffraction of light by a slit). If we close 1, we have the probability P_2. If both are open, we obtain the curve P_{12}. But now

$$P_{12} \neq P_1 + P_2. \tag{6.2}$$

When both slits are open, there is *interference* (a typical wave phenomenon) between the possible trajectories of the electron through each one of the holes. Which slit did the electron pass through? Through 1? Through 2? Through both? We can only state from the results of this experiment that the electron did not have a definite trajectory.

The description of the particles of the atomic world requires the inevitable use of objects that can be named in general *apparatus*, obeying the laws of classical mechanics, which are used to make the measurements of the physical quantities. The above-mentioned wall can be considered as one such piece of *apparatus* that *measures* through which hole the electron passes, and the more precise this measurement, the more uncertain the corresponding momentum, or the velocity.

This can be expressed as follows: quantum mechanics must be formulated in terms of principles essentially different from classical mechanics. For instance, in quantum mechanics there is no such concept as the path of a particle. This is a consequence of the uncertainty principle, formulated by Werner Heisenberg (1901–1976) in 1927.

Fig. 6.2 In the case of light or electrons, if we repeat experiment 1 by using an adequate wall, we find that $P_{12} \neq P_1 + P_2$.

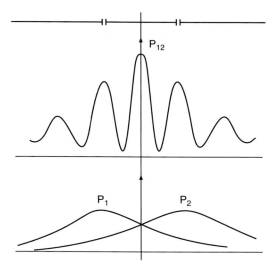

This principle also establishes that every measurement destroys part of our knowledge of a system that was obtained by previous measurements. As Heisenberg argued in his original paper, if Δx is the uncertainty in measuring the position along x, and Δp_x is the uncertainty in measuring simultaneously the conjugate momentum, their product is such that $\Delta p_x \Delta x \sim h$, where h is the Planck constant. This relation was the first quantitative expression of the uncertainty principle, although Heisenberg did not define precisely what the uncertainties meant. Later on, he refined the formula, but always preferred to think about the principle in heuristic terms. An exact expression was found in 1927 by Earle Hesse Kennard, by giving the uncertainties the statistical meaning of standard deviations (which is not really what Heisenberg meant):

$$\Delta p_x \Delta x \geq \hbar/2.$$

In some experiments, however, one can obtain information about both quantities: position and momentum or velocity, but each involving some indeterminacy. Then one could speak of the trajectory of the particle, but this has only an approximate meaning.

There exists a device to observe the trajectories of particles, called a Wilson cloud chamber, in honour of Charles Wilson (1869–1959), who was awarded the Nobel Prize in Physics in 1927 for this momentous invention. In such a chamber, the trajectory of a charged particle can be observed when it enters the chamber because it ionizes the medium, and water vapour condenses on the resulting ions to form small droplets. The width of the path described by such droplets, when observed, is large as compared with the atomic dimensions.

The process of measurement in quantum mechanics (which always requires macroscopic objects, or at least objects obeying the laws of classical physics) plays a central role in this science, since it always affects the quantum particle in such a way that, the more exact the measurement, the more significantly the motion of the particle is perturbed, according to the uncertainty principle.

The phenomenon of interference described in experiment 2 shows that the electron has wave properties. In fact, in order to describe a particle in quantum mechanics, a wave function $\Psi(x, t)$ is introduced, which is in general a complex function, the square of its modulus $|\Psi|^2$ being the probability density of localization of the particle in space, as we shall see later.

6.2   Evolution of the Concept of Atom

The Greek philosopher Democritus of Abdera (c.460–c.370 BCE) put forward the hypothesis that the Universe consists of empty space and an enormous number of indivisible particles, and that by joining and separating them, we get the creation and disappearance of bodies.

Approximately a century later, another Greek philosopher, Epicurus (341–270 BCE), gave the name *atoms* to these particles. More than twenty centuries after that, in 1738, Daniel Bernoulli was the first to attempt to construct a theory of gases based on the atomic structure model, using the calculus of probabilities, but his contemporaries did not pay very much attention to his work. At the beginning of the nineteenth century, John Dalton introduced the hypothesis of atomic structure once again, and Amedeo Avogadro contributed with the idea of integrated molecules (comprising several atoms each) and elementary molecules (as single atoms). Starting from the middle of the 19th century, the kinetic theory of gases was developed by James Prescott Joule, Rudolf Clausius, and James Clerk Maxwell, and subsequently by Ludwig Boltzmann, who based it on his statistical interpretation of the second law of thermodynamics. In 1881, Hermann von Helmholtz, after analyzing the work done by Michael Faraday on electrolysis, suggested the atomic nature of electricity, and later, George Johnstone Stoney proposed the term *electron* for the unit of electric charge.

In 1897, Joseph John Thomson, as a consequence of his studies of cathode rays, once again stated the atomic nature of electricity. Thomson is credited with the actual discovery of the electron, for which he was awarded the Nobel Prize in Physics in 1906. Later, Thomson proposed a model of the atom sometimes called the *plum pudding*, since it treated the atom as a positive charge in which the electrons were embedded like the plums in a pudding. The electrons were assumed to oscillate around their mean positions when emitting or absorbing radiation.

6.3   Rutherford's Experiment

In 1884, the Swiss mathematician Johann Jakob Balmer published the result of his investigations on the hydrogen spectrum data. When the radiation emitted by hydrogen gas is studied (for example, by producing electric arc sparks inside a jar containing the gas), the visible spectrum is found to consist of a series of discrete lines. Balmer gave an empirical formula for the frequencies of the different lines:

$$\nu = cR \left(\frac{1}{2^2} - \frac{1}{n^2} \right), \qquad n = 3, 4, 5, \ldots, \tag{6.3}$$

where $R = 1.09737 \times 10^5 \text{cm}^{-1}$ is the Rydberg constant and c is the speed of light. Later, similar empirical formulas for other spectral series of hydrogen were found.

In 1911, Ernest Rutherford (1871–1937) bombarded a thin sheet of gold with alpha particles (which are helium nuclei, having positive charge) and concluded that atoms comprise a massive positively charged nucleus around which the electrons move rather like in a very small planetary system. The nucleus was playing the role of the Sun, and the electrons were moving around it like the planets. The experimental data indicate that the nucleus contains more than 99.8% of the mass of atom but its radius is approximately 10^5 smaller than the atomic radius.

The basic idea behind Rutherford's experiment is very simple. Suppose that one has a bale of hay within which one wants to know whether a steel tool is hidden. We do not have at our disposal any other instrument than a shotgun with steel bullets, and we cannot undo the bale of hay. How can we find out whether the tool is hidden there? If we fire the shotgun and the bullets pass straight through, there is no tool in the bale. But if some bullets recoil, and if we measure their angles of deviation and the number of bullets emerging at each angle or in an interval of angles, we get much more information about the tool, i.e., we can determine whether it is large or small, etc.

Rutherford counted the alpha particles recoiling at different angles in his experiment, in which the "shotgun" was an emitter of alpha particles, and his "bale of hay" was a sheet of gold. The leading idea was that, since the charge of the alpha particles is positive, if there was a positively charged nucleus, the effect of the bombardment would give rise to collisions of alpha particles with the nuclei, similarly to the elastic collisions of the bullets with the tool hidden in the hay. This was the first reported experiment on particle scattering in physics.

6.4 Bohr's Atom

At this point, a contradiction appeared with the electromagnetic theory. The planetary model suggested that the electrons moved around the nucleus in elliptical orbits. But in this case, the electrons would be accelerated continuously and, according to electrodynamics, an accelerated charge should emit radiation, leading to a continuous loss of energy, so that the electron would eventually fall onto the nucleus. Furthermore, this emission of energy would give a continuous spectrum. However, spectroscopists had demonstrated that atoms do not emit energy continuously, but discretely, in the form of spectral lines (Fig. 6.3).

Fig. 6.3 The Danish
physicist Niels Bohr was one
of the most outstanding
physicists of the twentieth
century.

In 1913, a young Danish physicist, Niels Bohr (1885–1962), who had collaborated
with Rutherford, found a way to solve the crisis by exploiting the quantum ideas
introduced by Planck and Einstein. Bohr proposed two fundamental postulates:

1. *Out of all electron orbits, only those are permissible for which the angular
   momentum of the electron is an integer multiple of the reduced Planck constant
   ħ, and no energy is radiated while the electron remains in any one of these
   allowed orbits. There orbits are called stationary;*
2. *Whenever radiation energy is emitted or absorbed by an atom, this energy is
   emitted or absorbed in quanta that are integer multiples of $h\nu (= \hbar\omega)$, where ν
   is the frequency of the radiation, and the energy of the atom is changed by this
   amount.*

In other words, if E_i and E_f are the initial and final energies of the electron in the
atom emitting radiation, the condition

$$E_i - E_f = h\nu$$

is satisfied. For circular orbits, the quantization condition of Bohr (first postulate)
reads:

$$mvr = n\hbar, \quad n \text{ integer,} \tag{6.4}$$

where m is the mass of the electron, v is its velocity, and r is the radius of its orbit in the atom. On the other hand, the electron is kept on its orbit by the Coulomb force, which is equal in absolute value with the centripetal force:

$$\frac{mv^2}{r} = \frac{e^2}{r^2}. \tag{6.5}$$

From (6.4) and (6.5), we obtain easily the quantization of the radius and velocity:

$$r_n = \frac{n^2\hbar^2}{e^2m}, \quad v_n = \frac{e^2}{n\hbar}.$$

A simple calculation leads to the following expression for the energy of the electron in the hydrogen atom:

$$E_n = -\frac{me^4}{2\hbar^2}\frac{1}{n^2}, \quad n = 1, 2, 3\ldots \tag{6.6}$$

We observe that the constant coefficient in (6.6), if we multiply and divide by c^2, may be written as proportional to $mc^2\alpha^2$, where mc^2 is the rest energy of the electron according to Chap. 5, and the dimensionless constant $\alpha = e^2/\hbar c \simeq 1/137$, the so-called fine structure constant, characterizes the electromagnetic interactions in the atom. According to the second postulate, for the frequency of the spectral lines we obtain:

$$\nu = \frac{E_f - E_i}{h} = \frac{2\pi^2 me^4}{h^3}\left(\frac{1}{n_f^2} - \frac{1}{n_i^2}\right). \tag{6.7}$$

Here $n_f = 2$ yields the Balmer series (n_i is always greater than n_f), while $n_f = 1, 3, 4, 5$ yield the Lyman, Paschen, Brackett, and Pfund series, respectively. The Balmer series lies in the visible and near ultraviolet region. The Lyman series is in the ultraviolet, whereas the last three are in the infrared region. The number

$$\frac{2\pi^2 me^4}{ch^3} = 1.09740 \times 10^5 \text{cm}^{-1} \tag{6.8}$$

corresponds very well to the value of the Rydberg constant for hydrogen, named after Johannes Robert Rydberg (1854–1919). The reader may compare with (6.3), and observe that the value predicted by theory agrees very well with the experimental result.

The different spectral series result from the jumps of the electron from diverse excited states to a fixed final state. For instance, the Balmer series is produced by electron jumps from the initial levels $n_i = 3, 4, 5\ldots$, to the final level $n_f = 2$.

Bohr's theory concluded that atomic quantities must be discrete multiples of the quantum of action h. The system obeyed classical mechanics, except that the allowed motions satisfied the more general Bohr–Sommerfeld quantization rule

$$\oint p_i dq_i = n_i h,$$

where n is an integer and p_i is the momentum conjugated to the coordinate q_i, the integral being taken over the classical region, where $H(p,q) = E$. In the case of circular orbits as in (6.4), $p = mr\omega, dq = rd\theta, 0 \le \theta \le 2\pi$, where r, ω are constant.

Bohr's postulates led to an adequate model to explain the spectra of the hydrogen atom, but in the end they were replaced by a more complete quantum theory.

There is a historical analogy between the role of Bohr's quantum mechanics as regards the atom and Newton's mechanics as regards planetary motion. In Chap. 1, we noted that, starting with observational results, some empirical laws were formulated (the Kepler laws), and Newton subsequently constructed the *theory*: the second law of motion and the gravitational interaction law. This time there was also an experimental result (the discrete character of the emission spectra) and empirical laws (the Balmer series), and then a physical theory was formulated, based on Bohr's postulates, and from which the empirical laws could be deduced.

Newtonian mechanics and the theory of gravitation remained valid for more than two centuries, until it was demonstrated that they are limiting cases of more general theories, viz., Einstein's relativistic mechanics and theory of gravitation.

In contrast, Bohr's quantum mechanics (or the Bohr–Sommerfeld model of quantization) became obsolete after a much more shorter time. The model was unsatisfactory for describing more complicated atomic systems, such as the helium atom. It ignored the electron spin and the Pauli exclusion principle, and it contradicted the uncertainty principle, since it assumed classical orbits where position and momentum could be known simultaneously. Thus, after some twelve years, Bohr's model was substituted by the new quantum mechanics due to Erwin Schrödinger, Werner Heisenberg, Max Born, Paul Adrien Maurice Dirac, Pascual Jordan and others. Bohr himself had the privilege, not only of following this evolution of the quantum theory, but also of contributing significantly to its development.

6.5 Schrödinger's Equation

In 1924, Louis de Broglie (1892–1987), made the bold suggestion that, since radiation has dual behaviour as waves and particles, atomic particles like electrons, should also manifest wave properties. That is, if we have the relation

$$E = h\nu \tag{6.9}$$

between energy and frequency for a particle, then we must have the relation

$$p = mv = \frac{h}{\lambda} \tag{6.10}$$

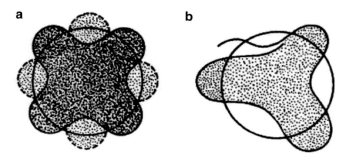

Fig. 6.4 a According to the de Broglie model, a stationary orbit should contain an integer number of wavelengths of the electron. **b** In the opposite case, there would be destructive interference and such an orbit would not exist for the electron.

between momentum and wavelength. This speculation by de Broglie was confirmed experimentally by George Paget Thomson (1892–1975), and independently by Clinton Joseph Davisson (1881–1958) and Lester Halbert Germer (1896–1971) in 1927, when they observed the phenomenon of electron diffraction in crystals. Louis de Broglie was awarded the Nobel Prize in 1929, and Clinton Davisson and George Thomson in 1937.

The de Broglie hypothesis can be shown to give rise to the Bohr stationary states. To have an electron in a stable orbit around the nucleus, the closed orbit must contain an integer number of wavelengths, otherwise the waves would interfere destructively (Fig. 6.4). Then, if r is the radius of the orbit, we must have

$$2\pi r = n\lambda. \tag{6.11}$$

However,

$$\lambda = h/mv, \tag{6.12}$$

whence

$$rmv = n\hbar, \tag{6.13}$$

which was the first Bohr postulate.

As already pointed out, Bohr's theory, developed subsequently by Arnold Sommerfeld (1868–1951) among others, could not account for the new atomic phenomena. Around 1925, Heisenberg, Jordan, and Born worked on a matrix form of mechanics which differed from Bohr's model and gave results compatible with experiment. For his work on matrix mechanics, Werner Heisenberg was awarded the Nobel Prize in 1932. But in 1926, a major step was taken by Erwin Schrödinger when he published his famous wave equation, which was the beginning of the new quantum mechanics, and for which he received the Nobel Prize in 1933. Schrödinger also showed the equivalence between his wave mechanics and the matrix mechanics of Heisenberg, which are thus different ways of expressing quantum mechanics.

The fundamental assumptions made by Schrödinger, which led him to his final equation, can be outlined as follows. There exists an analogy among the basic equations of classical mechanics and those of geometrical optics (we recall the analogy between Hamilton's principle of least action and Fermat's principle in optics, mentioned in Chaps. 1 and 4). Then, if atomic particles have wave properties, they should be governed by a wave mechanics that must bear the same relation to classical mechanics as wave optics bears with regard to geometrical optics, according to the following scheme:

$$\text{Wave Optics} \longrightarrow \text{Geometrical Optics}$$
$$\text{Wave Mechanics} \longrightarrow \text{Classical Mechanics}$$

In essence, the mathematical way to get the Schrödinger equation, is the following.
(a) First write down the classical expression for the energy of the system under investigation. The kinetic energy is expressed in terms of the linear momentum:

$$\frac{1}{2m}[p_x^2 + p_y^2 + p_z^2] + U(r) = E, \tag{6.14}$$

where $p^2/2m$, $U(r)$, and E are the kinetic, potential, and total energies, respectively. As an example, for the electron in the hydrogen atom, $U(r) = -e^2/r$.
(b) Classical quantities are replaced by operators, denoted by hatted letters, according to the following rules:

$$\begin{aligned}
p_x &\to \hat{p}_x = -i\hbar\tfrac{\partial}{\partial x}, \\
p_y &\to \hat{p}_y = -i\hbar\tfrac{\partial}{\partial y}, \\
p_z &\to \hat{p}_z = -i\hbar\tfrac{\partial}{\partial z}, \\
E &\to \hat{E} = i\hbar\tfrac{\partial}{\partial t}.
\end{aligned} \tag{6.15}$$

Coordinates are substituted by themselves as operators, $\hat{x} = x$, $\hat{y} = y$, $\hat{z} = z$.
(c) A differential equation is built for the *wave function*, using the substitutions (6.15) in (6.14), and applying the result to the wave function $\Psi(\mathbf{r}, t)$:

$$\left[-\frac{\hbar^2}{2m}\left(\frac{\partial^2}{\partial x^2} + \frac{\partial^2}{\partial y^2} + \frac{\partial^2}{\partial z^2} \right) + U(r) \right]\Psi = i\hbar\frac{\partial\Psi}{\partial t}, \tag{6.16}$$

where the quantity in squared brackets is named the Hamiltonian operator. In what follows it will be denoted by $\hat{H}$, so that (6.16) is written simply as

$$\hat{H}\Psi = i\hbar\partial\Psi/\partial t.$$

(d) In general, (6.16) is solved by imposing some simple conditions, viz., Ψ is periodic in time (as for any wave motion), vanishes at infinity, and is normalized so that $\int \Psi^*\Psi d^3x = 1$, where Ψ^* is the complex conjugate of Ψ. Assuming the separation of space and time variables, we write $\Psi(x, y, z, t) = u(x, y, z)\Theta(t)$. Then dividing the Schrödinger equation by $u(x, y, z)\Theta(t)$, we get

$$\left[-\frac{\hbar^2}{2m}\left(\frac{\partial^2}{\partial x^2} + \frac{\partial^2}{\partial y^2} + \frac{\partial^2}{\partial z^2}\right) + U(x, y, z)\right]u(x, y, z)/u(x, y, z) = i\hbar\frac{\partial\Theta(t)}{\partial t}/\Theta(t),$$

where the left-hand side depends only on $u(x, y, z)$ and the right-hand side only on $\Theta(t)$. Each side may thus be equated to a constant, which is the energy E, whence

$$\Theta(t) = e^{-i\frac{E}{\hbar}t}$$

and

$$\left(\frac{\partial^2}{\partial x^2} + \frac{\partial^2}{\partial y^2} + \frac{\partial^2}{\partial z^2}\right)u + \frac{2m}{\hbar^2}[E - U(x, y, z)]u = 0. \qquad (6.17)$$

For the hydrogen atom, Bohr's postulates and the energies of the stationary states follow as an immediate consequence:

$$E_n = -\frac{me^4}{2\hbar^2}\frac{1}{n^2}. \qquad (6.18)$$

Furthermore, the angular momentum is spatially quantized, that is, the inclination of the orbit of the electron can assume only a discrete set of values, depending on the value of n.

In order that the average value of quantum mechanical operators be real, the operators must be Hermitian or self-adjoint. This means that, if F is an operator, its average must satisfy $\langle F \rangle = \langle F \rangle^*$. This can be written explicitly as

$$\int\int d\mathbf{r}\varphi^* F\varphi = \int\int d\mathbf{r}\varphi F^*\varphi^*,$$

where φ is a function of x, y, z and $*$ indicates the complex conjugate. If the operator is a matrix M_{ij}, the condition is $M_{ij} = M_{ji}^*$, i.e., it must be the complex conjugate of its transpose (the transposed matrix is obtained by exchanging rows and columns). Denoting the transpose by the superscript t, the Hermitian conjugate of M_{ij} is $M_{ij}^{t*} = M_{ji}^*$. A unitary operator U is one whose Hermitian conjugate is its inverse, i.e., $U^{t*}U = 1$, or $U^{-1} = U^{t*}$. It is common to simplify the notation for Hermitian conjugation by putting $U^{t*} = U^\dagger$. In this notation the last two relations read $U^\dagger U = U^\dagger U$ and $U^{-1} = U^\dagger$ (Fig. 6.5).

Fig. 6.5 The Austrian physicist Erwin Schrödinger, whose famous equation is one of the cornerstones of modern physics. He worked in several other fields, particularly in statistical physics.

6.6 Wave Function

Schrödinger interpreted $\Psi(x, y, z, t)$ as a wave field, and from that one could assume that particles like electrons would be something like *wave packets*, similar to the pulse of radiation of Fig. 4.18. But this idea did not work. Among other problems, the *wave packet* would disperse and destroy itself in a very short time. However, the term 'wave packet' is frequently used to refer to quantum mechanical particles, with dual character.

Max Born (1882–1970) was the first to interpret the wave function amplitude as a quantity associated with the probability of locating the particle at a particular point. For this contribution, Max Born received the Nobel Prize in Physics in 1954. According to Born's interpretation, the square $|\Psi|^2$ of the modulus of the wave function is the probability density of finding the particle at a given point. If the wave function for the electron in the hydrogen atom is calculated, it is found that Ψ depends on three integers n, l, m. The first of these, n, determines the energy, while the second, l, is associated with the angular momentum, and the third, m, is associated with the component of the angular momentum along one of the coordinate axes. (A fourth number s would be necessary in order to characterize the spin or intrinsic angular momentum of the electron.)

The number n is a positive integer always greater than l, which is also a positive integer, whereas, for a given value of l, the number m can assume all integer values from $-l$ to l.

Table 6.1 Eigenvalues of the quantities determining the electron state in the atom.

General expressions	Particular values for $n = 2, l = 1$ $m = -1, m_s = -\frac{1}{2}$
Energy $= -\frac{me^4}{2\hbar^2}\frac{1}{n^2}$	$E = -\frac{me^4}{2\hbar^2}\frac{1}{4}$
Square of the angular momentum $L^2 = l(l+1)\hbar^2$	$L^2 = 2\hbar^2$
z-component of angular momentum $L_z = m\hbar$	$L_z = -\hbar$
z-component of spin $S_z = m_s\hbar$	$S_z = -\frac{1}{2}\hbar$

For each three numbers n, l, m, the wave function Ψ is a function of the coordinates x, y, z, and the time t, and it characterizes a quantum state for the electron in the hydrogen atom in which the energy, the angular momentum, and the component of angular momentum in a given direction are well-defined quantities.

Besides these three numbers, in order to completely characterize the state of the electron, it is necessary to specify its spin or intrinsic angular momentum. This is another quantum number s whose value is 1/2 in units of $\hbar$, but more interesting is the number specifying the orientation of the spin along an axis, m_s (the counterpart of the number m). This number m_s can take the values $\pm 1/2$. Then the state of the electron in the atom can be characterized by the set of four numbers n, l, m, m_s. For example, if $n = 2$, l may assume the values 0 and 1. In the first case, we can have only $m = 0$, and in the second m may take the values $-1, 0, 1$, whereas m_s may take the values $1/2, -1/2$. Then a particular state of the electron may be described, for instance, by the set of four numbers $(n, l, m, m_s) = (2, 1, -1, -1/2)$ (see Table 6.1) (Fig. 6.6).

In contrast with the classical case, the quantities energy, angular momentum, and z-component of the angular momentum assume discrete values. (Note that the spin is a particular case of angular momentum, intrinsic to the particle and independent of the orbital momentum angular l.) But there is another extremely important difference.

In the classical case, we could in principle know the particle position exactly, if the exact values of the energy and the angular momentum were also known. In the quantum case this is not so. The wave function Ψ characterizing the electron depends on the coordinates, that is, its value depends on the coordinates of the point of space we consider. These coordinates can be, for instance, the distance from the electron to the nucleus and two angles that would fix its position in space. Let Ψ_n correspond to an *eigenvalue* of the energy, say,

$$\hat{H}\Psi_n = E_n\Psi_n, \tag{6.19}$$

Fig. 6.6 The German
physicist Werner
Heisenberg, who formulated
the foundations of the new
quantum mechanics in
matrix form, independently
of Schrödinger.

where $n = 1, 2, \ldots$. This is a typical eigenvalue equation. In the atomic case, the function Ψ_n depends also on other quantum numbers, such as l, m, m_s, not written explicitly in (6.19). The state of minimal energy is called the ground state, and is non-degenerate in most cases. Other states are usually degenerate (more than one quantum state for the same energy eigenvalue).

In (6.19), the values of the coordinates would remain uncertain, and for a state of definite energy one can speak only about the probability that the electron be in some region. For the ground state $n = 1$, the probability $P(r)$ of finding the electron at a distance r from the nucleus is a curve of the form shown in Fig. 6.7. The maximum probability corresponds to the value $r_0 = \hbar^2/me^2$, that is, the radius corresponding to Bohr's theory for $n = 1$.

Our wave function must include the spin. To do this, one must multiply Ψ by a function ψ_s that does not depend on the coordinates, and accounts for the part of the wave function depending on the spin variables. We would then have a wave function describing the electron completely.

What would happen if we made a measurement to determine the position of the electron exactly? Then the electron would be in a state in which energy and angular momentum would be completely uncertain. That is, if the position is determined at some instant, the energy and the angular momentum do not have definite values at that instant.

Fig. 6.7 The probability distribution for the distance of the electron in the hydrogen atom, according to the new quantum mechanics, is depicted for the ground state, i.e., the state with minimum energy. The maximum probability corresponds to the Bohr radius, that is, the value of the radius of the orbit predicted by Bohr's model for $n = 1$.

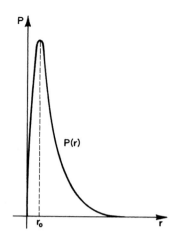

This is the consequence of Heisenberg's uncertainty principle, which we mentioned earlier. We wrote the uncertainty relations between position and momentum along the x-axis, viz.,

$$\Delta p_x \Delta x \sim h. \tag{6.20}$$

As pointed out earlier, the relation (6.20) is very similar to (4.33), typical of wave motion. Let us return to the diffraction problem in Chap. 4. A photon of wave number $k = \frac{2\pi}{\lambda}$ passes through the slit, of width D. Its uncertainty in position is given by the width of the slit along the y axis, $\Delta y = D$. The diffraction changes its momentum along the y axis in the amount $\Delta p = \hbar \Delta k_y$. We can estimate the quantity $\Delta k_y = k \sin \theta$ as indicated in Fig. 4.9. As θ is small, $\sin \theta \sim \tan \theta \sim \theta = y/R$. According to the plot in Fig. 4.10, $y \sim R\lambda/D$. Then, $\theta \sim \lambda/D$, from which $\Delta k_y \sim 2\pi/D$. This implies $\Delta y \Delta k_y \sim 2\pi$, and $\Delta y \Delta p_y \sim h$. Thus, we found the uncertainty relation for the conjugated variables y and p_y.

However, proceeding on more general grounds we may get a lower bound if we consider for instance the so-called minimum uncertainty wave packet. Let us start from the wave function

$$\psi(x) = A e^{\frac{i}{\hbar} p_0 x - \frac{\alpha x^2}{2\hbar}}, \tag{6.21}$$

which is a plane wave modulated by a Gaussian centred at the average value $x = \bar{x} = 0$, and A is a normalization constant. The density of probability for any value of x is

$$|\psi(x)|^2 = |A|^2 e^{-\alpha x^2/\hbar}, \tag{6.22}$$

which has to be normalized to one, i.e.

$$\int_{-\infty}^{\infty} |\psi(x)|^2 dx = 1.$$

This normalization condition gives a relation between the constants A and α, which is $A^2 = \sqrt{\frac{\alpha}{\pi\hbar}}$.

The uncertainty in the position is defined according to Kennard as the standard deviation, i.e. the root mean squared deviation of the position from its mean, whose expression is $\Delta x = \sqrt{\langle (x - \bar{x})^2 \rangle}$. Calculating the averages, one obtains $\Delta x = \sqrt{\hbar/2\alpha}$.

In momentum space, the wave function is the Fourier transform of (6.21), i.e.

$$\tilde{\psi}(p) = \frac{1}{\sqrt{2\pi\hbar}} \int dx\, \psi(x) e^{-\frac{i}{\hbar}px}.$$

One can show easily that its expression is

$$\tilde{\psi}(p) = \frac{A}{\sqrt{\alpha}} e^{-(p-p_0)^2/2\hbar\alpha}, \tag{6.23}$$

leading to the probability density in momentum space as $|\tilde{\psi}(p)|^2 = \frac{A^2}{\alpha} e^{-(p-p_0)^2/\hbar\alpha}$. The standard deviation, or the uncertainty, in p is $\Delta p = \sqrt{\langle (p - \bar{p})^2 \rangle} = \sqrt{\langle p^2 \rangle - \bar{p}^2}$. A short calculation of the averages gives $\Delta p = \sqrt{\hbar\alpha/2}$. Thus, we have found that

$$\Delta x\, \Delta p = \frac{1}{2}\hbar. \tag{6.24}$$

If we interpret the standard deviations as meaning uncertainties, the second term of (6.24) establishes the lower bound for the expression (6.20). The general formula $\Delta x\, \Delta p_x \geq \hbar/2$ is obtained by starting from the commutator of two noncommuting operators.

As in the classical case, for a particle moving in a central force field, like the electron in a hydrogen atom, it is more convenient in the quantum case, in place of the linear momentum, to use the angular momentum to characterize the state of motion, in addition to the energy. In that case, if the energy of a state is determined, it is only possible to know simultaneously the angular momentum and its component along one coordinate axis. Then the indeterminacy in the position of the electron with respect to the nucleus (which can only be known with a certain probability) is a consequence of the exact knowledge of the energy and angular momentum.

As we have already seen, quantum mechanical operators are associated with observable quantities. A quantum measurement of one of these observables leads to one of the eigenvalues of these operators. Two quantities p and q can be known simultaneously if the corresponding quantum operators $\hat{P}$ and $\hat{Q}$ commute, i.e., if

$$\hat{Q}\hat{P} - \hat{P}\hat{Q} = 0. \tag{6.25}$$

But if $\hat{Q}\hat{P} - \hat{P}\hat{Q} \neq 0$, it is not possible to know p and q simultaneously. For position and momentum, the two corresponding operators would be the position operator $\hat{Q} = x$ and the momentum operator $\hat{P} = -i\hbar\frac{\partial}{\partial x}$. Applied to a function of the coordinates

Fig. 6.8 The position of the electron in a system centred on the nucleus N, given in spherical coordinates r, θ, and φ.

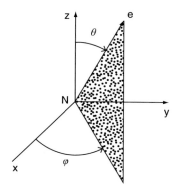

$f(x)$, it can be checked that

$$(\hat{Q}\hat{P} - \hat{P}\hat{Q})f(x) = i\hbar f(x). \tag{6.26}$$

The Hamiltonian operator $\hat{H}$ for the electron in the hydrogen atom has the form

$$\hat{H} = \frac{\hat{P}_r^2}{2m} + \frac{\hat{L}^2}{2mr^2} - \frac{e^2}{r}, \tag{6.27}$$

where $\hat{P}_r$ is the radial component of linear momentum, $\hat{L}$ is the total angular momentum, and r is the distance from the electron to the nucleus. This operator $\hat{H}$ does not commute with r, since it contains the momentum operator $\hat{P}_r$. And neither does it commute with $\hat{P}_r$. But it does commute with $\hat{L}^2$ and $\hat{L}_z$, even though it does not commute with the angles θ, φ that determine together with r the position of the electron (Fig. 6.8). From all this, it follows that, if the total energy is known, it is possible simultaneously to know the total angular momentum and its component in a given direction, but the distance r and the angles θ and φ, determining the position of the electron are not precisely known. However, it is possible to know their mean values. If $\Psi_a(\mathbf{r})$ is the wave function and $\Psi_a^*(\mathbf{r})$ is the complex-conjugate wave function in such a state, the average value of the position of the electron with respect to the nucleus is given by the expression

$$\langle r \rangle = \int \Psi_a^*(\mathbf{r}) \, r \, \Psi_a(\mathbf{r}) d\mathbf{r}, \tag{6.28}$$

where the integral extends over all space and the subscript a stands for the set of quantum numbers (n, l, m).

Similarly, we can find the average values of other quantities whose values cannot be known exactly in the quantum state Ψ_a. It can be shown that these average values satisfy the classical equations of motion, a result known as the *Ehrenfest theorem*, due to its discoverer, Paul Ehrenfest (1880–1933).

It should be noted that quantum mechanics can also be formulated in terms of path integrals, which generalizes the principle of least action of classical mechanics. To compute quantum amplitudes, it replaces the classical trajectory of a physical system in space and time, with a sum over an infinity of possible trajectories. The path integral method in quantum mechanics was developed by Richard Feynman (1918–1988), starting from an insightful observation made by Paul Dirac in 1933.

Note also the analogy between the diffusion equation (2.59) and the Schrödinger equation (6.16), which suggests that the Schrödinger equation is a diffusion equation with an imaginary diffusion coefficient. Incidentally, the path integral method was developed initially for the Brownian motion by the mathematician Norbert Wiener (1894–1964), in the beginning of the 1920s.

6.7   Operators and States in Quantum Mechanics

The mathematical formalism of quantum mechanics is essentially based on linear vector algebra. Below we shall briefly acquaint the reader to this formalism, at the same time introducing the widely used bra–ket notation proposed by Dirac in 1939.

The *state* of a microparticle is described in quantum mechanics by a function which we shall denote by $|\psi\rangle$ (a *ket* vector), and which is an element of a vector (or linear) space. The most familiar example of a vector space is the three-dimensional Euclidean space of radius-vectors. At an abstract level, all the properties of radius-vectors are fulfilled also by the state vectors of quantum mechanics. The dimension of the space of state vectors depends on the properties of the system; quite often, this dimension is infinite.

For a vector space, the dimension is the number of elements of a basis of the space, the basis being defined as a subset of linearly independent vectors, such that any other element of the vector space can be written in terms of the basis vectors. Mathematically, assuming a finite dimension n, the linear independence of the vectors $|\psi_1\rangle, |\psi_2\rangle, \ldots, |\psi_n\rangle$ is written as the condition

$$c_1|\psi_1\rangle + c_2|\psi_2\rangle + \cdots + c_n|\psi_n\rangle = 0, \tag{6.29}$$

valid if and only if all the coefficients vanish, $c_1 = c_2 = \cdots = c_n = 0$. An arbitrary element of the vector space, $|\psi\rangle$, is a linear combination of the basis vectors:

$$|\psi\rangle = \sum_{i=1}^{n} a_i |\psi_i\rangle, \tag{6.30}$$

where a_i, $i = 1, 2, \ldots, n$ are complex numbers.

The *bra* vector $\langle\psi|$ is defined as the Hermitian conjugate of the *ket* $|\psi\rangle$. The simplest way to express Hermitian conjugation is by adopting a matrix representation of the ket vectors, analogous to that of the radius-vector on an Euclidean space. Then Hermitian conjugation, denoted by †, means matrix transposition and complex conjugation:

$$\langle\psi| = |\psi\rangle^{\dagger} = (|\psi\rangle^*)^T. \tag{6.31}$$

In other words, if the ket (6.30) is written as the column vector

$$|\psi\rangle = \begin{pmatrix} a_1 \\ a_2 \\ \vdots \\ a_n \end{pmatrix},$$

the corresponding bra will be the row vector

$$\langle\psi| = \begin{pmatrix} a_1^* & a_2^* & \dots & a_n^* \end{pmatrix}.$$

One defines the *inner* product or the *scalar product* of the state vectors, $\langle\psi|\phi\rangle$, which is a complex number. The real quantity $\langle\psi|\psi\rangle$ has necessarily to be positive,

$$\langle\psi|\psi\rangle \geq 0,$$

and $\sqrt{\langle\psi|\psi\rangle}$ is called the *norm of the state vector* $|\psi\rangle$. Moreover, in quantum mechanics such norms have to be finite. The probability for the particle to be in a certain quantum state is $\langle\psi|\psi\rangle$.

There are some other assumption that the set of state vectors of a quantum mechanical system have to satisfy, but we shall not list them here, as they require more background knowledge in calculus. The space of state vectors is a particular example of a *Hilbert space*.

The basis of a vector space is not unique. Recall again, for illustration, the Euclidean space: one can choose a Cartesian coordinate system, or a spherical, or a cylindrical one. Each of these systems correspond to a different choice of basis. Of course, in each basis, the number of elements will be the same.

It is customary to consider the basis vectors to satisfy the property

$$\langle\psi_i|\psi_j\rangle = \delta_{ij}, \quad \text{for any} \quad i, j = 1, 2, \dots, n. \tag{6.32}$$

In this case we say that the basis is *orthonormal* (the norm of each basis vector is 1, and the vectors are orthogonal to each other, just like the directions of the coordinate axes in an Euclidean system). Then, if we have two state vectors,

$$|\psi\rangle = \sum_{i=1}^{n} a_i |\psi_i\rangle,$$

$$|\phi\rangle = \sum_{i=1}^{n} b_i |\psi_i\rangle,$$

using (6.31) and (6.32), we find their scalar product in terms of the expansion coefficients:

$$\langle \psi | \phi \rangle = \sum_{i=1}^{n} a_i^* \, b_i .$$ (6.33)

Note that the expansion coefficients are in effect scalar products of the state vector with the basis vectors:

$$a_i = \langle \psi_i | \psi \rangle, \quad i = 1, 2, \ldots, n.$$

Then, from

$$|\psi\rangle = \sum_{i=1}^{n} \langle \psi_i | \psi \rangle | \psi_i \rangle = \left(\sum_{i=1}^{n} |\psi_i\rangle \langle \psi_i| \right) |\psi\rangle,$$

we deduce that

$$\sum_{i=1}^{n} |\psi_i\rangle \langle \psi_i| = \mathbf{1},$$ (6.34)

where $\mathbf{1}$ is the unit operator. In the matrix representation, it is the unit $n \times n$ matrix.

The physical quantities, for example coordinate, momentum, energy, angular momentum etc., are represented in quantum mechanics by *linear operators* which act on the space of state vectors. Customarily, the operator associated to an observable A is denoted by $\hat{A}$. In contrast to classical mechanics, the measured values of the observables of a quantum system are selected by the equation

$$\hat{A} |\psi_i\rangle = A_i |\psi_i\rangle,$$ (6.35)

where $|\psi_i\rangle$ are called eigenstates or eigenvectors of the operator $\hat{A}$ and A_i are called eigenvalues. Finding the eigenstates and eigenvalues of an operator is equivalent to finding the *spectrum* of the operator. The bases of states used in quantum mechanics are sets of eigenvectors of various operators.

As the states are represented by column or row vectors, the operators are usually represented by matrices whose elements are complex numbers expressed by $\langle \psi_i | \hat{A} | \psi_j \rangle$.

The eigenvalues of an operator associated to an observable have to be real (since they are supposed to be the result of a measurement), which leads, by a theorem of linear algebra, to the requirement that the operators be Hermitian,

$$\hat{A} = \hat{A}^\dagger.$$

Let us assume that a quantum mechanical system is prepared to be in a state

$$|\psi\rangle = \sum_i c_i |\psi_i\rangle,$$

where $|\psi_i\rangle$, $i = 1, 2, \ldots, n$ form a complete and orthonormal set of eigenvectors of an operator $\hat{A}$. The system is with certainty (unit probability) in that state, therefore

$$\langle \psi | \psi \rangle = 1,$$

or, using (6.33),

$$\sum_i |c_i|^2 = 1. \tag{6.36}$$

The linear combination of basis vectors means, physically, that the system is actually in a *superposition* of eigenstates of $\hat{A}$, and the coefficients c_i tell us what is the contribution of the pure state $|\psi_i\rangle$ to the actual state of the system. The result (6.36) is interpreted as follows: with the system in the state $|\psi\rangle$, $|c_i|^2$ is the probability that, when testing the observable A, we measure the eigenvalue A_i. We can see that, according to the *principle of superposition*, while the quantum system is in a given state, the measurement of a certain observable can give, with specific probabilities, different values.

One very important result for quantum mechanics, expressed in a theorem, is that two operators, $\hat{A}$ and $\hat{B}$, which commute, $[\hat{A}, \hat{B}] = 0$, have a common set of eigenvectors. Physically, this means that one can measure simultaneously the two observables corresponding to the commuting operators. The famous commutation relation of Heisenberg, which we presented earlier, tells us that the coordinate operator $\hat{x}$ and the momentum operator $\hat{p}$ do not commute:

$$[\hat{x}(t), \hat{p}(t)] = i\hbar,$$

therefore they are not simultaneously measurable. This explains why the trajectory of a quantum mechanical particle is in principle not determinable. We wrote the commutation relation for a one-dimensional system and we shall continue with this simplification. The generalization to a space with more dimensions is straightforward. It should be stressed that the above commutation relation cannot be realized in terms of finite matrices, but in terms of the operators (6.15) introduced by Schrödinger.

One important set of eigenstates are those of the coordinate operator, $\hat{x}$, satisfying

$$\hat{x}|x\rangle = x|x\rangle, \tag{6.37}$$

and normalized as follows:

$$\langle x|x'\rangle = \delta(x - x'). \tag{6.38}$$

The coordinate operator has a continuous spectrum (therefore the sum over basis vectors becomes an integral), whose completeness is expressed as

$$\int_{-\infty}^{\infty} dx |x\rangle\langle x| = \mathbf{1}.$$

Similarly, we have the set of eigenstates of the momentum operator, $\hat{p}$, corresponding to the eigenvalue p:

$$\hat{p}|p\rangle = p|p\rangle, \quad \langle p|p'\rangle = \delta(p - p'), \tag{6.39}$$

for which

$$\int_{-\infty}^{\infty} dp \, |p\rangle\langle p| = \mathbf{1}. \tag{6.40}$$

What could be the meaning of the complex-valued wave function in this formalism? Let us expand the state vector of a system, $|\psi(t)\rangle$, in the basis of coordinate eigenstates:

$$|\psi(t)\rangle = \int dx \, |x\rangle\langle x|\psi(t)\rangle.$$

The coefficient of $|x\rangle$ in this expansion,

$$\Psi(x, t) = \langle x|\psi(t)\rangle, \tag{6.41}$$

is what we called previously the wave function. Here we find it as meaning the projection of a more abstract entity, the state vector $|\psi(t)\rangle$, on the coordinate eigenstates $|x\rangle$, and we call it wave function in *coordinate representation*. There is also a *momentum representation* wave function, defined analogously as

$$\tilde{\Psi}(p, t) = \langle p|\psi(t)\rangle. \tag{6.42}$$

The scalar product of the basis vectors in the coordinate and momentum representation is easily determined using (6.38) and (6.40):

$$\delta(x - x') = \langle x|x'\rangle = \int dp \, \langle x|p\rangle\langle p|x'\rangle,$$

and the Fourier expansion of the δ-function:

$$\delta(x - x') = \frac{1}{2\pi\hbar} \int dp \, e^{\frac{i}{\hbar} p(x - x')}.$$

Equating the integrands, we obtain:

$$\langle x|p\rangle = \frac{1}{\sqrt{2\pi\hbar}} e^{\frac{i}{\hbar} px},$$

$$\langle p|x'\rangle = \frac{1}{\sqrt{2\pi\hbar}} e^{-\frac{i}{\hbar} px'}. \tag{6.43}$$

Using (6.43), we find

$$\Psi(x, t) = \langle x | \psi(t) \rangle = \int dp \, \langle x | p \rangle \langle p | \psi(t) \rangle = \frac{1}{\sqrt{2\pi\hbar}} \int dp \, \tilde{\Psi}(p, t) e^{\frac{i}{\hbar} px}, \quad (6.44)$$

that is, the wave function in the momentum representation is the Fourier transform of the wave function in the coordinate representation. This relation is very important for solving the Schrödinger equation: the differential equation for $\Psi(x, t)$ becomes an algebraic equation for $\tilde{\Psi}(p, t)$.

6.8 One-Dimensional Systems in Quantum Mechanics

6.8.1 The Infinite Potential Well

The three-dimensional version of this model is called also "the particle in a box" and it is the simplest example that shows the differences between the quantum and classical mechanical systems. We present here the simplified one-dimensional case. Consider the motion of a particle of mass m. The potential is $U = 0$ for $0 < x < a$ and $U \to \infty$ for $x < 0$ and $x > a$. Classically, a particle colliding elastically with the walls would be expected to move continually between them with arbitrary speed or energy. Quantum mechanically, we may find the eigenfunctions $\psi_n(x, t)$ and energy eigenvalues E_n by solving the time-independent Schrödinger equation for the system. From (6.17) reduced to one spatial dimension, we have the time-independent equation for $0 < x < a$ as

$$\frac{d^2u}{dx^2} + \frac{2mE}{\hbar^2} u = 0, \quad (6.45)$$

since $U = 0$ inside the box and no forces act upon the particle, which means that the wave function inside the box oscillates and satisfies the condition $u(0) = u(a) = 0$.

The energy eigenvalues are given by

$$E_n = \frac{\hbar^2}{2m} \frac{\pi^2 n^2}{a^2}, \quad (6.46)$$

and the eigenfunctions are $\psi_n(x, t) = A_n e^{-i \frac{E_n}{\hbar} t} \sin \frac{n\pi x}{a}$, where $A_n = \sqrt{\frac{2}{a}}$ is the normalization constant. For each of these stationary states, the energy has a definite value given by (6.46); the average position of the particle is $\langle x \rangle = a/2$, and the probability densities are time-independent. Let us consider the states $n = 1, 2$. For $n = 1$ the probability density has a maximum at $x = a/2$ and a minimum at $x = 0$, $x = a$, where it vanishes. For $n = 2$, it has two maxima, the first at $x = a/4$ and the second at $x = 3a/4$, their average being $x = a/2$. The minima are located at the points $x = 0$, $x = a/2$, $x = a$, where the probability density is zero.

If we consider a state which is the superposition of the two eigenstates $n = 1$ and $n = 2$, the resulting state is not an energy eigenstate, and its wave function $\psi(x, t)$ is the linear combination of the two eigenfunctions $\psi_1(x, t)$ and $\psi_2(x, t)$. The probability density is

$$|\psi(x, t)|^2 = \frac{1}{a} \left[\sin^2 \frac{\pi x}{a} + \sin^2 \frac{2\pi x}{a} + 2 \cos \left(\frac{3\hbar\pi^2}{2ma^2} t \right) \sin \frac{\pi x}{a} \sin \frac{2\pi x}{a} \right].$$

This oscillates in time with the frequency $\omega = (E_2 - E_1)/\hbar = \frac{3\hbar\pi^2}{2ma^2}$. From $\psi(x, t)$, $|\psi(x, t)|^2$, we may find for instance the average energy $\langle E \rangle$ and show that the average position $\langle x(t) \rangle$ oscillates around the point $x = a/2$. This is proposed as a problem below.

6.8.2 Quantum Harmonic Oscillator

As another important example of a quantum system in one space dimension, we shall consider the harmonic oscillator. Recall from Chap. 1 its classical Hamiltonian.

$$H = \frac{p^2}{2m} + \frac{k}{2} x^2, \tag{6.47}$$

leading to the classical trajectory

$$x(t) = A \cos(\omega t + \beta), \quad \omega = \sqrt{\frac{k}{m}}. \tag{6.48}$$

The quantity β is some initial constant phase and A is the amplitude of the motion. Equation (6.48) describes the motion of a classical particle of mass m performing an oscillatory motion around the point $x = 0$. Such motion is typical of atoms in molecules and in solids, and in general occurs in any system whose more complicated potential is of the form $U(x) = U(0) + \frac{1}{2}U''(0)x^2 + \cdots$.

From (6.48) we get the velocity as $\dot{x}(t) = v(t) = -A\omega \sin(\omega t + \beta)$, and the classical energy as $E = \frac{1}{2}m\omega^2 A^2 = m\omega^2 \langle x^2 \rangle$, since $\langle \cos^2(\omega t + \beta) \rangle = \frac{1}{2}$. Thus, we find that $A = \sqrt{\frac{2E}{k}}$.

For further comparison with the quantum case, it is interesting to calculate the probability of finding the particle in the neighbourhood of some point x lying in the interval $[-A, A]$. The amount of time spent by the particle in the region dx, around the point x, is inversely proportional to the speed of the particle at x:

$$dt = \frac{dx}{v(x)}.$$

The probability of finding the particle in this region is the ratio of this time interval to $T/2$ – half of a period, which is the time needed to sweep the interval $[-A, A]$:

$$P(x)dx = \frac{2dt}{T} = \frac{2}{T}\frac{dx}{v(x)}.$$

From the expressions for $x(t)$ and $v(t)$, we easily obtain, for any moment of time, $x^2(t)/A^2 + v^2(t)/A^2\omega^2 = 1$, leading to

$$P(x)dx = \frac{dx}{\pi[A^2 - x^2]^{1/2}}, \quad \text{where} \quad \int_{-A}^{A} P(x)dx = 1. \tag{6.49}$$

We see that, since $P(x)$ is inversely proportional to the velocity of the particle, as the particle approaches the extremes of the trajectory $x = \pm A$, where the velocity vanishes, the probability density becomes infinite, $P(x \to \pm A) \to \infty$.

6.8.2.1 Schrödinger's Equation for the Harmonic Oscillator

Let us turn to the quantum mechanical description of the harmonic oscillator. We shall do a little mathematics to obtain the eigenvalues and eigenfunctions, since the oscillator is a simple, useful, and instructive example of a quantum system. Schrödinger's equation is

$$\hat{E}\Psi(x, t) = \hat{H}\Psi(x, t),$$

where $\hat{H}$ is the Hamiltonian (6.47) in which the momentum p is replaced by the operator $\hat{p}_x = -i\hbar\frac{\partial}{\partial x}$, and $\hat{E} = i\hbar\frac{\partial}{\partial t}$ is the energy operator. We seek a solution with separation of variables, and after introducing some adequate dimensionless variables,

$$\xi = x\sqrt{\frac{m\omega}{\hbar}} \quad \text{and} \quad \varepsilon = \frac{2E}{\hbar\omega}, \tag{6.50}$$

we write the total wave function as

$$\Psi(\xi, t) = Ce^{iEt/\hbar}\psi(\xi),$$

where C is a constant to be determined. By canceling the time-dependent exponential on both sides of the equation, one is left with a stationary Schrödinger equation (i.e. time-independent) whose solution has the form $\psi(\xi) = e^{-\xi^2/2}u(\xi)$, where $u(\xi)$ satisfies the new differential equation

$$\left(\frac{d^2}{d\xi^2} - 2\xi\frac{d}{d\xi} + (\varepsilon - 1)\right)u(\xi) = 0. \tag{6.51}$$

For
$$\varepsilon = 2n + 1, \tag{6.52}$$

where $n = 0, 1, 2, \ldots$, (6.51) coincides with the differential equation of Hermite polynomials, therefore $u(\xi) \sim H_n(\xi)$. These are polynomials of order n; for example, the first two of them are $H_0 = 1$, $H_1 = 2\xi$. The product $\psi_n(\xi) = N_n e^{-\xi^2/2} H_n(\xi)$ is named Hermite associated function, where N_n is a normalization constant. The details of the derivation can be found in any elementary quantum mechanics book. Plugging the condition (6.52) into the second relation of (6.50), we obtain the quantization of energy of the quantum harmonic oscillator:

$$E_n = \hbar\omega \left(n + \frac{1}{2}\right). \tag{6.53}$$

The total wave function in a state n is $\Psi_n(\xi, t) = e^{i E_n t/\hbar} \psi_n(\xi)$. Then, $|\Psi_n(\xi, t)|^2 = |\psi_n(\xi)|^2$ gives in the quantum mechanical approach the probability density $P(\xi)$ of locating the particle at the point ξ, for each n. It satisfies the natural normalization condition $\int P(\xi)d\xi = 1$.

In Fig. 6.9 are depicted both the classical and the quantum probability densities. The latter describes a fluctuation around the classical curve and follows on the average a path close to it. For very large n the quantum probability density approaches more and more the classical one.

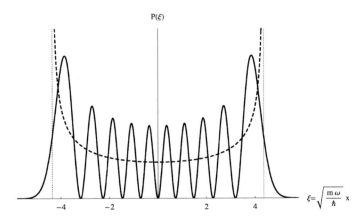

Fig. 6.9 The probability density curve for the quantum state $n = 9$ (*continuous line*) and the classical (*dashed line*) case for the one-dimensional linear oscillator. The quantum probability curve follows the classical curve on the average.

6.8.2.2 Ladder Operators

Finding the wave functions of quantum systems is an important step in the description of the system. However, in the case of the harmonic oscillator, there is a simplified method which can be used successfully to obtain essential information in a more intuitive way, avoiding to solve Schrödinger's equation. This is called the method of *ladder operators*, introduced by Dirac. In the following, we shall find the eigenstates of the Hamiltonian operator describing the harmonic oscillator.

Let us define linear combinations of the operators $\hat{x}$ and $\hat{p}$, as follows:

$$a = \sqrt{\tfrac{m\omega}{2\hbar}} \left(\hat{x} + \tfrac{i}{m\omega}\hat{p}\right),$$
$$a^\dagger = \sqrt{\tfrac{m\omega}{2\hbar}} \left(\hat{x} - \tfrac{i}{m\omega}\hat{p}\right), \tag{6.54}$$

as well as a *number operator*,

$$N = a^\dagger a. \tag{6.55}$$

From the canonical commutation relation

$$[\hat{x}(t), \hat{p}(t)] = i\hbar,$$

we obtain immediately the following commutators of the newly introduced operators:

$$[a, a^\dagger] = 1, \quad [a, a] = 0, \quad [a^\dagger, a^\dagger] = 0,$$

$$[N, a^\dagger] = a^\dagger, \quad [N, a] = -a. \tag{6.56}$$

The Hamiltonian operator $\hat{H} = \tfrac{\hat{p}^2}{2m} + \tfrac{k}{2}\hat{x}^2$ becomes

$$\hat{H} = \frac{1}{2}\hbar\omega(aa^\dagger + a^\dagger a) = \hbar\omega\left(N + \frac{1}{2}\right). \tag{6.57}$$

We construct now the space of states on which these operators act. With the experience gathered solving the Schrödinger equation for the harmonic oscillator, we require the states $|n\rangle$ to be eigenstates of the Hamiltonian operator, corresponding to the energy eigenvalues $E_n = \hbar\omega\,(n + 1/2)$, i.e.

$$\hat{H}|n\rangle = E_n|n\rangle.$$

This implies immediately that

$$N|n\rangle = n|n\rangle, \tag{6.58}$$

in other words the number operator really "counts" the levels of the harmonic oscillator.

Now let us see what is the role of the operators $a^\dagger$ and a. Computing

$$Na^\dagger|n\rangle = (a^\dagger N + [N, a^\dagger]) = (n + 1)a^\dagger|n\rangle, \tag{6.59}$$

where we have used (6.56), we deduce that $a^\dagger|n\rangle \sim |n + 1\rangle$, that is the operator $a^\dagger$ applied to a state raises the energy level by one. Similarly, one obtains that $a|n\rangle \sim |n - 1\rangle$, i.e. the lowering of the level by the action of a. The operator $a^\dagger$ is called *raising* operator, while a is called *lowering* operator, and together they are named *ladder operators*.

The lowest lying state is denoted by $|0\rangle$, it is called ground state or vacuum state, and it has the property that

$$a|0\rangle = 0.$$

Consequently, its energy is

$$E_0 = \hbar\omega/2,$$

named *vacuum or zero point energy* (see also Chap. 7).

Starting from the ground state, by successive applications of the raising operators, one can create all the states of the spectrum of the quantum harmonic oscillator, $\{|0\rangle, |1\rangle, \ldots, |n\rangle, \ldots\}$, infinite in number. The state $|n\rangle$ is chosen to be

$$|n\rangle = \frac{(a^\dagger)^n}{n!}|0\rangle,$$

so that we have the natural normalization $\langle n|n'\rangle = \delta_{nn'}$ for any n and n'.

The generalization to a system of noninteracting harmonic oscillators, with the angular frequencies ω_k, $k = 1, 2, \ldots$ is straightforward. The total number operator will be written as $N = \sum_k a_k^\dagger a_k$ and the Hamiltonian $\hat{H} = \sum_k \hbar\omega_k(a_k^\dagger a_k + 1/2)$, where we sum over all the oscillators k. The commutation relations $[a_k, a_j^\dagger] = \delta_{kj}$, $[a_k, a_j] = 0$, and $[a_k^\dagger, a_j^\dagger] = 0$ hold for any oscillators k and j.

The importance of the ladder operator method in the case of the harmonic oscillator is hard to overestimate, as it lies at the basis of the theory of quantized fields, and all the modern particle physics theories are quantum field theories.

6.8.3 Charged Particle in a Constant Magnetic Field

As a direct application of the quantum harmonic oscillator problem, we present the motion of a charged particle in a constant external magnetic field (see Chap. 5). The canonical momentum vector is $\mathbf{P} = \mathbf{p} - \frac{e}{c}\mathbf{A}$, where $\mathbf{A}$ is the vector potential. For a constant magnetic field $\mathbf{B}$ along the z axis, it may be taken as $\mathbf{A} = (-By, 0, 0)$, although the expression for $\mathbf{A}$ is not unique, due to gauge freedom (see Chap. 5). The Hamiltonian operator is

$$\hat{H} = \frac{1}{2m}\left[\left(-i\hbar\frac{\partial}{\partial x} + \frac{e}{c}By\right)^2 + \left(-i\hbar\frac{\partial}{\partial y}\right)^2 + \left(-i\hbar\frac{\partial}{\partial z}\right)^2\right] - \boldsymbol{\mu}\cdot\mathbf{B}. \quad (6.60)$$

To solve the Schrödinger equation, we assume a separation of the variables of the form $\Psi = \psi(\mathbf{r})\theta(t)$. We then note that, in the resulting equation $\hat{H}\psi = E\psi$, the x and z coordinates do not appear in the Hamiltonian, whence the generalized momenta p_x, p_z are conserved. We thus assume a wave function of the form $\psi(\mathbf{r}) = e^{\frac{i}{\hbar}(p_x x + p_z z)}\psi(y)$. After substituting it in (6.60), we get an equation for $\psi(y)$:

$$\frac{d^2\psi}{dy^2} + \frac{2m}{\hbar^2}\left[E + \boldsymbol{\mu}\cdot\mathbf{B} - \frac{p_z^2}{2m} - \frac{m}{2}\left(\frac{eB}{mc}\right)(y - y_0)^2\right]\psi(y) = 0, \quad (6.61)$$

where $y_0 = -\frac{cp_x}{eB}$. Equation (6.61) is similar to the Schrödinger equation for the oscillator (6.47)–(6.51), and setting $x = y - y_0$, it can be written as

$$\frac{d^2\psi}{dx^2} + \frac{2m}{\hbar^2}(E' - \frac{1}{2}m\omega^2 x^2)\psi = 0,$$

where $E' = E + \boldsymbol{\mu}\cdot\mathbf{B} - \frac{p_z^2}{2m}$. We take $\boldsymbol{\mu}\cdot\mathbf{B} = \pm\mu B$, where the electron magnetic moment is $\mu = e\hbar\sigma/2mc$ and $\sigma = \pm 1$ is due to the spin of the particle. We may write the energy eigenvalues as

$$E = \frac{p_z^2}{2m} + \left(n + \frac{1}{2} \mp \frac{1}{2}\right)\frac{eB\hbar}{mc}.$$

We observe that the energy depends on the magnetic field B and is quantized by the integers $n = 0, 1, 2, \ldots$, called the Landau quantum numbers. Along the z axis, it behaves as a free particle. As $\sigma = \pm 1$, after the (non-degenerate) ground state $n = 0$, $\sigma = -1$, there is a two-fold degeneracy ($n = 0$, $\sigma = 1$, and $n = 1$, $\sigma = -1$, and so on). But E is also independent of p_y, that is, it is also degenerate with respect to y_0, which is interpreted as the coordinate of the centre of the orbit. This is especially important in the quantum Hall effect, leading to the degeneracy term $eB/\hbar c$, whence there is a number of states per Landau energy level which grows in proportion to B (see Chap. 8, expression (8.39)). The reader may compare the quantum problem of motion of a particle in a magnetic field to the classical case, discussed in Chap. 3.

6.9   Emission and Absorption of Radiation

One of the phenomena leading to the discovery of the quantum nature of atomic processes was the discrete nature of the emission and absorption spectra of several substances. In the case of hydrogen, as already mentioned, it was possible to explain the origin of the already known spectral series by starting from Bohr's postulates.

Identical results were obtained by starting from the new quantum mechanics, but it was also possible to go farther, obtaining selection rules, that is, conditions that must be satisfied by the set of quantum numbers of the initial and final states of the electron so that the emission or absorption of the radiation can occur.

When the electron in the atom emits or absorbs radiation, it suffers a change in the quantum numbers n, l, m. Since the electromagnetic radiation has angular momentum, both l and m can change. The radiation can be considered as composed of particles (photons), and the processes of emission and absorption – as interactions between particles (electrons and photons). However, this treatment of the problem must be made in the framework of quantum electrodynamics.

According to quantum mechanics, if atoms are made to interact with radiation of several frequencies, these atoms will absorb certain frequencies and jump to excited states. In order for a frequency ν to be absorbed by an atom in a state of energy E_i, there must be another permissible state for the electron, of energy E_f, such that the condition $h\nu = E_f - E_i$ is satisfied. If this condition is not satisfied, the radiation cannot be absorbed. The electrons in the atom, by absorbing radiation, are excited to higher energy states.

If we irradiate the excited atoms with a beam of radiation of frequency ν, which can be absorbed, the excited atoms will emit photons of frequency ν, and the emitted photons will be in phase with those of the incident beam and going in the same direction. The probability of emission is higher than if we leave the excited atoms unperturbed. Thus, the incident beam is augmented. This is called stimulated emission.

If we do not induce the atoms initially excited to emit radiation, that is, if we leave them unperturbed, they will emit radiation spontaneously with some non-vanishing probability. They will tend to return to the ground state by emitting photons of frequency equal to the initially absorbed photons, but in random directions and with random initial phases. This is called spontaneous emission.

Why does spontaneous emission occur? Its existence indicates that the excited states are not really stationary and that the tendency of the electron is to fall to the ground state. Spontaneous emission is explained within the framework of quantum electrodynamics: it is due to the interaction of the electron with the virtual photons of the quantum vacuum in the presence of the electric field of the electron–nucleus system.

6.10 Stimulated Emission and Lasers

If an atom interacting with radiation is able to emit a photon in a quantum state of frequency ν and momentum $\mathbf{k}$, the probability of emission is proportional to the number N of photons of frequency ν and momentum $\mathbf{k}$ already existing in the system. That is, if N is increased, the probability of emission by an excited atom of a photon of frequency ν and momentum $\mathbf{k}$ will also be increased.

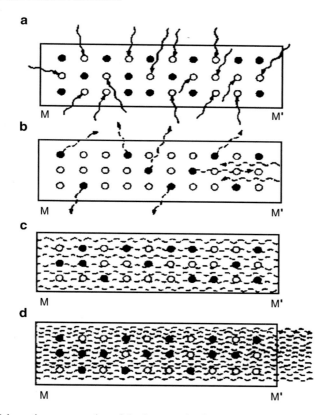

Fig. 6.10 Schematic representation of the laser mechanism. **a** The atoms of some substance are excited from level 1 (*white*) to level 2 (*black*), due to absorption of radiation (photons) coming from a lamp. **b** In less than a millionth of a second, the atoms, by spontaneous emission, pass from level 2 to level 1, and emit radiation in random directions, part of it perpendicular to the system of mirrors, *M* which is silvered (completely reflectant) and *M'* which is half-silvered. **c** The radiation perpendicular to the mirrors begins to oscillate between them. It has the characteristics of being monochromatic and coherent. **d** This radiation induces the atoms (excited again by the effect of the external radiation) to emit photons of the same frequency and phase. In this process, the intensity of coherent radiation increases inside the resonant cavity formed by the mirrors *M* and *M'*, and becomes large enough to cross the half-silvered mirror *M'* and emerge as laser light.

Let us consider for simplicity a gas composed of atoms with two energy levels $E_1 < E_2$, inside a resonant cavity comprising two mirrors between which a beam of radiation of frequency

$$\nu = \frac{E_2 - E_1}{h} \tag{6.62}$$

can be reflected and kept oscillating. A lamp emits radiation that is absorbed by the atoms of the gas, whence they pass from level E_1 to E_2. After a very short time, these atoms emit spontaneously radiation of frequency ν in all directions. A large number of photons is lost, except those which are directed perpendicularly to the mirrors in the resonant cavity. By means of this mechanism, the population of photons oscillating inside the cavity begins to increase, all of them having frequency ν, momentum $\mathbf{k}$, and identical phase and polarization. This induces the atoms to emit new photons in the same state, with greater probability than any other, all of which will be in the same quantum state. Their number increases exponentially to reach such a density that the light passes through one of the mirrors and emerges from the system as coherent light. This is the laser light (Fig. 6.10).

6.11 Tunnel Effect

Let us assume a particle in a potential $V(x)$, as shown in Fig. 6.11. If the energy E is lower than the height of the barrier B and the particle is in region I, then in classical physics, the particle cannot pass to region II. From the quantum point of view, the situation is different. The probability density, that is the square of the modulus of the wave function, has its maximum in region I, but it is not zero inside the barrier B and still exists inside region II, after the barrier. This means that the particle can pass from region I to region II. When this happens, we call this phenomenon the tunnel effect or penetration of a potential barrier.

The probability of penetration diminishes if the height of the barrier increases. Quantum tunneling has great importance in physics, since it provides explanations for several phenomena, e.g., the alpha disintegration of atomic nuclei and it is essential also in the functioning of various modern electrotechnical devices, like integrated circuits.

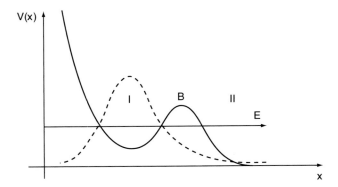

Fig. 6.11 Potential of a particle with a barrier of finite height B. If the particle is located initially in region I, before the barrier, it can escape through the barrier even though its energy E is lower than the height of the barrier. The dotted line represents the probability density.

6.12 Indistinguishability and Pauli's Principle

Consider an atom with two electrons, for instance, the helium atom. We neglect the mutual interaction between the electrons. Then each of them will be characterized by wave functions similar to those of the electron in the hydrogen atom. Let us denote one of the electrons by 1 and assume that its state is described by a set of quantum numbers that will be called a. Similarly, the other electron is denoted by 2 and b will be its set of quantum numbers. The wave functions of the electrons are denoted by $\Psi_1(a)$ and $\Psi_2(b)$, respectively. Then the wave function of the system of electrons will be

$$\Psi_I = \Psi_1(a)\Psi_2(b). \tag{6.63}$$

But since these wave functions extend in space, they may overlap in some region, and the state of the system can actually correspond, without having any possibility to distinguish among them, to another state in which the two particles are exchanged:

$$\Psi_{II} = \Psi_1(b)\Psi_2(a). \tag{6.64}$$

Which is the true wave function? It cannot be determined.

The real wave function of the system should be a linear combination of Ψ_I and Ψ_{II}, in such a manner that the physical properties do not change if the two particles are interchanged. Then, we have the two possibilities:

$$\Psi_S(1, 2) = \frac{1}{\sqrt{2}}[(\Psi_1(a)\Psi_2(b) + \Psi_1(b)\Psi_2(a)], \tag{6.65}$$

or

$$\Psi_A(1, 2) = \frac{1}{\sqrt{2}}[\Psi_1(a)\Psi_2(b) - \Psi_1(b)\Psi_2(a)]. \tag{6.66}$$

The function $\Psi_S(1, 2)$ is symmetric since it does not change if the particles are exchanged, while the function $\Psi_A(1, 2)$ is antisymmetric, since it changes its sign if the particles are exchanged.

The same situation occurs whatever the number of particles: the wave function of the complete system should be symmetric or antisymmetric. This property is a consequence of the indistinguishability of identical particles in quantum systems.

In which case is the wave function symmetric and in which case antisymmetric? If the identical particles have half-integer spin (like electrons, protons, neutrinos, etc.), the wave function of the system is antisymmetric. Such particles are said to obey the Fermi–Dirac statistics, and for that reason are called fermions. If the particles have integer spin (like photons, π mesons, atoms of ^{4}He, etc.), the wave function describing a set of identical particles of this kind is symmetric. Such particles obey the Bose–Einstein statistics and are called bosons.

Two fermions cannot be in the same state, since in that case the antisymmetric wave function of the system vanishes. This is the Pauli exclusion principle. This is easily seen in the case of two particles. If the set of quantum numbers a and b are identical, then $\Psi_A = 0$. The Pauli principle determines the distribution of the electrons in atoms and plays a fundamental role in the theory of the chemical bond.

On the other hand, one can have an arbitrary number of bosons in the same quantum state. For instance, a beam of laser radiation contains millions of photons in the same quantum state. This property of bosons also manifests itself in phenomena like superconductivity and superfluidity.

6.13 Exchange Interaction

Consider an atom with two electrons, such as a helium atom. The complete wave function Ψ of the system is the product of a function S_{12} depending on the spins and another function depending on the positions of the electrons, $\Phi(\mathbf{r}_1, \mathbf{r}_2)$, the latter being a solution of the Schrödinger equation. The total wave function is

$$\Psi = S_{12}\Phi(\mathbf{r}_1, \mathbf{r}_2). \tag{6.67}$$

The total wave function should be antisymmetric. If S_{12} is symmetric (which occurs if the spins are parallel, that is, if they have the same direction), then $\Phi(\mathbf{r}_1, \mathbf{r}_2)$ must be antisymmetric. If S_{12} is antisymmetric (antiparallel spins), then $\Phi(\mathbf{r}_1, \mathbf{r}_2)$ must be symmetric. So the energy eigenvalues corresponding to the symmetric solutions of the Schrödinger equation are taken when the spins are antiparallel (total spin equal to zero) and the energy eigenvalues corresponding to the antisymmetric solutions of the Schrödinger equation are taken when the spins are parallel (total spin equal to 1).

As the value of the energy is different in these two cases, we see that in a system of several electrons the possible values of the energy depend on the total value of the spin. Thus, we may consider that there is an interaction between the particles depending on their spin, and we call this the *exchange interaction*.

Let $\mathbf{r}_1$ and $\mathbf{r}_2$ be the position vectors, so that $\Phi_1(\mathbf{r}_1)$ and $\Phi_2(\mathbf{r}_2)$ correspond to the wave functions of electron 1 at the position $\mathbf{r}_1$ and electron 2 at the position $\mathbf{r}_2$, respectively. On the other hand, $\Phi_1(\mathbf{r}_2)$ and $\Phi_2(\mathbf{r}_1)$ are the wave functions of electron 1 at the position $\mathbf{r}_2$ and electron 2 at the position $\mathbf{r}_1$, that is, when the two electrons are exchanged. Let the potential between the electrons be $V(\mathbf{r}_1, \mathbf{r}_2)$. We define the exchange integral by

$$J = \int \Phi_2^*(\mathbf{r}_1)\Phi_1^*(\mathbf{r}_2)V(\mathbf{r}_1, \mathbf{r}_2)\Phi_1(\mathbf{r}_1)\Phi_2(\mathbf{r}_2)d\mathbf{r}_1 d\mathbf{r}_2. \tag{6.68}$$

E denotes the energy of the two-electron system in the field of the nucleus, with the exchange effects switched-off, i.e. $E = \varepsilon_1 + \varepsilon_2 + Q$, where ε_1, ε_2 are hydrogen-like energies and $Q = \int \Phi_1^2(\mathbf{r}_1)\Phi_2^2(\mathbf{r}_2)V d\mathbf{r}_1 d\mathbf{r}_2$ is the average Coulomb interaction energy between the electrons. The total energy, including the exchange term has two possible values:

$$E_s = E + J \quad \text{and} \quad E_A = E - J. \tag{6.69}$$

The first corresponds to the case in which the two spins are antiparallel (total spin zero and $\Psi(\mathbf{r}_1, \mathbf{r}_2)$ symmetric), and the second to the case in which the spins are parallel (total spin 1 and $\Psi(\mathbf{r}_1, \mathbf{r}_2)$ antisymmetric). In the excited helium atom, we call the first case parahelium and the second case orthohelium. Orthohelium has lower energy and is a metastable state of helium (it does not decay to the ground state with spin 0, since a flip of spin would be involved and this cannot happen spontaneously).

6.14 Exchange Energy and Ferromagnetism

In a system with many electrons, the exchange energy is determined by the quantity

$$U = J \sum_{i,j} \mathbf{S}_i \cdot \mathbf{S}_j, \tag{6.70}$$

where J is the exchange integral and $\mathbf{S}_i \cdot \mathbf{S}_j$ is the scalar product of the spins of the particles i, j (the spins have been taken as operators).

In a ferromagnetic substance, this exchange energy plays an essential role. The ferromagnetic properties are due to the coupling between the magnetic dipoles of the electrons, associated with the exchange interaction between them. As the quantity J is due to a purely quantum effect, we see that ferromagnetism has a quantum origin.

The exchange interaction determines that the energy is minimum when the spins (and in consequence, the magnetic dipoles) are parallel. The effect is equivalent to an external field aligning the spins. For that reason one can speak of an *exchange field*.

This ordering tendency, of quantum origin, competes with the disordering tendency due to thermal agitation, so that when we heat a ferromagnetic substance, there is a critical temperature called the Curie temperature (T_c) at which the ferromagnetic effect disappears and the material becomes paramagnetic for $T > T_c$.

For some substances, said to be antiferromagnetic, the exchange integral is negative, determining an antiparallel ordering of spins (see Chap. 3).

6.15 Distribution of Electrons in the Atom

We have already mentioned that the energy of the electron in the hydrogen atom is determined by a positive integer. The angular momentum is specified by another integer, l that takes any value between zero and $n - 1$, i.e.,

Table 6.2 Electronic configurations for $n = 2$ (atomic shell L).

Shell	L	
Orbitals	$2s$	$2p$
Maximum number of electrons	2	6

Table 6.3 Electronic configurations for the fist ten elements of the periodic table.

Atomic number	Symbol	Shell Orbital	K $1s$	L $2s$	$2p$
1	H		1		
2	He		2		
3	Li		2	1	
4	Be		2	2	
5	B		2	2	1
6	C		2	2	2
7	N		2	2	3
8	O		2	2	4
9	F		2	2	5
10	Ne		2	2	6

$$0 \leq l \leq n - 1. \tag{6.71}$$

This number is associated with the modulus of the angular momentum vector. There is another integer, m, that characterizes the direction of the component of the angular momentum along a given axis:

$$- l \leq m \leq l. \tag{6.72}$$

Finally, for each set of values n, l, m, there are two possible values of the spin, $m_s = \pm 1/2$. These rules have a general validity, and starting from such rules, one can get the distribution of electrons in any atom, bearing in mind that the Pauli principle permits at most one electron in each quantum state.

For instance, for $n = 1$, we have the so-called K shell. The number l has to be zero, and there can be a maximum of two electrons in such a shell, with opposite spins. In the hydrogen atom, there is only one, and in the helium atom, two electrons.

For $n = 2$, the shell is called L. Then l can take the values 0 and 1. The spatial distributions corresponding to different values of l are called orbitals, and in this case are symbolized by $2s$ $2p$ ($1s$ is the orbital corresponding to the shell K). For $l = 0$, the only acceptable value is $m = 0$, as in the previous case, and for $l = 1$, the number m can take 3 values, viz., $-1, 0, 1$. Then for $n = 2$, one can have the electronic configurations shown in Table 6.2.

Six electrons occupy the orbital $2p$ due to the three possible orientations of the orbital angular momentum, $m = -1, 0, +1$, to each of which correspond two electrons with opposite spins. In total there would be 8 electrons in the L shell. This configuration corresponds to the neon atom.

The electronic configurations of the first ten elements are given in the Table 6.3. In the same way, one can continue with the higher shells. For $n = 3$, the shell is called M, and has orbitals $3s$, $3p$, $3d$ to which correspond a maximum of 2, 6, and 10 electrons, respectively, while for $n = 4$, the shell is called N, with orbitals $4p$, $4d$, $4f$, etc.

The chemical and physical properties of substances and the location of the elements in Mendeleev's periodic table are determined by the electronic configurations, which, as we have seen, are direct consequences of quantum laws such as Schrödinger's equation and Pauli's exclusion principle.

6.16 Quantum Measurement

Quantum mechanics is tested experimentally through the interaction of quantum objects with apparata obeying classical physics. For example, the photons composing spectral lines, which are emitted when electrons jump from one energy level to another inside atoms, are observed using a spectrometer; the motion of a particle is observed in the Wilson cloud chamber through the formation of droplets of vapour, which are classical objects; in scattering experiments, angles, energies, and momenta of scattered particles are usually measured, after their interaction with macroscopic devices. Classical mechanics thus plays a double role, as both the limiting case, and also a necessary basis for formulating quantum measurements.

One essential feature of a quantum measurement of a physical quantity is this: *unless the particle is in an eigenstate of the operator corresponding to the observable quantity measured by the apparatus, in general, the measurement destroys the initial wave function, leading to a final state which is an eigenfunction of the operator associated with the measured quantity.*

Suppose we send toward a detector an electron in an eigenstate of momentum, described by the wave function $\Psi(x, t) = Ae^{i(px-Et)/\hbar}$. The wave function is obviously a plane wave extending from $-\infty$ to ∞, like the case we considered in Chap. 4. At the instant t_0, the electron enters the detector, which registers it at the position x_0. We can say that the electron wave function at time t_0 is proportional to $\delta(x - x_0) = (2\pi\hbar)^{-1} \int_{-\infty}^{\infty} e^{ip(x-x_0)/\hbar} dp$, i.e., a pulse of zero width, since the Dirac δ function (here we are speaking about the one-dimensional delta function), is not a standard function in the purely mathematical sense. In particular, it vanishes everywhere except at the point x_0, where it has infinite value. As mentioned earlier, two basic properties of the δ function are:

$$\int_{-\infty}^{\infty} \delta(x - x_0)dx = 1 \qquad (6.73)$$

and

$$\int_{-\infty}^{\infty} f(x)\delta(x - x_0)dx = f(x_0). \tag{6.74}$$

The initial plane wave can be written as a linear superposition of δ functions:

$$Ae^{i(px-Et)/\hbar} = Ae^{-iEt/\hbar} \int_{-\infty}^{\infty} e^{i(px')/\hbar}\delta(x' - x)dx'. \tag{6.75}$$

Thus, as a result of the measurement, only *one* specific component of the infinite superposition survives, namely the one corresponding to $x' = x_0$. Perhaps this is a more transparent example than the one in (6.30), where we have an arbitrary wave function expanded in terms of eigenfunctions of a given operator. The coefficients of the expansion a_i are the probability amplitudes, i.e., $|a_i|^2$ are the probabilities of finding the system in the eigenstate i after a measurement of the corresponding observable. We emphasize that a measurement in quantum mechanics generally modifies the wave function unless the quantum system is in an *eigenstate* of a certain quantity, and only this quantity is measured. For instance, if a measurement of the energy is made on a quantum system, it gives as a result the energy E_n, and the wave function is Ψ_n. Then, if the energy is measured again immediately, the wave function does not change, and the new measurement gives the same result as the previous one.

The very notion of *destroying* the wave function implies *changing or reducing the wave function instantaneously throughout physical space* (according to a clock fixed on the measuring apparatus). This in turn implies a non-local action on the wave function.

We conclude that the process of measurement has new ingredients not present in the basic postulates of quantum mechanics. It does not satisfy the superposition principle, and as a consequence it is essentially *nonlinear* (the measurement of a state described by a linear combination of wave functions is not generally the linear combination of the measurements). In this way we see two different properties or procedures merging as basic ingredients of quantum theory, which we discuss below.

6.16.1 U and R Evolution Procedures

John von Neumann (1903–1957) pointed out the need to distinguish the role of two basic evolution procedures in quantum mechanics. This idea has been completed by Roger Penrose (b. 1931), who called these procedures U and R. The usually more clearly recognized procedure is U, the unitary time evolution of the quantum state $\Psi(x, t)$, which is governed by the Schrödinger equation, as a deterministic process, which does not violate reversibility or time inversion. This means that $\Psi(x, t \pm \Delta t)$ can be predicted with certainty to be the wave function at the times $t \pm \Delta t$. In other words, by solving $i\hbar\frac{\partial\Psi}{\partial t} = \hat{H}\Psi$, one can get some evolution operator U able to effect

Table 6.4 Characteristics of U and R evolution procedures.

U (Unitary transformation)	R (Reduction of the wavepacket)
Deterministic	Probabilistic
Preserves superposition	Violates superposition
Acts continuously	Acts essentially discontinuously

the transformation $\Psi(x, t) \rightarrow \Psi(x, t \pm \Delta t)$ by going either forward or backward in time (in the latter case, by taking $t \rightarrow -t$).

The process R is associated with measurements of some observable. It *magnifies quantum aspects to the classical level*, forming squares of the modulus of the amplitudes to yield classical probabilities. R leads to a reduction of the wave function to an eigenstate of the measured observable or to the *collapse* of the wave function. It distinguishes the future from the past, that is, it is asymmetric with respect to time. It is the procedure R, and only R, that introduces uncertainties and probabilities in quantum theory. The two procedures are compared in Table 6.4.

Both U and R procedures are necessary for the correspondence between quantum effects and observations.

The genesis of what was later named U and R procedures, and the problem which led to the uncertainty principle, were subject of debate. Important contributions were made by Bohr and Heisenberg, but several facts remained unclear for some of the founders of quantum theory, and for instance, at least for some time, Schrödinger did not accepted the "quantum jumps." Einstein, although disagreeing with the mainstream of quantum mechanical way of thinking, contributed with several remarkable comments and objections to the advance of very basic quantum ideas.

6.16.2 On Theory and Observable Quantities

Einstein raised some objections to Heisenberg's presentation in Berlin in 1926 concerning the initial interpretations of observable quantities in quantum mechanics. Einstein, in a private talk, told him (in Heisenberg's words): *Whether you can observe a thing or not depends on the theory which you use. It is the theory which decides what can be observed.* His argument was like this: *Observation means that we construct some connection between a phenomenon and our realization of the phenomenon. There is something happening in the atom, the light is emitted, the light hits the photographic plate, we see the photographic plate and so on and so on. In this whole course of events between the atom and your eye and your consciousness you must assume that everything works as in the old physics. If you changed the theory concerning this sequence of events then of course the observation would be altered.*

Einstein remarked that *it is really dangerous to say that one should only speak about observable quantities, because every reasonable theory will, besides all things which one can immediately observe, also give the possibility of observing other things more indirectly.* In Heisenberg's words, *Einstein had pointed out to me that it is really dangerous to say that one should only speak about observable quantities... In quantum theory it meant, for instance, that when you have quantum mechanics then you cannot only observe frequencies and amplitudes, but for instance, also probability amplitudes, probability waves and so on, and these, of course, are quite different objects.*

Later, Bohr and Heisenberg discussed a problem: if in an atom one must abandon the concept of trajectory, what about the cloud chamber, where the electron could be seen to have moved along a track? Was this a trajectory or not? Heisenberg remembered Einstein's statement, and turned around the question of *how to represent in quantum mechanics the orbit of the electron in the cloud chamber?* He transformed the question in: *Is it not true that only such situations occur in Nature, even in a cloud chamber, which can be described by the formalism of quantum mechanics?* Quantum theory stated that it was not possible to measure at the same time the exact position and exact velocity of a particle, that is, the uncertainty principle. In the cloud chamber there was a trajectory, but the path had a significant width – there was uncertainty in both quantities, velocity and position. When Heisenberg met Bohr, he found that Bohr had also elaborated an answer, based on his famous complementarity principle. (Complementarity means that quantum systems can be observed either as having particle or wave behaviour, but not both simultaneously). Finally they agreed to have understood quantum theory.

However, Einstein was sceptical about the uncertainty principle, and during the Solvay Conference in 1927, he discussed with Bohr every day, suggesting *Gedanken* experiments as disproving examples, which Bohr found the way to refute. At the end, Bohr succeeded and Einstein did not raise any more objections.

6.17   Paradoxes in Quantum Mechanics

6.17.1   De Broglie's Paradox

Consider a box with walls able to reflect electrons inside in such a way that the electrons are not absorbed. Suppose we place one electron inside. The box has a partition which also has perfect reflecting properties. If the partition is drawn, the box is separated into two half-boxes. We may keep the two half-boxes joined, or even separate them. For instance, we may take the half-box 1 in a plane to Havana, while the half-box 2 remains in Helsinki.

Imagine that we make a measurement in box 2 in Helsinki at time *t* and it is found that the electron is inside it.

If we wonder where the electron was at the moment $t - \epsilon$, we can adopt different attitudes. A 'realistic' physicist, who believes in determinism, would say:

The electron was already in 2 at $t - \epsilon$. Standard quantum mechanics does not give suffi-
cient information; it is necessary to have additional parameters (hidden variables). A more
complete theory would indicate where the electron was at $t - \epsilon$.

A physicist interpreting standard quantum mechanics in the traditional way would say:

It is meaningless to ask where the electron was before doing the measurement.

Let us denote the coordinates in 1 by x_1 and in 2 by x_2. The wave functions describing the electron inside box 1 and box 2 would be $\Psi_1(x_1, t - \epsilon)$ and $\Psi_2(x_2, t - \epsilon)$, respectively. The total wave function describing the electron inside the box, according to the U procedure, is

$$\Psi(x_1, x_2, t - \epsilon) = a\left[\Psi_1(x_1, t - \epsilon) + \Psi_2(x_2, t - \epsilon)\right], \tag{6.76}$$

where a is a normalization constant. If one integrates $|\Psi(x_1, x_2, t - \epsilon)|^2$ over the whole volume of the box, one obtains unity. Let us assume for simplicity that the wave function Ψ is non-zero only inside the volume V, while Ψ_1 and Ψ_2 are nonzero only in the left and right halves of V, respectively, so that if we write

$$|\Psi|^2 = a^2\left(|\Psi_1|^2 + |\Psi_2|^2 + |\Psi_1^*\Psi_2 + \Psi_2^*\Psi_1|\right), \tag{6.77}$$

and integrate (6.77) over the whole volume V, the third term on the right vanishes, i.e., we exclude interference for simplicity. We take then $a^2 = 1/2$. Integrating (6.77) over the whole volume V we obtain 1 on the left, while the integral of $\frac{1}{2}|\Psi_1(x_1, t - \epsilon)|^2$ over the half-box 1, as well as the integral of $\frac{1}{2}|\Psi_2(x_2, t - \epsilon)|$ over 2, are equal to $\frac{1}{2}$. Thus, at the moment $t - \epsilon$ the probabilities are $P_1 = 1/2$ that the electron is in Havana and $P_2 = 1/2$ that it is in Helsinki.

But when we carry out the measurement, i.e., the operation R, the wave packet is reduced, and the wave function is, say, $\Psi = \Psi_2(x, t)$ in box 2 and zero in box 1. We conclude that some non-local property of quantum mechanics is involved in the measurement process.

6.17.2   Schrödinger's Cat Paradox

Let us consider a simplified version of the original Schrödinger's cat paradox. A cat is inside a box, and a gun is activated by a photocell in such a way that, if it receives a photon polarized vertically (V), the gun shoots the cat, while if the polarization is horizontal (H), there is no shot, and the cat remains alive. Photons enter the photocell after passing through a polarizer, say, a calcite crystal, with H or V polarizations. We have an external source sending photons to the polarizer. Suppose it sends a photon polarized at $45°$ with respect to the vertical. The photon wave function is

$$\Psi = \frac{1}{\sqrt{2}}\Psi_V + \frac{1}{\sqrt{2}}\Psi_H, \tag{6.78}$$

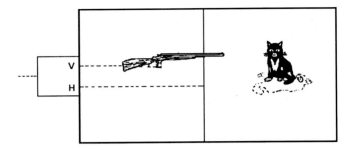

Fig. 6.12 Schrödinger's cat.

where Ψ_V and Ψ_H are the wave functions corresponding to vertical and horizontal photon polarizations.

If we open the box and the cat is alive, what was the state of the cat before opening the box (Fig. 6.12)? Since

$$\Psi^C = \frac{1}{\sqrt{2}}\Psi_D^C + \frac{1}{\sqrt{2}}\Psi_A^C, \tag{6.79}$$

we conclude that it was a linear combination of states in which it was dead, Ψ_D^C, and it was alive, Ψ_A^C. We would be forced to admit that the cat would be partially alive and partially dead, or in a state of 'suspended animation'. This is a very strange idea, and it may be argued that we cannot apply the quantum mechanical rules naively to an extremely complex system like a living organism.

6.17.3 Toward the EPR Paradox

In 1935, Albert Einstein, Boris Podolsky (1896–1966), and Nathan Rosen (1909–1995) wrote a famous paper in which they put forward strong arguments to show that quantum mechanics did not give a complete description of reality. The ideas expressed in that paper were later to become known as the *EPR paradox*, and were the subject of a long controversy for nearly 50 years. This paradox states problems very similar to the de Broglie box and Schrödinger's cat paradoxes. At the present time, many physicists consider that the EPR paradox is, if not solved, at least better understood, as a result of the works by John S. Bell (1928–1990) in 1967 and the experiments done by Alain Aspect (b. 1947) in 1982, to which we refer below. In this section, we shall try to use some simple examples to describe the EPR paradox.

In some substances, the excited atoms decay into two photons of different wavelengths and orthogonal polarizations in such a way that, whenever the left photon is V, the right is H, and whenever the left photon is H, the right is V. Or equivalently, if the left photon emerges polarized at an angle of $+45°$ with respect to the vertical,

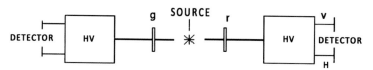

Fig. 6.13 A source of pairs of orthogonally polarized photons is placed at the centre, between two filters of green (*left*) and red (*right*) photons. The photons enter the polarizers *HV*, which contain photon detectors, in such a way that, if the left photon is found to be *V* by a detector, the right photon is found to be *H*, and whenever the left photon is *H*, the right is *V*.

Fig. 6.14 The same system as in the previous figure, except that the polarizer on the right has been removed.

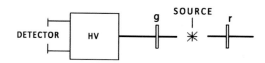

the right photon is polarized at $-45°$ with respect to the vertical. Suppose we arrange a device as shown in the Fig. 6.13, so that one can read the orthogonal polarizations in the detectors whenever a pair of photons is emitted.

Now consider the same system, but without the polarizer and the detectors at the right, as in Fig. 6.14. Suppose the polarization of a photon is measured on the left. Since it is orthogonal to the polarization of the photon on the right, we can be sure what is the polarization of the right photon, without making a measurement on it. In other words, measuring the photon polarization on the left determines the polarization of the photon on the right.

But a quantum measurement changes the measured system. We do not know what the polarization of the left photon actually is before measuring it. The polarizer rotated the direction of polarization of the left photon through some angle, so that it is now *H* or *V*, so the polarizer could not have changed the polarization of the right photon, being at the other end of the laboratory.

Similarly, we can consider a system of two electrons described by a wave function Ψ. Assume that the total spin of the two electrons is $s_1 + s_2 = 0$, that is, the state of the system was prepared so that its total spin is zero. Suppose now that the spin of one of them, say, the left one, is measured in some arbitrary direction, and the result is, for instance, $s_1 = 1/2\hbar$. According to quantum mechanics, this determines the value of the spin of the *right electron* as $s_2 = -1/2\hbar$, in the *same arbitrarily chosen direction*, independently of how far apart the two electrons may be.

If we reject the idea that the measuring apparatus in either of these two examples is able to influence the distant photon or electron, how can we explain the previous results?

The idea of EPR is that, if the left-hand apparatus could measure a property of the right-hand particle without perturbing it, this would reflect the incompleteness of standard quantum mechanics in describing the real world. In Einstein's words:

> If, without in any way perturbing a system, we can predict with certainty (i.e., with probability equal to unity) the value of a physical quantity, then there exists an element of physical reality corresponding to this physical quantity.

Fig. 6.15 Rotation of the polarizer on the right-hand side, through an angle ϕ. The direction ϕ_- is orthogonal to ϕ_+.

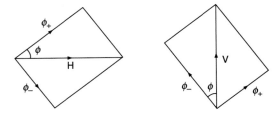

A possible way of completing quantum mechanics, as suggested later by David Bohm, would be by including in the theory a *hidden variable* determining the result of the measurement.

Thus we may conclude at this point that either the measurement affects the distant particle or there is a deterministic theory of hidden variables. There would then be at least two possible versions of quantum mechanics: the conventional one, and another one based on hidden variables.

6.17.4 A Hidden Variable Model and Bell's Theorem

Let us assume the following properties in the case of the experiment with photons:

1. The photons are emitted plane polarized in random directions;
2. The photons in each pair have perpendicular polarizations;
3. The photon emerges through the nearest channel (H or V), i.e., if its polarization is at an angle of less than 45° with the vertical, it will emerge through the V channel, while if its polarization is at an angle of less than 45° with the horizontal, it will emerge through the H channel;
4. The results of the experiments at each point of observation depend only on the values of the physical quantities at these points, and not on measured properties of other, distant, particles. In other words, there is locality.

Observe that at this point we are assuming very special *particle* properties for the pair of photons, but nothing connected with non-locality.

Let us suppose that the polarizer on the right is rotated through an angle ϕ with respect to some axis H chosen as the horizontal (see Fig. 6.15). Then let us denote by $n(V, \phi_+)$ the average number of left photons whose right partner emerges through an angle ϕ with respect to the horizontal axis. It is assumed that every time a photon with polarization V is detected on the left, a photon with polarization H would be detected on the right if a suitable apparatus were placed there.

Consider three experiments. In the first experiment, the right apparatus has been rotated through an angle ϕ with respect to the horizontal axis. We note that photons emerging on the right with polarization $\phi_+ = \phi$ are those horizontal with respect to the polarizer rotated through an angle ϕ. The direction ϕ_- is orthogonal to ϕ_+. We have

Table 6.5 Pattern generated by assigning to each symbol H, ϕ, and θ either a $+$ or a $-$, for a large number of repetitions.

H	ϕ	θ
$+$	$+$	$-$
$+$	$-$	$+$
$-$	$+$	$-$
$+$	$-$	$+$
$-$	$+$	$+$
...	...	...

$$n(V, \phi_+) = n(H = +, \phi = +), \qquad (6.80)$$

which means that the number of photons arriving on the left with polarization V and partner photons on the right emerging polarized through an angle ϕ_+ is equal to the fraction of photons that would be H (if the polarizer were not rotated) on the right, but emerge polarized through an angle ϕ.

In the second experiment, we have a similar situation but now the right polarizer has been rotated through an angle θ with respect to the H direction:

$$n(V, \theta_+) = n(H = +, \theta = +). \qquad (6.81)$$

In the third setup, the left polarizer has been rotated through an angle ϕ and the right polarizer through an angle θ. We have in this case

$$n(\phi_+, \theta_+) = n(\phi = -, \theta = +). \qquad (6.82)$$

Consider the Table 6.5. It can be checked that

$$n(H = +, \phi = +) + n(\phi = -, \theta = +) \geq n(H = +, \theta = +), \qquad (6.83)$$

as a simple consequence of Boolean algebra, if we compare the three sets formed by columns (H, ϕ), (ϕ, θ), and (H, θ). Under the previous assumptions, we can in principle apply the counting of $\pm$ signs in Table 6.5 to the problem of counting photons in our systems of polarizers. Then, from the previous equations we obtain

$$n(V, \phi_+) + n(\phi_+, \theta_+) \geq n(V, \theta_+), \qquad (6.84)$$

under the assumption that the results of the experiments are determined by the properties of the measured photons (encoded in some hidden variables), and not by the configuration of a distant apparatus.

This is the content of the Bell theorem, also known as the Bell inequality.

6.17.5 Bell Inequality and Conventional Quantum Mechanics

Now comes another fundamental step. It is the connection between the number of photons and the energy of the electromagnetic wave. According to the conventional quantum theory, if N is the total number of emitted photons ($N/2$ on each side), then since this number is proportional to the energy intensity, we have $N \sim \int d^3x E^2$, where E is the electric field and the integral extends over spatial coordinates. The number of photons emerging on the right in the direction ϕ_+ must be $n = N \cos^2 \phi \sim \int d^3x E^2 \cos^2 \phi$. Then we have for the photon density

$$n(V, \phi_+) = \frac{1}{2} N \cos^2 \phi,$$

$$n(V, \theta_+) = \frac{1}{2} N \cos^2 \theta,$$

$$n(\phi_+, \theta_+) = n(H, (\theta - \phi)_+) = \frac{1}{2} N \sin^2(\theta - \phi),$$

or, using the Bell inequality,

$$\cos^2 \phi + \sin^2(\theta - \phi) \geq \cos^2 \theta,$$

for any θ and ϕ.

We can choose, for instance, $\phi = 3\theta$. Then we have

$$F(\theta) = \cos^2 3\theta + \sin^2 2\theta - \cos^2 \theta \geq 0. \tag{6.85}$$

But for $0 \leq \theta \leq 30°$, the Bell inequality is violated (Fig. 6.16).

The experiments done by Alain Aspect, starting from 1982, confirmed the violation of the Bell inequalities and were in full agreement with the predictions of quantum mechanics. This is interpreted by most physicists as the final proof of the validity of conventional quantum mechanics.

These results can be stated in a weaker form as follows: no quantum theory based on the preservation of locality is consistent with experiment.

6.17.6 EPR Paradox: Quantum Mechanics Versus Special Relativity

Does the photon whose polarization is measured somehow *warn* the other photon about it? Does it emit a signal carrying information at a velocity $V > c$? There are strong arguments against that. EPR pairs cannot be used to send messages at $V > c$.

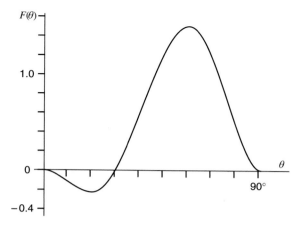

Fig. 6.16 Bell's inequality is violated for $\phi = 3\theta$, in the interval $0 \le \theta \le 30°$, where the function $F(\theta)$ is negative.

There is no conflict between EPR and causality, but with the *spirit* of relativity in our description of physical reality, associated with the problem of non-locality. The pair of photons moves as a single non-local entity. When the polarization of one of the photons is measured, the wave function jumps, in such a way that the photon that was not measured has a definite polarization.

Before the measurement, the state is, for instance,

$$\Psi = H_G V_R + V_G H_R,$$

where H_G, V_R denotes a horizontally polarized green photon on the left *and* a vertically polarized red photon on the right, whereas V_G, H_R describes the same situation but exchanging horizontal and vertical.

If by rotating through an angle θ in a plane perpendicular to the photon propagation, we conceive another state

$$\Psi' = H'_G V'_R + V'_G H'_R,$$

where

$$H'_G = H_G \cos\theta + V_G \sin\theta, \qquad V'_G = -H_G \sin\theta + V_G \cos\theta,$$

and

$$V'_R = H_R \sin\theta + V_R \cos\theta, \qquad H'_R = H_R \cos\theta - V_R \sin\theta,$$

it is easy to show that $\Psi' = \Psi$. These states are now called *entangled states*. (The symmetric and antisymmetric states seen earlier in Sect. 6.12 can also be considered as entangled states.)

If a measurement is made in an arbitrary direction α, the wave function is reduced, say, to

$$\Psi_\alpha = H_G V_R$$

or

$$\Psi_\alpha = V_G H_R,$$

where H_G, V_G refer to one point in space, namely $(-x_0, t_0)$, and H_R, V_R refer to another, far from the first. Let us call it (x_0, t_0).

Due to the relativity of simultaneity, if we now consider an observer moving at velocity V toward the right, he would conclude that the second photon is measured before the first one, whereas another observer moving toward the left would find the opposite, i.e., the second is measured after the first.

Referring to the Lorentz transformation, let us write this for the observer moving to the right. He would find that the measurements were made at times t_1, t_2 given by

$$t_2 = \frac{t_0 - V x_0/c^2}{\sqrt{1 - V^2/c^2}} < \frac{t_0 + V x_0/c^2}{\sqrt{1 - V^2/c^2}} = t_1. \tag{6.86}$$

If we make simultaneous measurements, in the system at rest, on the left and right photons, the moving observer would find that the left part of the wave function *jumps* before being measured!

Thus we have a conflict: different moving observers would arrive at different conclusions about wave function reduction (the R operation) in the EPR experiment.

This fact implies a subtle contradiction between the current formulations of quantum theory and special relativity. The statement regarding the violation of the Bell inequalities involved some possible loopholes relating to the required spacelike separation between measured events, the limited efficiency of the detectors, and a reliance on the randomness of spontaneous emission events. The three loopholes have been closed by several fine test experiments performed between 1998 and 2017. We mention for instance an international collaborative effort to close the 'freedom-of-choice loophole' using human free will instead of random number generators. In an experiment conducted in 2016, over 100 000 volunteers participated in an online video game that used random human decisions to provide sufficient input for the experiments to be statistically significant. Other important work was done in 2017 by the group led by David Kaiser of the Massachusetts Institute of Technology and Anton Zeilinger of the Institute for Quantum Optics and Quantum Information and University of Vienna. They performed experiments that "produced results consistent with nonlocality" by measuring starlight that had taken 600 years to travel to Earth.

6.18   Quantum Computation and Teleportation

The EPR particle pairs play an important role in quantum computation, a theory of communication based on quantum concepts that could materialize in the future in much faster computers than conventional ones. First of all, there is a new concept called the *qubit*, the quantum partner of the classical *bit* that we saw in Chap. 2. The bit is based on the two logical alternatives 0, 1 of the Boolean algebra. The quantum alternatives of two base states $|0\rangle$, $|1\rangle$ are larger, because besides the $0 = |\updownarrow\rangle$ and $1 = |\leftrightarrow\rangle$, we can have qubits in states like $|\searrow\rangle = (|0\rangle + |1\rangle)/\sqrt{2}$ and $|\nearrow\rangle = (|0\rangle - |1\rangle)/\sqrt{2}$. Two qubits can also be entangled in states like $\Psi = (|01\rangle - |10\rangle)/\sqrt{2}$, where neither of the two-component qubits alone has a definite state.

Quantum computers are still a long way from becoming a reality. But if they could be built, they should be able to solve problems like the factorization of very large numbers (which is essential in cryptography) very quickly. This would allow super-fast Fourier analysis, and would open the way to solving various other interesting problems.

Using EPR pairs, it is possible in principle *to teleport* a particle in a state A, from which part of the information about its state was scanned, by causing the remaining unscanned part to pass to another particle in state C which does not interact with A. This is accomplished by making A interact with another particle B which is a partner of C in an EPR pair (B and C interacted before A interacted with B, so obviously, when A interacts with B, some effect is produced on C). By using the initial scanned data from A, plus the EPR information gleaned from C, the exact initial state of A is reproduced in the particle which was in state C. The original particle in state A passes to another state which is of no interest. In 1997, quantum teleportation was achieved experimentally for the first time by a team led by Sandu Popescu (b. 1956) and independently by another team led by Anton Zeilinger (b. 1945). In 2004, researchers at the University of Innsbruck and the US National Institute of Standards and Technology (NIST) teleported one atom state to another with the help of a third auxiliary atom. Teleportation of some other particle states has also been reported. Recent experiments performed in 2017 by scientists from the Chinese Academy of Sciences, the Austrian Academy of Sciences and Vienna University through the Project QUESS (Quantum Experiments at a Space Scale), achieved quantum teleportation at distances of 7600 Km.

6.19   Classical vs. Quantum Logic

Quantum logic is a set of rules (for reasoning about propositions), taking into account the principles of quantum theory. The name originated in work carried out in 1936 by Garrett Birkhoff (1911–1996) and John von Neumann, who were attempting to reconcile the apparent inconsistency of classical logic with quantum facts.

Classical logic satisfies Boolean algebra, and propositions satisfy a distributive law. We consider an example in the classical case. We use the symbol $+$ to represent the operation "or" and the symbol $\cdot$ to represent "and", and consider a ball of mass m carrying a momentum P toward a stick divided by a line into two parts, upper a and lower b. Let p be the proposition "the ball hits the stick at a point", q the proposition "it is a point of a", and r the proposition "it is a point of b". Obviously, the proposition $p \cdot (q + r) = (p \cdot q) + (p \cdot r)$ satisfies the distributive law, since

The ball hits the stick at a point **and** (it is a point of a **or** it is a point of b) =(the ball hits the stick at a point **and** it is a point of a) **or** (the ball hits the stick at a point **and** it is a point of b).

But some properties of quantum logic drastically distinguish it from classical logic. One example is the failure of the distributive law of propositions. Consider the propositions p = the particle has momentum in the interval $[0, +1/8]$, q = the particle is in the interval $[-1, 1]$, and r = the particle is in the interval $[1, 4]$. Now consider the statement

$$p \cdot (q + r) = (p \cdot q) + (p \cdot r), \tag{6.87}$$

where the symbols p, q, and r are the propositional variables. To illustrate why the distributive law fails, consider a particle moving along a line, and setting momentarily Planck's constant $\hbar = 1$, note that $p \cdot (q + r) = true$, since the uncertainty in the momentum is $\Delta p = 1/8$, while the uncertainty in the coordinates is $\Delta x = 5$. Thus $\Delta p \Delta x = 5/8 > 1/2$. On the other hand, $(p \cdot q) + (p \cdot r) = false$, since the proposition $p \cdot q$ implies $\Delta p \Delta x = 1/4 < 1/2$ and $p \cdot r$ implies $\Delta p \Delta x = 3/8 < 1/2$, whence these are both excluded, since they impose tighter restrictions on simultaneous values of uncertainties in position and momentum than is allowed by the Heisenberg principle. Note that (6.87) is a linear relation, and the fact that it fails in logical reasoning (which connects theory with observable facts) means that new logical relations are needed, specific to quantum theory.

Problems

Problem 6.1 Classical concepts are valid in the Bohr model. Thus, we may speak of the average speed of an electron in the first Bohr orbit of an atom of atomic number Z. Is it (a) $Z^{1/2}c$, (b) Zc, or (c) $Zc\alpha$, where α is the fine structure constant.

Problem 6.2 In the one-dimensional infinite potential well, assume a state which is the superposition of the first two eigenstates $n = 1$ and $n = 2$. The resulting state is not an energy eigenstate, and its wave function $\psi(x, t)$ is a linear combination of the two eigenfunctions $\psi_1(x, t)$ and $\psi_2(x, t)$. (a) Calculate the probability density, the average energy $\langle E \rangle$, the average squared energy $\langle E^2 \rangle$, and the standard deviation energy $\sigma(E) = \sqrt{\langle E^2 \rangle - \langle E \rangle^2}$. (b) Show that the average position $\langle x(t) \rangle$ oscillates around the point $x = a/2$ and calculate the oscillation frequency. (c) Calculate the period of oscillation T and its relation with $\sigma(E)$.

Problem 6.3 In the atomic nucleus, the potential binding protons and neutrons is sometimes approximated by an infinite potential well. Assume a proton is confined in an infinite square well of width $a = 2 \times 10^{-12}$ cm (which is 20 fm, where fm stands for the fermi or femtometer, with 1 fm $\equiv 10^{-13}$ cm). Calculate the energy E_γ of the photon emitted by the proton under a transition from the excited state $n = 2$ to the ground state $n = 1$.

Problem 6.4 The neutron and proton masses are $=1.6749275 \times 10^{-24}$ g and $= 1.6726219 \times 10^{-24}$ kg, respectively. Express their masses in units MeV/c^2.

Literature

1. R.P. Feynman, *The Feynman Lectures on Physics*, vol. 3. (Addison Wesley, Reading, 1965). The basic principles of quantum mechanics are presented and applied with unique originality
2. R.H. Dicke, J.P. Wittke, *Introduction to Quantum Mechanics* (Addison Wesley, Reading, 1965). An excellent introductory textbook. The problem of quantum measurement is discussed clearly
3. L.D. Landau, E.M. Lifschitz, *Quantum Mechanics (Non-Relativistic Theory)* (Pergamon, London, 1981). An excellent advanced textbook
4. A.S. Davidov, *Quantum Mechanics* (Pergamon, Oxford, 1965). This is a comprehensive textbook, with clear presentation of the topics
5. M. Chaichian, R. Hagedorn, *Symmetries in Quantum Mechanics: From Angular Momentum to Supersymmetry* (IOP, Bristol, UK, 1997). This is especially recommended for readers interested in the topic of symmetries in quantum mechanics
6. M. Chaichian, A. Demichev, *Path Integrals in Physics. Volume 1: Stochastic Processes and Quantum Mechanics* (IOP, Bristol, UK, 2001). Readers may find in this book a comprehensive treatment of quantum mechanics in terms of path integrals
7. A. Salam (ed.), *From a Life of Physics*. Evening Lectures at the International Centre for Theoretical Physics (Supplement of the IAEA Bulletin, Trieste, Italy, June 1968). The evening lecture *Theory, Criticism and a Philosophy* given by Werner Heisenberg contains very interesting historical anecdotes
8. A.I.M. Rae, *Quantum Physics, Illusion or Reality?* (Cambridge University Press, Cambridge, 1986). A monograph devoted to the quantum mechanical paradoxes
9. M.A. Nielsen, I.L. Chuang, *Quantum Computation and Quantum Information* (Cambridge University Press, Cambridge, 2000). A basic, widely read book on the topic of quantum information and quantum computation

Chapter 7
Quantum Electrodynamics

The formulation of a relativistic quantum equation for the electron by Paul Dirac, as well as the works of Max Born, Werner Heisenberg, and Pascual Jordan on the quantization of the electromagnetic field as a system of harmonic oscillators—an idea anticipated earlier by Paul Ehrenfest—established the basis for the development of *quantum electrodynamics*—a relativistic theory describing the interaction of the quantized electromagnetic field with the electron–positron field.

The first steps were taken by Einstein in 1905, with his theory of the photoelectric effect, and subsequently, in 1917, in his works on the emission and absorption of radiation by an atom. Einstein was the first to suggest that the electromagnetic interaction exists and it is emitted and absorbed in *quanta of radiation*, which were later named *photons*.

Other outstanding physicists of the twentieth century also participated in later developments of quantum electrodynamics: Niels Bohr, Freeman J. Dyson, Enrico Fermi, Richard P. Feynman, Vladimir A. Fock, Wolfgang Pauli, Julian Schwinger, Sin-Itiro Tomonaga, Victor Weisskopf, and many others. But a particularly remarkable step was the formulation of the relativistic equation for the electron, the Dirac equation.

7.1 Dirac Equation

7.1.1 The Spin of the Electron

In 1922, Otto Stern (1888–1969) and Walther Gerlach (1889–1979) performed an experiment on the deflection of an electrically neutral atom through an inhomogeneous magnetic field, thus discovering the existence of an intrinsic angular momentum for those particles, with certain quantized values. This was the first experimental

© The Author(s), under exclusive license to Springer-Verlag GmbH, DE, part of Springer Nature 2021
M. Chaichian et al., *Basic Concepts in Physics*, Undergraduate Lecture Notes in Physics, https://doi.org/10.1007/978-3-662-62313-8_7

observation of what is known as *spin*. In 1943, the Nobel Prize in Physics was awarded solely to Otto Stern. In 1925, by the theoretical work of George Uhlenbeck (1900–1988) and Samuel Goudsmit (1902–1978), it was known that the electrons possessed an intrinsic angular momentum, named spin. Earlier that year, Wolfgang Pauli had proposed the *exclusion principle*, according to which at most one electron can exist in a given quantum state. He also included the spin in Schrödinger's equation in 1927. Pauli's non-relativistic equation for the wave function reads:

$$i\hbar \frac{\partial}{\partial t} \phi(\mathbf{r}, t) = \left[-\hbar^2 \frac{\Delta}{2m} + \frac{e}{2m} \left(\hat{\mathbf{L}} + g_s \hat{\mathbf{S}} \right) \cdot \mathbf{B} \right] \phi(\mathbf{r}, t), \tag{7.1}$$

where

$$\phi(\mathbf{r}, t) = \begin{pmatrix} \phi_1(\mathbf{r}, t) \\ \phi_2(\mathbf{r}, t) \end{pmatrix}$$

is a two-component wave function, e and m are the charge and the mass of the particle, respectively, $\hat{\mathbf{L}}$ is the angular momentum operator, $\mathbf{B}$ is an external magnetic field, and $\hat{\mathbf{S}} = \frac{\hbar}{2}\boldsymbol{\sigma}$ is the spin operator, whose components are three 2×2 matrices, called Pauli's matrices:

$$\sigma_1 = \begin{pmatrix} 0 & 1 \\ 1 & 0 \end{pmatrix}, \quad \sigma_2 = \begin{pmatrix} 0 & -i \\ i & 0 \end{pmatrix}, \quad \sigma_3 = \begin{pmatrix} 1 & 0 \\ 0 & -1 \end{pmatrix}. \tag{7.2}$$

These matrices satisfy the relations $\sigma_1^2 = \sigma_2^2 = \sigma_3^2 = -i\sigma_1\sigma_2\sigma_3 = \mathbf{1}$, where

$$\mathbf{1} = \begin{pmatrix} 1 & 0 \\ 0 & 1 \end{pmatrix}.$$

The orbital angular momentum operator is obtained from the classical formula $\mathbf{L} = \mathbf{r} \times \mathbf{p}$, in which we replace, according to formula (6.15), $\mathbf{p}$ by $-i\hbar\nabla$,

$$\hat{\mathbf{L}} = -i\hbar\, \mathbf{r} \times \nabla. \tag{7.3}$$

The coefficient g_s is called gyromagnetic ratio and for the electron it is equal to 2. This number had been obtained in spectroscopic experiments determining the ratio between the intrinsic magnetic moment and the orbital magnetic moment of the electron. Nobody knew why the ratio should have been 2, but so it was, and Pauli included it by hand in his equation. This was, however, not a fundamental theory of spin. The spin was to appear naturally only when quantum mechanics was formulated in a relativistically invariant way.

From the relativistic expression

$$E^2 = \mathbf{p}^2 c^2 + m^2 c^4 \tag{7.4}$$

for the energy of the electron in terms of its momentum seen in Chap. 5, it was quite natural to derive a quantum mechanical equation, which would be the generalization

of the Schrödinger equation to the relativistic case. All one had to do was to replace E and $\mathbf{p}$ by the corresponding operators, according to (6.15). However, the equation obtained was not adequate to describe the motion of the relativistic electron, considered either as a free particle or as part of an atom. Among other difficulties, it was not possible to introduce the spin in the relativistic equation, and besides, negative probabilities arose, which was absurd. All these problems were revealed in the works of Walter Gordon, Oskar Klein, and Vladimir Fock during 1926–1927. Actually, it is known that Schrödinger himself wrote first a relativistic quantum mechanical equation of the type (7.4), but he obtained for the spectrum of the hydrogen atom energy levels which did not correspond to the experimental findings. He then abandoned the relativistic equation and settled for the non-relativistic one, which reproduced with remarkable accuracy the data. The reason for the lack of success of extending (7.4) to the quantum mechanical treatment of the hydrogen atom was the fact that the equation described a particle with spin zero, while the electron was not such a particle. This difficulty was to be solved in 1928 by Paul Adrien Maurice Dirac (1902–1984), and for this theory he was awarded the Nobel Prize in Physics in 1933, jointly with Erwin Schrödinger (Fig. 7.1).

Referring to the times when the general principles of quantum mechanics were formulated, we quote some paragraphs from a talk given by Dirac at the International Centre for Theoretical Physics of Trieste in 1968: *In order to understand the atmosphere in which theoretical physicists were then working, one must appreciate the enormous influence of relativity. As relativity was then understood, all relativistic theories had to be expressible in tensor form. On this basis we could not do better than the Klein–Gordon theory. Most physicists were content with the Klein–Gordon theory as the best possible relativistic quantum theory for an electron, but I was always dissatisfied with the discrepancy between it and general principles, and continually worried over it till I found the solution. Tensors were inadequate and one had to get away from them, introducing two-valued quantities, now called spinors. Those people who were too familiar with tensors were not fitted to get away from them and think up something more general, and I was able to do so only because I was more attached to the general principles of quantum mechanics than to tensors. One should always guard against getting too attached to one particular line of thought. The introduction of spinors provided a relativistic theory in agreement with the general principles of quantum mechanics, and also accounted for the spin of the electron, although this was not the original intention of the work. But then a new problem appeared, that of negative energies. The theory gives symmetry between positive and negative energies, while only positive energies occur in Nature. As frequently happens with the mathematical procedure in research, the solving of one difficulty leads to another. The difficulty is removed by the assumption that in the vacuum all the negative energy states are filled. One is led to a theory of positrons together with electrons. But again a new difficulty appears, this time connected with the interaction between an electron and the electromagnetic field. One gets divergent integrals for quantities that ought to be finite. Again, this difficulty was really present all the time, lying dormant in the theory, and only now becoming the dominant one.*

Dirac obtained an equation linear in energy and momentum, while in (7.4) the dependence is quadratic. In other words, Dirac linearized the square-root

$$\hat{E} = \sqrt{\hat{\mathbf{p}}^2 c^2 + m^2 c^4},$$

in which $\hat{E}$ and $\hat{\mathbf{p}}$ are *operators*. How can such a linearization be achieved? Observe that (7.4) can be written in terms of operators as:

$$\hat{E}^2 = (\hat{p}_1^2 + \hat{p}_2^2 + \hat{p}_3^2)c^2 + m^2 c^4, \tag{7.5}$$

where $\hat{p}_1$, $\hat{p}_2$, $\hat{p}_3$ are the components of the total momentum, that is, they satisfy the relation $\hat{\mathbf{p}}^2 = \hat{p}_1^2 + \hat{p}_2^2 + \hat{p}_3^2$. Dirac's ingenious idea was to write:

$$\hat{E} = c(\alpha_1 \hat{p}_1 + \alpha_2 \hat{p}_2 + \alpha_3 \hat{p}_3) + \beta m c^2, \tag{7.6}$$

where α_1, α_2, α_3 are coefficients to be determined. Then, in order to get (7.5) by squaring (7.6), one must satisfy the relations

$$\alpha_1^2 = \alpha_2^2 = \alpha_3^2 = \beta^2 = 1, \tag{7.7}$$

and also

$$\begin{aligned}
\alpha_1\alpha_2 + \alpha_2\alpha_1 &= 0, & \alpha_1\beta + \beta\alpha_1 &= 0, \\
\alpha_1\alpha_3 + \alpha_3\alpha_1 &= 0, & \alpha_2\beta + \beta\alpha_2 &= 0, \\
\alpha_2\alpha_3 + \alpha_3\alpha_2 &= 0, & \alpha_3\beta + \beta\alpha_3 &= 0.
\end{aligned} \tag{7.8}$$

Obviously, the quantities α_1, α_2, α_3, and β cannot be numbers, because they do not commute. The set of properties (7.7) and (7.8) can be satisfied only by other mathematical entities, viz. matrices, with the special property of being anticommuting, that is, the product of two of them in a certain order is equal but of opposite sign to the product when their order is reversed. It turns out that the minimum number of dimensions in which one can find four independent matrices satisfying the properties (7.7) and (7.8) is four. Thus, the matrices α_1, α_2, α_3, and β are 4×4 matrices; consequently, the wave function has to have four components. The matrices α_1, α_2, α_3, and β can be expressed in terms of Pauli's matrices (7.2) and the unit matrix as follows:

$$\alpha_1 = \begin{pmatrix} 0 & \sigma_1 \\ \sigma_1 & 0 \end{pmatrix}, \quad \alpha_2 = \begin{pmatrix} 0 & \sigma_2 \\ \sigma_2 & 0 \end{pmatrix}, \quad \alpha_3 = \begin{pmatrix} 0 & \sigma_3 \\ \sigma_3 & 0 \end{pmatrix},$$

$$\beta = \begin{pmatrix} 1 & 0 \\ 0 & -1 \end{pmatrix}. \tag{7.9}$$

The resulting quantum mechanical equation of Dirac, based on the above properties, describes a free electron:

$$i\hbar \frac{\partial \Psi}{\partial t} = -ic\hbar \left(\alpha_1 \frac{\partial \Psi}{\partial x} + \alpha_2 \frac{\partial \Psi}{\partial y} + \alpha_3 \frac{\partial \Psi}{\partial z} \right) + \beta mc^2 \Psi, \tag{7.10}$$

where $\Psi(\mathbf{r}, t)$ is a complex wave function having four components:

$$\Psi(\mathbf{r}, t) = \begin{pmatrix} \Psi_1 \\ \Psi_2 \\ \Psi_3 \\ \Psi_4 \end{pmatrix}. \tag{7.11}$$

Still, we cannot call this entity a four-vector, because upon a Lorentz transformation it does not have the transformation properties of the four-vectors. One finds how $\Psi(\mathbf{r}, t)$ transforms by requiring that Dirac's equation does not change its form (is *covariant*) when passing from one inertial reference frame to another. The transformation law thus obtained in a way corresponds to that of the *square root* of a four-vector. We call it a *spinor*, because it describes a particle with nontrivial spin, equal to $\hbar/2$.

To see that this is indeed the case, we have to make a small calculation. We know that spin is intrinsic angular momentum. Since the electron has a nontrivial spin, its total angular momentum will be the sum of its usual, or orbital, angular momentum and its spin. Total angular momentum has to be conserved in Dirac's theory, since the theory is relativistically covariant, and rotations in the three-dimensional space – which are related by Noether's theorem (see Chap. 1) to angular momentum conservation—are part of the Lorentz transformations which we discussed in Chap. 5. Now, if we wish to check the conservation of an operator $\hat{O}$ in quantum mechanics, we do this by calculating its commutator with the Hamiltonian operator $\hat{H}$, since by Heisenberg's equation this commutator gives the change in time of the operator:

$$i\hbar \frac{d\hat{O}}{dt} = [\hat{O}, \hat{H}].$$

We have still to identify the Hamiltonian operator for a free electron, according to Dirac's theory. This is very easy—we read it off from the right-hand side of (7.10):

$$\hat{H} = -ic\hbar \left(\alpha_1 \frac{\partial}{\partial x} + \alpha_2 \frac{\partial}{\partial y} + \alpha_3 \frac{\partial}{\partial z} \right) + \beta mc^2 = -ic\hbar\boldsymbol{\alpha} \cdot \nabla + \beta mc^2 . \quad (7.12)$$

Now we can easily calculate the commutator of $\hat{\mathbf{L}}$ with $\hat{H}$, using the properties (7.7) and (7.8) and the action of the differential operator ∇ on $\mathbf{r} = (x, y, z)$. The result, as expected, is not zero, but

$$[\hat{\mathbf{L}}, \hat{H}] = \frac{c\hbar}{i}\boldsymbol{\alpha} \cdot \mathbf{p}. \quad (7.13)$$

Consequently, $\hat{\mathbf{L}}$ changes in time, $d\hat{\mathbf{L}}/dt \neq 0$. This confirms that the orbital angular momentum itself is not conserved in Dirac's theory. In order to achieve conservation, we have to add to $\hat{\mathbf{L}}$ another term, whose commutator with the Hamiltonian will compensate exactly the right-hand side of (7.13). This extra term is the spin operator $\hat{\mathbf{S}}$, which has the expression

$$\hat{\mathbf{S}} = \frac{\hbar}{2}\boldsymbol{\Sigma}, \quad (7.14)$$

where the components of the matrix $\boldsymbol{\Sigma}$ are

$$\Sigma_j = -i\epsilon_{jkl}\alpha_k\alpha_l = \begin{pmatrix} \sigma_j & 0 \\ 0 & \sigma_j \end{pmatrix}. \quad (7.15)$$

With the addition of spin, the total angular momentum

$$\hat{\mathbf{J}} = \hat{\mathbf{L}} + \hat{\mathbf{S}} \quad (7.16)$$

is conserved:

$$i\hbar \frac{d\hat{\mathbf{J}}}{dt} = [\hat{\mathbf{J}}, \hat{H}] = 0.$$

From the expression of $\hat{\mathbf{S}}$ one finds straightforwardly that its eigenvalues are $\pm\hbar/2$, in other words, the spin of the Dirac particle is 1/2 in units of $\hbar$.

Massive particles with spin $s = 1/2$ have two possible orientations of the spin vector with respect to their momentum: parallel and antiparallel. For a general spin s, the number of projections of the spin on the direction of momentum is $2s + 1$.

Another success of Dirac's equation was that it *predicted* the gyromagnetic ratio of the electron, $g_s = 2$. If one takes the non-relativistic limit of (7.10), by considering the velocity of the particle much less than c, one obtains exactly Pauli's wave equation (7.1), for a two-dimensional spinor.

The conclusion which imposes itself is that spin is a relativistic quantum number (the fundamental equations describing particles with spin have to be relativistically covariant). Later, in the 1930s, Eugene Paul Wigner (1902–1995) showed that the relativistic particle states are classified by their *mass* and *spin*, and the wave functions describing such particles are the so-called representations of the Lorentz group.

7.1.2 Hydrogen Atom in Dirac's Theory

Dirac extended (7.10) to describe an electron in a hydrogen atom, and to achieve it he had to include a potential term describing the Coulomb interaction between the electron and the nucleus:

$$i\hbar\frac{\partial\Psi}{\partial t} = -ic\,\hbar\left(\alpha_1\frac{\partial\Psi}{\partial x} + \alpha_2\frac{\partial\Psi}{\partial y} + \alpha_3\frac{\partial\Psi}{\partial z}\right) + \beta mc^2\Psi - \frac{Ze^2}{r}\Psi, \tag{7.17}$$

where Z is the atomic number of the nucleus, which is equal to 1 for hydrogen.

By solving the above equation, he obtained the energy levels of the electron in the hydrogen and other atoms with great accuracy, predicting also its *fine structure*: each energy level of non-relativistic quantum mechanics is in fact split into a set of finer energy levels and these are in excellent agreement with the experimental data.

The energy spectrum of the electron in the field of a nucleus of atomic number Z is given by the following expression:

$$E = mc^2\left[1 + \left(\frac{Z\alpha}{n - (j + 1/2) + \sqrt{(j + 1/2)^2 - Z^2\alpha^2}}\right)^2\right]^{-1/2}, \tag{7.18}$$

where $n = 1, 2, \ldots$ is a positive integer quantum number, and j is connected to the total angular momentum operator (7.16), whose eigenvalues range from 0 to $j + 1/2 \leq n$. Here, $\alpha = e^2/\hbar c \approx 1/137$ is the fine structure constant (see also Sect. 6.4). The expression (7.18) can be approximated by expanding it in powers of $(Z\alpha)^2$, with the result:

$$E = mc^2 \left[1 - \frac{(Z\alpha)^2}{2n^2} - \frac{(Z\alpha)^4}{2n^4} \left(\frac{n}{j+1/2} - \frac{3}{4} \right) + \mathcal{O}\left((Z\alpha)^6 \right) \right]. \qquad (7.19)$$

The first term, mc^2, represents the rest energy of the electron, the second gives the non-relativistic spectrum, obtained from Bohr's model (6.6) The third term produces the *fine structure corrections* to the previous term, i.e. a further split of the spectral lines, due to the effects of the spin.

Dirac theory and the Bohr–Sommerfeld atomic model. We must state at this point that an equation of form similar to (7.18) was actually obtained in 1915–1916, before Dirac's work, by Arnold Sommerfeld, as a relativistic generalization of the Bohr model which is known as the Bohr–Sommerfeld model. Sommerfeld concluded that the theory of relativity should be applied to electrons orbiting with high velocity v around the nucleus, taking into account the fact that their mass was not constant, but given by

$$m = \frac{m_0}{\sqrt{1 - \frac{v^2}{c^2}}}. \qquad (7.20)$$

In the frame of the Bohr model, this implied changes in the elliptic orbits of the electrons but, more fundamentally, corrected values for the energy levels given by the two quantum numbers n and k, i.e., the principal and azimuthal quantum numbers, respectively, where $k = 0, 1, \ldots, (n-1)$. The energy formula for a one-electron system, i.e., the hydrogen atom, as given by Sommerfeld in 1916, is similar to (7.18), if one makes the replacement $(j + 1/2) \rightarrow k$ in the latter. If expanded in powers of α, the result is also similar to the fine structure expression (7.19).

For some time it was believed that the relativistic problem was solved and even Einstein congratulated Sommerfeld on his results. But the discovery of spin raised a new problem. In 1926, Heisenberg and Pascual Jordan succeeded in accounting for the hydrogen fine structure spectral lines by adding to the usual Hamiltonian a perturbation term describing the relativistic correction plus a magnetic term referring to the electron spin. With their approximation, they reproduced the fine-structure Sommerfeld term, with k replaced by $j + 1/2$. The Heisenberg–Jordan theory was not genuinely relativistic, and the spin effect was introduced "by hand". The full and consistent quantum-relativistic theory was formulated by Dirac two years later. The Bohr–Sommerfeld theory remained as its quasi-classical approximation.

7.1.3 Hole Theory and Positrons

Dirac's equation was, for these reasons, a great success. But it went even further in predicting the existence of "positive electrons". Let us see how they came out.

The particles described by the free Dirac equation (7.10) have energies which satisfy (7.4). There are actually two solutions for the energy as a function of the momentum, after taking the square root:

$$E = \pm c\sqrt{\mathbf{p}^2 + m^2c^2},\tag{7.21}$$

that is, two possible values for the energy of the particle. Negative-energy solutions cannot be simply discarded as non-physical due to the requirement of the completeness of the set of solutions in quantum mechanics. For free particles, this is not a problem, since free particles do not suffer any change in their momentum or energy. However, as soon as we allow for electromagnetic interactions, transitions between positive- and negative-energy states are possible, i.e. an electron at rest can tumble down into a negative-energy state by emitting radiation. For an electron in the ground state of the hydrogen atom, such transitions were calculated and it was found that the rate for an electron to make a transition between $-mc^2$ and $-2mc^2$ was $10^8\,\mathrm{s}^{-1}$, and the rate would blow up if all negative-energy states were considered! This is indeed not the way an electron behaves, so negative-energy states had to be interpreted in a way that would make sense physically.

To this end, Dirac proposed in 1930 his *hole theory*, which incorporated Pauli's exclusion principle into the picture. According to this theory, the ground state is the state in which all the negative-energy states are filled with electrons, thus being unobservable, and all the positive-energy states are unoccupied. As a result, the electrons bound in the hydrogen atom cannot make anymore transitions to lower energy states, because Pauli's exclusion principle forbids the existence of two electrons with the same quantum numbers.

The *negative-energy sea* can loose one electron of energy $-E$ if this absorbs a quantum of radiation with energy at least $E + mc^2$ (Fig. 7.2). Then, the electron will have positive energy and will become observable. Still more interesting is that its place in the negative-energy sea will become an observable *hole*. An observer will interpret, relative to the vacuum, the *absence* of an electron with energy $-E$ and charge $-|e|$ as the *presence* of a particle with energy E and charge $|e|$, which is called a *positron*. Thus, holes in the negative-energy sea are meaningful objects, assimilated to positive-energy particles with charge opposite to the electrons. The phenomenon described above is the production of an electron–positron pair from radiation. The opposite phenomenon, pair annihilation into radiation, is explained by the trapping of a positive-energy electron into a hole, with emission of photons.

When Dirac proposed the hole theory, the only known subatomic positively charged particles with charge $|e|$ where the protons, and Dirac initially considered that the holes were protons. Later it was shown by Robert Oppenheimer (1904–1967) that they had to be particles with the same mass as the electron, but with positive charge. In 1931, Dirac predicted that such antielectrons are real, not fictitious, particles and one year later, Carl Anderson (1905–1991) indeed found them in his experiments with cosmic rays. They were the first observed manifestation of *antimatter*. Anderson was awarded the Nobel Prize in 1936.

Dirac's hole theory was historically an interesting and useful model in explaining the existence of particle–antiparticle pairs (as the electron–positron). It certainly had its drawbacks, for example the fact that the sea of negative-energy electrons should have an infinite electric charge and infinite mass, which would have a tremendous

Fig. 7.2 A photon of energy equal or greater than $2mc^2$ can excite an electron occupying a state of negative energy in the Dirac sea. The electron becomes observable, and a vacant state is left, a positively charged hole, which is the positron or antielectron.

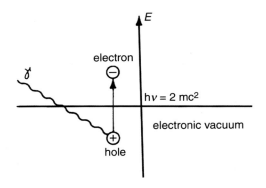

influence both electromagnetically, as well as gravitationally. It also created a breach in the quantum mechanical interpretation of the wave function: in the non-relativistic case, the number of particles of any quantum mechanical system is fixed and given by the integral over space of the square of the wave function. The hole theory implies variable numbers of particles, and there is no way to reconcile this with the probabilistic interpretation of the wave function.

The hole theory becomes obsolete in the quantum theory of fields, although the ideas of polarization of vacuum and the creation and annihilation of virtual electron–positron pairs have some ingredients of it. At present, the hole theory is used in condensed matter physics, where there exists a similar situation concerning the creation of electron–hole pairs, in the electronic bands of semiconductors. These holes are named quasi-particles, since they do not exist outside matter, but are important to explain the phenomenon of electric conduction in semiconductors.

The particle–antiparticle symmetry exists not only for electrons and positrons, but for all the particles of the subatomic world, as protons, neutrons, neutrinos, π mesons, muons, etc. In particular, the photon is its own antiparticle, as it is neutral with respect to all quantum numbers.

In 1955, Emilio Segrè (1905–1989) and Owen Chamberlain (1920–2006) discovered the antiproton, with negative charge. They received the Nobel Prize in 1959 for their discovery. A positron bounded to an antiproton forms an antihydrogen atom. Atoms of antimatter have been produced in high energy laboratories.

The existence of antiatoms led to the idea that in our Universe there may exist large portions of antimatter (including stars and galaxies). If so, they would have to be far from bodies formed from ordinary matter, since otherwise they would mutually annihilate, producing a large amount of radiation. Stars of antimatter would not be distinguishable from ordinary stars by the spectroscopic analysis of the light they emit, since the emission and absorption of radiation is similar for atoms and antiatoms. However, such anti-stars are not observed. It seems that the observable Universe contains only matter, whose nuclei are formed by protons and neutrons. Thus, there is a *baryon asymmetry* in the Universe (see Chap. 11), whose origin is still a topical subject of investigation.

7.2 Intermezzo: Natural Units and the Metric Used in Particle Physics

The reader has probably observed that all the formulas in quantum mechanics contain Planck's constant $\hbar$, and in the relativistic case they contain usually also the speed of light in vacuum, c. The formulas would be much less heavy if we were to write them without these constants, but being able to revert to the correct dimensions in the end of the calculation.

In the system of *natural units*, the fundamental quantities are mass, action and speed. The unit of action is $\hbar$, which is set to 1, as is also the unit of speed, c:

$$\hbar = c = 1. \tag{7.22}$$

This is the system of units adopted in particle physics. The only dimensionfull fundamental quantity is the mass, and all the other dimensions are expressed in powers of mass dimension, $[M]^n$. Clearly, action and speed have mass dimension zero, $n = 0$; energy and momentum have the same dimension as the mass itself, $n = 1$; time and length have inverse mass dimension, $n = -1$. Lagrangian and Hamiltonian densities, in a three-dimensional space, have mass dimension $n = 4$. The fine structure constant is dimensionless in CGS, therefore the electric charge has mass dimension $n = 0$ in natural units. The mass dimensions of all the quantities are easily found from the basic formulas which relate them to each other.

After performing calculations in natural units, one reverts to CGS units by multiplying the result with the appropriate powers of $\hbar$ and c to obtain the correct dimension for the calculated quantity.

Before starting our discussion of quantized fields, we recall the redefinition of four-vectors which we introduced in the end of Sect. 5.8. We shall use it in what follows. To fix the essentials, we remind the reader that a contravariant four-vector is defined as $a^\mu = (a^0, a^1, a^2, a^3) = (a^0, \mathbf{a})$. The corresponding covariant vector is $a_\mu = (a_0, a_1, a_2, a_3)$, with $a^0 = a_0$ and $a^i = -a_i$, for $i = 1, 2, 3$. In particular, the coordinate four-vector is denoted generically by x or $x^\mu = (t, x^1, x^2, x^3) = (t, \mathbf{x})$, where we have employed also natural units.

The scalar product of two four-vectors, a and b, is

$$ab = a^\mu b_\mu = a^0 b^0 - \mathbf{a} \cdot \mathbf{b},$$

where Einstein's summation convention is used.

The metric tensor $g^{\mu\nu}$ by which the raising and lowering of the indices is achieved is the diagonal 4×4 matrix

$$g^{\mu\nu} = \begin{pmatrix} 1 & 0 & 0 & 0 \\ 0 & -1 & 0 & 0 \\ 0 & 0 & -1 & 0 \\ 0 & 0 & 0 & -1 \end{pmatrix}. \tag{7.23}$$

Its inverse is $g_{\mu\nu}$, which is identical to $g^{\mu\nu}$. From a contravariant vector we obtain the covariant one by $a_\mu = g_{\mu\nu}a^\nu$, and viceversa, $a^\mu = g^{\mu\nu}a_\nu$.

The matrix $g^{\mu\nu}$ in this representation has the signature (or trace) -2, and sometimes the metric of the Minkowski space is identified by its signature. This metric is most commonly used in particle physics.

7.3 Quantized Fields and Particles

The reconciliation between quantum theory and special relativity was achieved by the formulation of the *relativistic quantum theory of fields*. In this theory, unlike the non-relativistic quantum mechanics, the role of dynamical variables is taken by *fields*, which become operators.

The prototype for a field is the electromagnetic field, which has also a classical manifestation and is therefore more appealing to intuition. In the free space without electric or magnetic sources, the electromagnetic field propagates as a wave which is self-sustainable, the curl of the electric field $\mathbf{E}\,(\mathbf{r},\,t)$ producing the time-variation of the magnetic field $\mathbf{B}\,(\mathbf{r},\,t)$ and vice-versa, according to Maxwell's equations (4.3) and (4.4), which are the equations of motion for the electromagnetic field. Remark that the radius-vector $\mathbf{r}$ is used here as a label for the space point at which the field is considered. The change in time of $\mathbf{r}$ is not the object of the study, as it were in the mechanics of particles or bodies, but the change in time of $\mathbf{E}$ and $\mathbf{B}$ is. For this reason we call the fields dynamical variables. Another essential remark is that the value of the field at one point is independent of its value at any other point, therefore for each value of $\mathbf{r}$ we have one degree of freedom for each polarization of the wave. Since $\mathbf{r}$ varies continuously, the number of degrees of freedom of a field is infinite.

When a field is quantized, it becomes an operator which acts on a space of states. Those states represent particles, which can be created or destroyed by the quantum field. For illustration, let us consider a non-interacting real scalar field, $\hat{\Phi}(x)$. Its equation of motion is identical to the quantum mechanical equation of a scalar particle, the so-called Klein–Gordon equation,

$$\left(\Box + m^2\right)\hat{\Phi}(x) = 0, \tag{7.24}$$

where $\Box = \partial^2/\partial^2 t - \partial^2/\partial^2 x - \partial^2/\partial^2 y - \partial^2/\partial^2 z$ is the d'Alembertian operator (four-dimensional generalization on the Minkowski space of the Laplacian operator) and m is a positive parameter, which will turn out to be the mass of the particles associated to the field. The argument x, labeling the degrees of freedom, is actually the four-vector x^μ.

The general solution of this equation, taking a finite volume V for the system, can be written as a Fourier series, whose coefficients are operators:

$$\hat{Phi}(x) = \sum_k \left(\frac{1}{2V\omega_k}\right)^{1/2} \hat{a}(\mathbf{k})e^{-ikx} + \sum_k \left(\frac{1}{2V\omega_k}\right)^{1/2} \hat{a}^\dagger(\mathbf{k})e^{ikx}, \quad (7.25)$$

where k is the wave four-vector of a relativistic particle with mass m, momentum $\hbar\mathbf{k}$ and energy (in natural units)

$$E_\mathbf{k} = \omega_\mathbf{k} = \sqrt{m^2 + \mathbf{k}^2}.$$

In the exponents of (7.25), $kx = k^\mu x_\mu = \omega_\mathbf{k} t - \mathbf{k} \cdot \mathbf{x}$. The operators $\hat{a}(\mathbf{k})$ and $\hat{a}^\dagger(\mathbf{k})$ satisfy the commutation relations:

$$[\hat{a}(\mathbf{k}), \hat{a}^\dagger(\mathbf{k}')] = \delta_{\mathbf{k}\mathbf{k}'}, \quad (7.26)$$
$$[\hat{a}(\mathbf{k}), \hat{a}(\mathbf{k}')] = [\hat{a}^\dagger(\mathbf{k}), \hat{a}^\dagger(\mathbf{k}')] = 0,$$

which are analogous to the commutation relations of the raising and lowering operators of the quantum harmonic oscillator, which we have encountered in Chap. 6, (6.56).

The space of states for these operators is constructed in much the same way as for the harmonic oscillator. One defines the ground state, or *vacuum state*, denoted by $|0\rangle$, with the property that, for any $\mathbf{k}$,

$$\hat{a}(\mathbf{k})|0\rangle = 0. \quad (7.27)$$

The particle states are created from the vacuum state by successive applications of the operators $\hat{a}^\dagger(\mathbf{k})$. For example, a state with one particle with momentum $\mathbf{k}$ will be described by

$$|\mathbf{k}\rangle = \hat{a}^\dagger(\mathbf{k})|0\rangle,$$

while a state with n particles with different momenta will be

$$|\mathbf{k}_1, \mathbf{k}_2, \ldots, \mathbf{k}_n\rangle = \hat{a}^\dagger(\mathbf{k}_1)\hat{a}^\dagger(\mathbf{k}_2)\ldots\hat{a}^\dagger(\mathbf{k}_n)|0\rangle.$$

If a state contains a particle with momentum $\mathbf{k}$, then by applying the operator $\hat{a}(\mathbf{k})$ to that state, the particle is destroyed:

$$\hat{a}(\mathbf{k})|\mathbf{k}\rangle = \hat{a}(\mathbf{k})\hat{a}^\dagger(\mathbf{k})|0\rangle = \left(\hat{a}^\dagger(\mathbf{k})\hat{a}(\mathbf{k}) + 1\right)|0\rangle = |0\rangle,$$

where the commutation relations (7.26) were used.

Consequently, the operators $\hat{a}^\dagger(\mathbf{k})$ are called *creation* operators, while $\hat{a}(\mathbf{k})$ are called *annihilation* operators. The particle states are labeled by their momenta (since for free particles the momentum does not change) and they behave like harmonic oscillators of a given relativistic energy, corresponding to that momentum.

Speaking of quantum mechanical particles, which are indistinguishable, we have to make sure that they satisfy also the correct statistics. The particles being scalar, they have spin zero, meaning that they have to obey the Bose–Einstein statistics.

Indeed, we see that a state with any number of particles with the same momentum (same quantum number) can exist, i.e.

$$|\mathbf{k}, \mathbf{k}, \ldots, \mathbf{k}\rangle = \frac{1}{\sqrt{n!}} \hat{a}^\dagger(\mathbf{k}) \hat{a}^\dagger(\mathbf{k}) \ldots \hat{a}^\dagger(\mathbf{k}) |0\rangle \tag{7.28}$$

(the factor $1/\sqrt{n!}$ is just a normalization factor, where n is the number of particles in the state). Moreover, any multiparticle state is symmetrical under the interchange of any two of its particles. For example, for a two-particle state, using (7.26), we have:

$$|\mathbf{k}_1, \mathbf{k}_2\rangle = \hat{a}^\dagger(\mathbf{k}_1) \hat{a}^\dagger(\mathbf{k}_2) |0\rangle = \hat{a}^\dagger(\mathbf{k}_2) \hat{a}^\dagger(\mathbf{k}_1) |0\rangle = |\mathbf{k}_2, \mathbf{k}_1\rangle. \tag{7.29}$$

With some modifications, this quantization scheme applies to all integer spin fields, in particular to the electromagnetic field, which is a vector field $\hat{A}_\mu(x)$, and has spin 1. We adopt for the electromagnetic field the vector-potential description because we need to develop a relativistic theory, and the space-vectors $\mathbf{E}$ and $\mathbf{B}$ are not suitable for this purpose. The photons, which are the particles corresponding to the quantized electromagnetic field, have no electric charge, and this is mathematically encoded in the fact that $\hat{A}_\mu(x)$ is a real field.

If we wish to describe electrons in the theory of quantized fields, we have to incorporate the two spin degrees of freedom, as well as Pauli's exclusion principle, according to which in a given quantum state one can have at most one electron. These requirements are fulfilled if we take as the equation of motion for the free electron field the Dirac equation:

$$\left(i\gamma^\mu \partial_\mu - m \right) \hat{\Psi}(x) = 0, \quad \gamma_0 = \beta, \quad \gamma_i = \beta\alpha_i, \quad i = 1, 2, 3, \tag{7.30}$$

where we denoted $\partial_\mu = \partial/\partial x^\mu$, and we modify appropriately the commutation relations analogous to (7.26). Let us see what modifications would suit our purpose.

The solution of Dirac's equation (7.30) can be written as:

$$\hat{\Psi}(x) = \sum_{\mathbf{p}} \sum_{s=1,2} \left(\frac{m}{V E_\mathbf{p}} \right)^{1/2} [\hat{c}_s(\mathbf{p}) u_s(\mathbf{p}) e^{-ipx} + \hat{d}_s^\dagger(\mathbf{p}) v_s(\mathbf{p}) e^{ipx}], \tag{7.31}$$

where $E_\mathbf{p} = \sqrt{m^2 + \mathbf{p}^2}$.

Since the field $\hat{\Psi}(x)$ is complex, its so-called Dirac conjugated, defined by the relation $\hat{\bar{\Psi}}(x) = \hat{\Psi}^\dagger \gamma_0$, has the expansion:

$$\hat{\bar{\Psi}}(x) = \sum_{\mathbf{p}} \sum_{s=1,2} \left(\frac{m}{V E_\mathbf{p}} \right)^{1/2} [\hat{d}_s(\mathbf{p}) \bar{v}_s(\mathbf{p}) e^{-ipx} + \hat{c}_s^\dagger(\mathbf{p}) \bar{u}_s(\mathbf{p}) e^{ipx}]. \tag{7.32}$$

Note that the summation is performed also over the spin orientations, besides the summation over the momenta. The factors $u_s(\mathbf{p})$ and $v_s(\mathbf{p})$ are spinors (in other words, column matrices with four components), depending on $\mathbf{p}$ and m. For our discussion, the interesting factors are the creation and annihilation operators, $\hat{c}_s^\dagger$, $\hat{c}_s$ and $\hat{d}_s^\dagger$, $\hat{d}_s$. The field being complex, we obtain two sets of operators for each polarization, and they indeed correspond to electrons (particles) and positrons (antiparticles), respectively.

What about Pauli's exclusion principle? Suppose that $\hat{c}_s^\dagger(\mathbf{p})$ creates an electron with momentum $\mathbf{p}$, energy $E_\mathbf{p}$ and spin polarization s,

$$|e^-(\mathbf{p}, s)\rangle = \hat{c}_s^\dagger(\mathbf{p})|0\rangle.$$

Then the state with two electrons with the same quantum numbers has to vanish, i.e.

$$|e^-(\mathbf{p}, s), e^-(\mathbf{p}, s)\rangle = \hat{c}_s^\dagger(\mathbf{p})\hat{c}_s^\dagger(\mathbf{p})|0\rangle = 0,$$

which implies

$$\hat{c}_s^\dagger(\mathbf{p})\hat{c}_s^\dagger(\mathbf{p}) = 0.$$

This happens indeed if instead of commutation relations, we impose *anticommutation relations* between creation and annihilation operators between the creation and annihilation operators of the spinor field:

$$\{\hat{c}_s(\mathbf{k}), \hat{c}_{s'}^\dagger(\mathbf{k}')\} = \delta_{ss'}\delta_{\mathbf{k}\mathbf{k}'}, \tag{7.33}$$
$$\{\hat{c}_s(\mathbf{k}), \hat{c}_{s'}(\mathbf{k}')\} = \{\hat{c}_s^\dagger(\mathbf{k}), \hat{c}_{s'}^\dagger(\mathbf{k}')\} = 0,$$

and

$$\{\hat{d}_s(\mathbf{k}), \hat{d}_{s'}^\dagger(\mathbf{k}')\} = \delta_{ss'}\delta_{\mathbf{k}\mathbf{k}'}, \tag{7.34}$$
$$\{\hat{d}_s(\mathbf{k}), \hat{d}_{s'}(\mathbf{k}')\} = \{\hat{d}_s^\dagger(\mathbf{k}), \hat{d}_{s'}^\dagger(\mathbf{k}')\} = 0.$$

The curly brackets are used to denote the *anticommutator* of two operators:

$$\{\mathcal{O}_1, \mathcal{O}_2\} = \mathcal{O}_1\mathcal{O}_2 + \mathcal{O}_2\mathcal{O}_1.$$

The above relations will lead also to antisymmetric multiparticle electron or positron states. Thus, the particles created and destroyed by the field $\hat{\Psi}(x)$ satisfy the Fermi–Dirac statistics.

The quantum harmonic oscillator is the basis of discussion for all free fields, and the theory of quantized fields is thus solidly anchored in the usual quantum mechanics. For bosonic fields, quantized by commutation relations, we have a direct generalization of the non-relativistic quantum mechanics, while for fermionic fields we take into account Pauli's exclusion principle by using anticommutation relations for quantization.

Since *observation* implies interaction, the next step is, therefore, to quantize fields in interaction.

7.4   Quantum Electrodynamics (QED)

The interactions of photons with electrons and positrons were the first to be explained in the framework of the theory of quantized fields. The energies at which these interactions show their quantum peculiarities are relatively low, and the results of various experiments in the 1940s could not be explained using just quantum mechanics. In general, when speaking about quantum electrodynamics, it is understood the interaction between photons and electrically charged particles.

For the formulation of quantum electrodynamics it is very useful to use a Lagrangian function that, as in mechanics and classical electrodynamics, allows us to obtain the equations of motion. For instance, the Maxwell equations discussed in Chap. 3 are the equations of motion for the electromagnetic field, and can be obtained by starting from a certain Lagrangian function. The relativistic quantum field theoretical description of the interactions of photons with charged particles is a task of some mathematical complexity.

For the purposes of the present book, we shall avoid this task. Here it will be enough to point out that the Lagrangian density of quantum electrodynamics contains three terms: the first describes the free electron–positron field, the second—the free electromagnetic field, and the third—the interaction of the electron–positron field with the electromagnetic field. The latter term is written as:

$$\hat{\mathcal{L}}_{int} = e\hat{\bar{\Psi}}\gamma^\mu \hat{A}_\mu \hat{\Psi}. \tag{7.35}$$

where e is the electric charge of the electron, and the other factors are the same as in the previous section. According to (7.31) and (7.32), $\hat{\Psi}$ annihilates electrons and creates positrons, and $\hat{\bar{\Psi}}$ creates electrons and annihilates positrons.

The equations of motion for the interacting fields $\hat{\Psi}(x)$ and $\hat{A}_\mu(x)$ are more complicated than for free fields: they are non-homogeneous differential equations, which are not exactly solvable. Considering that the energy of the electromagnetic interaction is small compared to the other terms in the Lagrangian density (due to the smallness of e in (7.35)), the most convenient procedure to treat quantum electrodynamics is in a perturbative expansion. The role of expansion parameter is played by the coupling constant α, characterizing the strength of the electromagnetic interactions:

$$\alpha = e^2/\hbar c \simeq \frac{1}{137}, \tag{7.36}$$

which is present, for example, in the formula for the energy of the electron in the hydrogen atom, as mentioned in Sect. 7.1.2. We see that in the constant α enter the square of the electron charge e^2, the Planck constant $\hbar$, and the speed of light c. The resulting number is dimensionless, a true universal constant, whose value does not depend on the system of units used.

7.4.1 Unitarity in Quantum Electrodynamics

In quantum electrodynamics, as in any other quantum field theory, the scattering of particles is described by a unitary evolution operator called the S-matrix. The S-matrix is expanded in a perturbative series in the fine structure constant α and each term of the expansion can be expressed in terms of Feynman diagrams. This expansion is an essential tool in the calculation of scattering probabilities and it is due to Freeman Dyson (1923–2020), one of the pioneers of quantum electrodynamics.

The total Hamiltonian of quantum electrodynamics is separated into two parts: one describing the free fields (corresponding to the kinetic energy of particles), $\hat{H}_0$, and another describing the interaction (corresponding to the potential energy of particles), $\hat{H}_{int}$:

$$\hat{H} = \hat{H}_0 + \hat{H}_{int},$$

such that the interaction term, whose contribution to the energy is much smaller than that of the first term, is treated as a perturbation. The Hamiltonian of interaction (or the potential term) in quantum electrodynamics is

$$\hat{H}_{int} = \int d^3x \, \hat{\mathcal{H}}_{int}, \quad \text{with} \quad \hat{\mathcal{H}}_{int} = -\hat{\mathcal{L}}_{int} = -e\hat{\bar{\Psi}}\gamma^\mu \hat{A}_\mu \hat{\Psi}.$$

This separation leads naturally to the so-called *interaction picture*, in which the operators (e.g., the fields) evolve in time as dictated by the free Hamiltonian, while the states of the system, generally denoted by $|\Psi(t)\rangle$, evolve according to a Schrödinger-like equation driven by the Hamiltonian of interaction:

$$i\frac{d}{dt}|\Psi(t)\rangle = \hat{H}_{int}|\Psi(t)\rangle, \tag{7.37}$$

which is written here as a particular case of the general Tomonaga–Schwinger equation. Then, formally, the solution of this equation can be written as

$$|\Psi(t)\rangle = U(t, t_0)|\Psi(t_0)\rangle, \tag{7.38}$$

with the evolution operator satisfying the differential equation with boundary condition

$$i\frac{d}{dt}U(t, t_0) = \hat{H}_{int}U(t, t_0), \quad U(t_0, t_0) = \mathbf{1}, \tag{7.39}$$

or, alternatively, satisfying the integral equation

$$U(t, t_0) = \mathbf{1} - i\int_{t_0}^{t} d\tau \, \hat{H}_{int}(\tau)U(\tau, t_0). \tag{7.40}$$

This equation can be solved only iteratively, the result being an expansion in the *coupling constant*, which is the electric charge e. Solving the equation is beyond the scope of this book, however it is the evolution operator which we intend to interpret physically.

In a scattering process, we are not interested in the intermediate states when the particles are actually interacting, but rather in the *initial state*, in which we prepare the system in an experiment, and in the *final state*, in which we observe the system after the interaction. According to the adiabatic approximation, one considers the initial state as the state in which the particles are in the distant past ($t_0 \to -\infty$) and sufficiently far apart not to interact:

$$|\Psi(-\infty)\rangle = |i\rangle.$$

In the final state, taken in the distant future ($t \to \infty$), the particles behave again as if they were free. The interaction takes place during a finite period of time in between, as if the Hamiltonian of interaction were "switched on" at the time $-T$ and then "switched off" at the time T.

The S-matrix relates the initial state $|\Psi(-\infty)\rangle$ to the final state $|\Psi(\infty)\rangle$, which means that, according to (7.38),

$$S = \lim_{t \to \infty} \lim_{t_0 \to -\infty} U(t, t_0).$$

The initial state is fixed by the observer as a state with a definite number of particles, with given properties (like momenta, spin polarizations, etc.), and far apart, so that they do not interact. The system is with *certainty* (unit probability) in the initial state at the time $t_0 \to -\infty$, therefore the state is normalized to unity:

$$\langle i|i \rangle = \langle \Psi(-\infty)|\Psi(-\infty)\rangle = 1.$$

The probability is conserved, since the S-matrix contains all the possibilities for a final state, i.e.

$$\langle \Psi(\infty)|\Psi(\infty)\rangle = 1.$$

But when we observe the system, we look for *particular final states*, $|f\rangle$, whose probability to be detected upon the scattering is

$$|\langle f|\Psi(\infty)\rangle|^2.$$

Consequently, the *probability amplitude* for the transition from $|i\rangle$ to $|f\rangle$ can be written as an element of the S-matrix:

$$\langle f|\Psi(\infty)\rangle = \langle f|S|i\rangle = S_{fi}. \tag{7.41}$$

Considering that all possible final states form a complete orthonormal set, we can expand the state $|\Psi(\infty)\rangle$:

$$|\Psi(\infty)\rangle = \sum_f |f\rangle\langle f|\Psi(\infty)\rangle = \sum_f S_{fi}|f\rangle,$$

which leads to the unitarity of the S-matrix expressed in the form:

$$\sum_f |S_{fi}|^2 = 1. \tag{7.42}$$

This expression of the conservation of probability in quantum field theory has a higher generality than the conservation of particles in non-relativistic quantum mechanics, which is not anymore valid, since particles can now be created and annihilated.

The unitarity of the S-matrix is a powerful tool in quantum field theory and an indication whether a theory is plausible or not. Studies of unitarity have led to impressive theoretical advances, one of the farthest-reaching for particle physics being the introduction of the so-called Faddeev–Popov ghosts in gauge field theories in 1967, by Ludvig Faddeev (1934–2017) and Victor Popov (1937–1994). However these topics, as well as the expression of the conservation of probability by the *optical theorem*, are beyond the scope of this book.

7.4.2 Feynman Diagrams

The processes of quantum electrodynamics or any other quantum field theory are represented by diagrams, of which those in Fig. 7.3a, b are special cases. They are called *Feynman diagrams* and represent a customary technical tool in the study of the interactions of particles. To each term of an S-matrix element corresponds a Feynman diagram.

The diagrams illustrating the interaction Lagrangian (7.35) involve two fermions (e.g. electron–positron, or electron–electron, or positron–positron) and a photon. For each factor in the Lagrangian (7.35) there will be one line on the diagram, as shown in Fig. 7.3. All the processes appearing in any order in the perturbation expansion will be made out of these basic diagrams, which correspond to the so-called *vertex*, or interaction point. When connecting vertices among themselves, the theory requires one to impose the conservation of energy, momentum and electric charge at each interaction point.

The diagrams in Fig. 7.3 would correspond to the following processes:

1. The scattering of a photon by an electron: the incoming electron and the photon are annihilated and an electron is created. Classically, we would say that the electron absorbs a photon and increases its energy and momentum. The time-reversed process of emission of a photon by an electron is also conceivable.

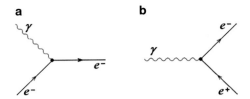

Fig. 7.3 Vertex diagrams representing the interaction of a photon (*wavy line*) with the electron–positron field (*straight lines*). The initial state particles are drawn customarily to the left and the final state particles to the right. The arrows represent the flow of negative charge. In the case of electrons (particles), the arrows are in the direction of the momentum, while in the case of positrons (antiparticles), the arrows are in the opposite direction to the momentum.

2. Another fundamental process is the creation of an electron–positron pair, from a photon. The time-reversed process would be the annihilation of the pair, creating a photon. None of these processes has a classical counterpart.

There is a problem, however, with these processes, which is common to all processes in the first order in the perturbation expansion of QED: they are forbidden due to the requirement of energy-momentum conservation, if the particles involved are real, i.e. observable. Observable electrons and positrons have to satisfy the energy-momentum dispersion relation $E_e^2 = m^2 c^4 + \mathbf{p}^2 c^2$, while photons, being massless, have to satisfy $E_\gamma^2 = \mathbf{k}^2 c^2$. The reader may check that energy-momentum conservation with such particles cannot be fulfilled in the above two processes. (However, if the electromagnetic field is treated classically as an external source, while the electron–positron field is a quantum field, the processes are allowed. They can actually happen, for example, in the Coulomb field present in the neighbourhood of an atomic nucleus, or in the presence of a magnetic field.)

7.4.3 Virtual Particles

If the first order in the perturbative expansion does not provide any physical process, is it possible that higher orders do? Actually, this is indeed the case. If for one of the particles at the vertex we do not require the energy-momentum dispersion relation to hold, then energy and momentum can be conserved. But what about the particle which makes this possible? That particle will not be physical, but what we call a *virtual particle*. Such a particle cannot be observed, but its effect can be. Virtual particles are allowed by Heisenberg's uncertainty principle. Since the uncertainties in energy and time for a quantum particle are in the relation $\Delta E \, \Delta t \sim h$, it means that for very short times one can have very big fluctuations in energy. According to special relativity, energy is equivalent to mass, therefore those energy fluctuations can be treated as particles. Their life-time is too short to allow observation, therefore they are *not* constrained by the energy-momentum dispersion relation, $E = c\sqrt{\mathbf{p}^2 + m^2 c^2}$. Virtual particles appear in intermediate states only, connecting two vertices in the higher orders of the perturbation expansion. The mathematical functions charac-

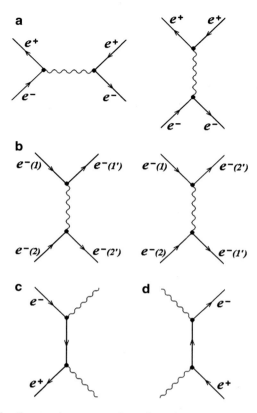

Fig. 7.4 Lowest order diagrams in quantum electrodynamics. The direction of the arrows shows the flow of the negative charge. **a** Bhabha scattering, $e^+ + e^- \to e^+ + e^-$. There are two diagrams contributing to this process, the first being a purely quantum process, while the second has a classical limit. **b** Møller scattering, $e^- + e^- \to e^- + e^-$. Again there are two diagrams, due to the indistinguishability of the electrons in the final state - it is impossible to know that the electron (1′) is created at the vertex where the electron (1) or electron (2) is annihilated, therefore the two diagrams have to be summed up. The same process can take place also with positrons. **c** Pair annihilation, $e^+ + e^- \to \gamma + \gamma$. While the pair annihilation into one real (not virtual) photon is forbidden by the energy-momentum conservation, the process with two photons in the final state is allowed. **d** Pair creation, $\gamma + \gamma \to e^+ + e^-$.

terizing virtual particles in intermediate states are called propagators and they are essential for the Lorentz-covariant treatment of any relativistic quantum field theory; the higher-order corrections to basic processes are due only to virtual particles.

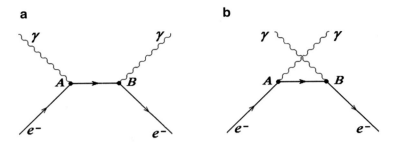

Fig. 7.5 Feynman diagrams of the Compton scattering. Both diagrams contribute to the process.

7.4.4   Compton Scattering

If we go to the next order in the perturbative series, we can describe various physical processes. The second order terms are proportional to the fine structure constant α. For example, an electron–positron pair can be created from two photons, if their total energy is higher than the double of the rest energy of an electron (see Fig. 7.4d).

A two-photon process occurs also in the *Compton effect*, in which a high-energy photon (i.e. a photon whose energy is comparable with the rest energy of the electron) is scattered by an electron (Fig. 7.5). In the final state we find a "recoil electron" and a photon with an increased wavelength and a different direction compared to the initial one. The effect was discovered in 1923 by Arthur Compton (1892–1962), who was awarded the Nobel Prize in 1927. The great importance of the discovery at that time was that it gave further support to the hypothesis that light behaves as a particle in certain phenomena. Classically, the scattering of lower frequency electromagnetic radiation by electrons had been explained by J. J. Thomson, the discoverer of the electron. According to his explanation, the electron absorbs electromagnetic radiation, which causes it to accelerate. As a result of the accelerated motion, the charged particle emits radiation of the same frequency as the absorbed one, but of lower intensity. The difference in the frequency between the incident and the scattered beams observed by Compton could not be explained classically. In QED, any process

$$\gamma + e^- \rightarrow \gamma + e^-$$

is called Compton scattering, irrespective of the energy of the incident photons. The low-energy limit of the Compton scattering is called the Thomson limit.

The relativistic kinematics of Compton scattering is easy to understand, assuming that the electromagnetic radiation behaves like particles and the energy and momentum are conserved (Fig. 7.6). In this section we keep explicit the h and c factors. The natural reference system to treat this process is the rest frame of the target electron. In this frame, the electron has zero momentum, $\mathbf{p} = 0$, and the rest energy $E = mc^2$. The incident photon's energy is $E_i = h\nu_i$ and the relativistic momentum $|\mathbf{p}_i| = h\nu_i/c$. After the interaction, the recoil electron gains the momentum $\mathbf{p}'$, such that its total energy becomes $E' = \sqrt{\mathbf{p}'^2 c^2 + m^2 c^4}$. The photon scattered at an angle

Fig. 7.6 Kinematics of the
Compton scattering.

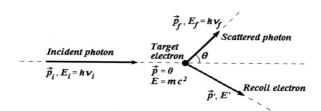

θ with respect to the direction of the incident photon has the energy $E_f = h\nu_f$ and
the momentum $|\mathbf{p}_f| = h\nu_f/c$. The energy conservation is expressed by the equality

$$mc^2 + h\nu_i = \sqrt{(\mathbf{p}'c)^2 + (mc^2)^2} + h\nu_f,$$

which gives, upon squaring,

$$h^2(\nu_i - \nu_f)^2 + 2mc^2 h(\nu_i - \nu_f) = \mathbf{p}'^2 c^2. \tag{7.43}$$

The momentum conservation can be written in the form:

$$\mathbf{p}_i - \mathbf{p}_f = \mathbf{p}',$$

which we plug into (7.43), obtaining:

$$h^2(\nu_i - \nu_f)^2 + 2mc^2 h(\nu_i - \nu_f) = (\mathbf{p}_i - \mathbf{p}_f)^2 c^2. \tag{7.44}$$

Inspecting the Fig. 7.6, we can write easily the right-hand side of (7.44):

$$\begin{aligned}
(\mathbf{p}_i - \mathbf{p}_f)^2 c^2 &= \mathbf{p}_i^2 c^2 + \mathbf{p}_f^2 c^2 - 2\mathbf{p}_i \cdot \mathbf{p}_f c^2 \\
&= \mathbf{p}_i^2 c^2 + \mathbf{p}_f^2 c^2 - 2|\mathbf{p}_i| \cdot |\mathbf{p}_f| c^2 \cos\theta \\
&= h^2(\nu_i^2 + \nu_f^2 - 2\nu_i\nu_f \cos\theta).
\end{aligned}$$

Using this last result, we put (7.44) into the form

$$\frac{1}{h\nu_f} - \frac{1}{h\nu_i} = \frac{1 - \cos\theta}{mc^2},$$

or, in terms of the wavelengths of the incident and scattered beams:

$$\lambda_f - \lambda_i = \frac{h}{mc}(1 - \cos\theta). \tag{7.45}$$

This calculated wavelength shift was in full agreement with Compton's measurements in scattering experiments with X rays and gamma rays. In the conclusion of

his 1923 paper presenting the effect, Compton noted: *This remarkable agreement between our formulas and the experiments can leave but little doubt that the scattering of X-rays is a quantum phenomenon.* With the explanation of the Compton scattering, a new constant was introduced, the *Compton wavelength* of a particle of mass m, representing the wavelength of a photon whose energy is equal to the rest energy of that particle, $\lambda_c = h/mc$.

In quantum electrodynamics, the Compton scattering is explained as follows: the incident photon is annihilated, together with the target electron, at the point A, where a virtual electron is created (Fig. 7.5a). This electron propagates for an extraordinarily short time from A to B. At the point B, this virtual electron is annihilated, simultaneously with the creation of the recoil electron and of the scattered photon. To this effect contributes also the process depicted in Fig. 7.5b, in which at the point A where the initial electron is annihilated, the scattered photon is created, together with a virtual electron. At B, after a brief propagation of the virtual electron, the physical recoil electron is created, while the incident photon is destroyed. Remark that in this explanation we have never used the expression "the electron absorbs a photon" or "the electron emits a photon", although later in this chapter we shall use this terminology, with the warning that it has its roots in the classical intuition. (This way of speaking—absorption and emission of the photon by charged particles—is widely used. We are permanently facing classical concepts "at the border" of quantum theory.) An electron which would be classically interpreted to absorb a photon and continue its passage with a different energy and momentum, in quantum field theory is interpreted as being destroyed, while another electron is created. Whether the electron is the same or another, we have no way of knowing for sure, because the electrons (as well as any other elementary particles) cannot be in principle distinguished from one another. What is sure, however, is that the calculation of the scattering probability according to the rules of quantum electrodynamics is in perfect agreement with the observation. Moreover, for low-energy incident photons, the energy of the recoil electron is negligible and in this limit, the result of quantum electrodynamics reduces to the classical Thomson scattering formula.

The attribute of virtual is assigned also to those non-observable photons responsible for the electrostatic interactions. The electrostatic field (which is exerted between two charges kept at rest, for instance, and can be attractive or repulsive) is polarized longitudinally. This means that it is along the direction of propagation, and although it is formally considered as composed of quantum particles, that is, longitudinal photons, these are not observable in vacuum (see Sect. 4.5). In a medium where free charges exist, for instance in a plasma, the electrostatic interactions can propagate as longitudinal waves and the corpuscular character of these longitudinal oscillations may become observable.

7.4.5 Electron Self-energy and Vacuum Polarization

As a consequence of the fundamental processes mentioned in Sect. 7.4.2, we should understand the electron as being in permanent interaction with a cloud of virtual

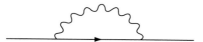

Fig. 7.7 One-loop diagram of the electron self-energy, called also mass operator. It represents the emission and absorption of a virtual photon by an electron (or a positron). It is proportional to the fine structure constant α (since it contains two vertex factors, each of them contributing one factor e).

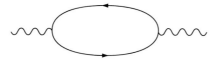

Fig. 7.8 Vacuum polarization diagram.

photons, which it "emits" and "absorbs" all the time. This is represented by the *loop diagram* in Fig. 7.7, illustrating the emission and absorption of a virtual photon by an electron. The interaction of the electron with its own electromagnetic cloud changes the energy of the electron, consequently it modifies the mass of the electron. The diagram is said to represent the *electron self-energy* and it gives the difference δm between the mass m of an interacting, or *physical electron*, and the mass $m_0 = m - \delta m$ of a non-interacting electron, which is sometimes called *bare*.

The *polarization of vacuum* is another important effect of quantum electrodynamics. In this case, a propagating photon creates a virtual electron–positron pair, which subsequently annihilates again, reproducing the original photon which continues propagating (Fig. 7.8). Since the electron–positron pair which is created and then annihilated has virtual nature (they are not observable particles), it is not necessary that the photon has high energy (or frequency), and the process may occur for photons of any energy. In the presence of a charge producing a background electric field, the distribution of these virtual pairs will be modified and it would appear as if the vacuum were polarized. For example, an electron would attract the virtual positrons and repel the virtual electrons. Thus, the vacuum behaves like a dielectric, polarizable, medium (the situation is similar to the one depicted in Fig. 3.20). Then, if a small test charge is brought at a relatively large distance from the electron, it would experience a force corresponding to a smaller charge of the electron, due to the screening effect of the vacuum polarization. This effect is indeed observable, as we shall see in Sect. 7.4.6.

What would happen to a photon? Would it get some non-zero mass due to its interaction with virtual particles in the vacuum? The answer is negative. Calculations show that the effect of the vacuum polarization diagram in quantum electrodynamics is such that it keeps observable free photons on the light cone. This result is expected, since QED is a relativistic theory, and it is also gauge invariant, as we will discuss in Sect. 7.6, so photons must remain massless.

There is still one diagram which gets modified in the higher orders in α, and that is the vertex correction represented in Fig. 7.9. The latter diagram contributes to the observed intrinsic magnetic moment of the electron. Recall that the magnetic

Fig. 7.9 Vertex correction
diagram.

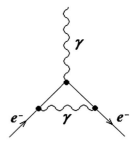

moment due to the spin of the electron involved the gyromagnetic factor $g_s = 2$ (see Sect. 7.1). However, when the measurements were refined, it turned out that there was a slight departure from this integer number (*anomalous magnetic moment*), and the actual g_s-factor was higher:

$$g_s = 2(1 + a_e). \tag{7.46}$$

The anomaly a_e was first measured by Polykarp Kusch and Henry Foley in 1947–1948 to be 0.00119 ± 0.00005. Julian Schwinger derived the correction a_e in 1948 from the process depicted in the diagram in Fig. 7.9 and found it to be exactly $\alpha/2\pi = 0.00116$. Over the years, the measurement of the anomalous magnetic moment of the electron became one of the highest precision tests of quantum electrodynamics, while the theoretical value has been calculated up to the order α^4. The agreement between theory and experiment is staggering, making the anomaly of the magnetic moment of the electron the most accurately verified prediction in the history of physics. As of 2012, the value given by the Particle Data Group is

$$a_e = 0.00115965218076 \pm 0.00000000000027.$$

In 1947, Willis Lamb discovered a spectral line of 1,058 MHz in the hydrogen atom spectrum, corresponding to the energy difference between the levels $2S_{1/2}$ and $2P_{1/2}$, i.e. the two levels with $n = 2$ and $j = 1/2$, but corresponding to different angular momentum quantum numbers, $l = 0$ (denoted in spectroscopy by S) and $l = 1$ (denoted by P). Dirac's theory for the hydrogen atom spectrum predicts degeneracy, that is, the same energy for both levels, as can be seen from the formula (7.18), which does not depend at all on l. Hans Bethe (1906–2005) made a non-relativistic estimation of the effect of the electron's self-energy and found a surprisingly good agreement with the experimentally observed value (1,040 MHz). Actually, besides the electron self-energy, also the vacuum polarization and the vertex correction slightly contribute to the Lamb shift (Fig. 7.10).

In the end of the 1940s, quantum electrodynamics was in the process of being elaborated. The relativistic calculation of the Lamb shift was the great challenge of the day. Many theoretical physicists were attempting it, but finally the credit went to Sin-Itiro Tomonaga (1906–1979), Julian Schwinger (1918–1994), and Richard

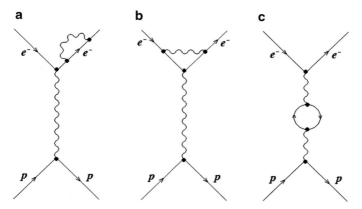

Fig. 7.10 The diagrams contributing to the Lamb shift—the hydrogen spectral line corresponding to the transition $2S_{1/2} - 2P_{1/2}$. (**a**) The electron self-energy contributes 1,017 MHz, (**b**) the anomalous magnetic moment contributes 68 MHz, and (**c**) the vacuum polarization diagram has a small negative contribution of -27 MHz. Altogether they give the experimentally measured 1,058 MHz.

Feynman, (1918–1988), who developed the full relativistic theory while seeking the answer to this concrete problem. They were awarded jointly the Nobel Prize in 1965, "for their fundamental work in quantum electrodynamics, with deep-ploughing consequences for the physics of elementary particles". Already in 1955, the Nobel Prize had been awarded to Willis Lamb (1913–2008) "for his discoveries concerning the fine structure of the hydrogen spectrum" and to Polykarp Kusch (1911–1993) "for his precision determination of the magnetic moment of the electron".

7.4.6 *Renormalization and Running Coupling Constant*

Why was it, after all, so difficult to build quantum electrodynamics? In the discussion of the previous section we intentionally omitted to mention the reason. The first order in α diagrams give finite results, in excellent agreement with the experiment. When one calculates higher-order corrections, like for example those in Fig. 7.10, one would expect them to be much smaller than the first-order result, i.e. of the order of α^2. Surprisingly, the calculation of the loop diagrams involve divergent integrals, and the corrections seem to be infinite! The great achievement of Tomonaga, Schwinger, and Feynman, was to elaborate a consistent formalism to subtract the divergences, leading to finite quantities, comparable with experiments. The procedure is generally called *renormalization* and it consists of three main steps.

The first step is to regularize the integrals, i.e. to make them finite. The divergences are named *ultraviolet divergences*, since they are due to the integration up to infinite momenta of the virtual particles which run in the loops. One intuitive method of regularization is to cut off the integration limit to a very large, but finite momentum.

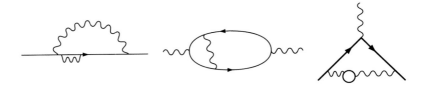

Fig. 7.11 Two-loop diagrams contributing to the electron self-energy, vacuum polarization and vertex correction. We may conceive easily higher order diagrams. Each new loop introduces a factor α.

The second step consists of relating the quantities describing bare (i.e. non-interacting) particles to those describing the physical particles. For example, one has to find the relation between the mass of a bare electron, m_0, and that of a physical electron, m, or between the charge of a bare electron, e_0, and the charge of a physical electron, e. All the predictions of the theory (which are mainly scattering probabilities, expressed as scattering cross sections) have to be written in terms of e and m, which are set at this point to their experimental values, measured at a specific energy, which gives the renormalization scale.

In order to revert to the original theory, we have to remove the regularization (in the cut-off method, we let the momenta of the loop go to infinity again). However, due to the renormalization done at the second step, the infinities reappear only in the relation between the bare and the physical quantities. In a way, we claim that the bare mass m_0 and the bare charge e_0 of an electron are infinite—but this is no problem anymore, since bare particles cannot be observed. As all the observables have been expressed at the second step in terms of the finite (physical) e and m, the corrections will be in their turn finite and proportional to α, if we work at one-loop level, or to α^2, if we work at two-loop level etc. (Fig. 7.11).

A result of special significance is the effective value of the electric charge when we change the energy scale at which we make the measurement. Recall that we have mentioned in Sect. 7.4.5 that the vacuum polarization effect has as a consequence the screening of the charge of an electron, such that from a large distance, a test charge "feels" it as being smaller. If we bring the test charge closer to the one we wish to measure, then the screening effect obviously diminishes, and the effective charge appears as being larger. The closer we bring the test charge, the larger the charge of the electron appears to be. To bring the test charge close, we need to give it more and more energy, therefore short distance is equivalent to high energy. Instead of saying that the electric charge depends on the distance to the test charge, we can say that it depends on the energy scale μ of the measurement, $\alpha = \alpha(\mu)$. The Coulomb law has to be modified, because *the coupling constant runs.* This means that the interaction potential between two charges q and q' should be written at very short distances as

$$V(r) \simeq \frac{qq'}{r}\left[1 + \frac{2\alpha(\mu)}{3\pi}\left(\ln\frac{h}{mcr} - \text{const.}\right)\right], \qquad (7.47)$$

where const. $= 5/6 + \ln \gamma$ and $\gamma = 1.781 \ldots$, that is, the dependence of the potential on the distance becomes of the form $[1 + A \ln(r_0/r)]/r$, where $A = \frac{2\alpha}{3\pi}$ and $r_0 = h/mc$ is the Compton wavelength of the electron, of the order 10^{-2} Å.

The effect is indeed observable. With utmost present-day precision, we can say that the fine structure constant has the value $\alpha = 1/137.03599911(46)$ at an energy equal to the rest energy of the electron, $\mu_0 = mc^2 = 0.510998918(44)$ MeV (low-energy limit, called also Thomson limit). However, measurements of the fine structure constants have been regularly reported for higher energies. It has been found that the effective value for the fine structure constant, at a value of the energy equal to the mass of the Z boson, $m_Z = 91.1876 \pm 0.0021$ GeV, is $\alpha(m_Z) = 1/(128.937 \pm 0.047)$, in excellent agreement with the theoretical prediction of $\alpha(m_Z) = 1/(128.940 \pm 0.048)$, and significantly different from the value $\alpha \simeq 1/137$ at low energies.

For large values of $\ln \frac{\mu^2}{\mu_0^2}$, the dependence on the energy scale of the coupling constant $\alpha(\mu)$, to all orders in perturbation, is

$$\alpha(\mu) = \frac{\alpha(\mu_0)}{1 - \dfrac{\alpha(\mu_0)}{3\pi} \ln\left(\dfrac{\mu^2}{\mu_0^2}\right)}. \tag{7.48}$$

At the nowadays attainable energies in high energy physics laboratories, the increase in α is not big, although observable. However, at the energy scale $\mu = mc^2 e^{3\pi/2\alpha(\mu_0)}$, which is a giant value, 10^{283} keV, but still finite, the denominator vanishes, and $\alpha(\mu)$ becomes infinite. The only way to solve this inconsistency is to assume that $\alpha(\mu_0)$ vanishes, but then $\alpha(\mu)$ would be zero at any energy, i.e. QED would be a non-interacting theory, which is absurd. This problem is known as the *Landau pole*, or sometimes the *Moscow zero*. It is a limitation of quantum electrodynamics, and it is absent in more complete theories, like the Standard Model or Grand Unified Theories (see Chap. 11).

7.5 Quantum Vacuum and Casimir Effect

In quantum electrodynamics the properties of vacuum are interpreted as the properties of the *ground state*. For instance, the energy spectrum of the free electromagnetic field is given by its quantum Hamiltonian operator, whose expression is

$$\hat{H} = \sum_{\mathbf{k}} \sum_{\lambda=1,2} \hbar\omega_{\mathbf{k}} \left(\hat{a}_\lambda^\dagger(\mathbf{k})\hat{a}_\lambda(\mathbf{k}) + \frac{1}{2} \right),$$

where $\hat{a}_\lambda^\dagger(\mathbf{k})$ and $\hat{a}_\lambda(\mathbf{k})$ are the creation and annihilation operators of a free photon of momentum $\mathbf{k}$ and polarization λ (free real photons can have only transverse polarizations, therefore the index λ can take only two values). When we apply this

Hamiltonian to the vacuum state $|0\rangle$, the first term gives no contribution due to the fact that all annihilation operators destroy the vacuum (see (7.27)). The second term gives the energy of the vacuum, which is obviously infinite, due to the summation over all possible values of $\mathbf{k}$:

$$E_0 = \frac{1}{2} \sum_{\mathbf{k}} \sum_{\lambda} \hbar \omega_{\mathbf{k}}.$$

This energy is associated to the virtual modes of vacuum which are infinite in number. As we have seen earlier, quantum vacuum is by no means equivalent to simple empty space but instead it contains unobservable particles (electrons, positrons, photons, etc.)—the virtual particles created and annihilated continuously, in quantum fluctuation processes.

When we calculate the energy of any particle state, this vacuum energy will always give an infinite contribution (sometimes called *zero-point energy*), which is usually subtracted in a renormalization process, arguing that the vacuum energy is unobservable, and the only thing that we can in principle measure is the difference between the ground state and the particle states energies.

However, the vacuum energy of the electromagnetic field has actually a macroscopic effect—the *Casimir effect*—discovered and explained in 1948 by Hendrik Casimir (1909–2000). It consists in the following: if one places two uncharged metallic plates a few micrometer apart in vacuum, the only waves allowed to propagate in between the plates are those which correspond to a vanishing transverse electric field on the metallic plates. Consequently, the waves have to be such that half wavelengths fit exactly within the distance d, i.e. $\lambda_n = 2d/n$, where $n = 1, 2, \ldots$, corresponding to the angular frequencies $\omega_n = \pi n c / d$. On the outside of the plates, however, all the possible frequencies still exist (Fig. 7.12). As a result, a smaller *radiation pressure* arises from the modes inside the cavity than from those outside it, leading to an attractive force between the plates. The attractive force per unit area between the plates (or the corresponding pressure) is given by the expression

$$F/S \equiv p = -\frac{\pi^2 \hbar c}{240 d^4}. \tag{7.49}$$

For two metallic plates with the area of $1\,\text{cm}^2$, separated by a distance $d = 1\,\mu\text{m}$, the force is about 1.3×10^{-2} dyne $= 1.3 \times 10^{-7}\,\text{N}$. The first experiment in which the Casimir effect was observed was performed by Marcus Sparnaay in 1957, but while the results did not infirm the theory, the errors were very large. In 1997 started the era of more accurate measurement of the Casimir force, with the experiments of Steve Lamoreaux, and Umar Mohideen and Anushree Roy.

Fig. 7.12 Inside the plates, only those electromagnetic modes are allowed whose frequencies are integer multiples of $\omega = \pi c/d$.

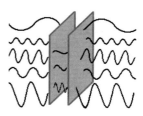

7.6 Principle of Gauge Invariance

In Chap. 5 we discussed the gauge invariance of the electromagnetic field components **E** and **B**. When considering the interaction of the electromagnetic field with the electron–positron field, gauge invariance becomes a more general property.

As we have mentioned in Chap. 5, in order to describe electromagnetic theory in a manifestly relativistic way, one has to re-cast it in the language of the vector potential A_μ, and start from the relativistically invariant Lagrangian density of the field, whose expression is:

$$\mathcal{L} = -\frac{1}{4} F^{\mu\nu} F_{\mu\nu}, \qquad F_{\mu\nu} = \partial_\mu A_\nu - \partial_\nu A_\mu. \tag{7.50}$$

The components of $F_{\mu\nu}$ are shown in (5.37), i.e. $E_k = F_{0k}$ and $B_k = \frac{1}{2}\epsilon_{ijk} F_{ij}$. Applying the least action principle which we discussed for particle mechanics in Chap. 1, one obtains the equations of motion for the field, which in this case are the Maxwell equations. Under the gauge transformations

$$A_\mu(x) \rightarrow A'_\mu(x) = A_\mu(x) + \partial_\mu f(x), \tag{7.51}$$

where $f(x)$ is an arbitrary differentiable real function, the field strength tensor $F_{\mu\nu}$ is invariant, which means that the action itself is invariant, as well as all the observables of the theory.

What looked like a simple peculiarity of the reformulation of electromagnetism in terms of the vector potential, turns up to be a manifestation of the *gauge invariance principle*, which underlies all the modern relativistic theories of particle physics. Let us sketch the argumentation leading to this principle. In the discussion we shall refer to classical fields and Lagrangian densities.

The relativistic theory of the interaction of charged particles with the electromagnetic field starts with the formulation of a relativistically invariant action. The Lagrangian density which leads to the free Dirac equation (7.30) has the form:

$$\mathcal{L}_{Dirac}(x) = \bar{\Psi}(x)(i\gamma^\mu \partial_\mu - m)\Psi(x). \tag{7.52}$$

This function remains invariant if we make a transformation on the fields which does not affect the spacetime, namely:

$$\Psi(x) \rightarrow \Psi'(x) = e^{-i\epsilon}\Psi(x),$$
$$\bar{\Psi}(x) \rightarrow \bar{\Psi}'(x) = e^{i\epsilon}\bar{\Psi}(x), \tag{7.53}$$

where ϵ is any real number. Such transformations are unitary, in the sense that the inverse of a transformation is its Hermitian conjugate. They also form a group, denoted by $U(1)$, which is the unitary group with one parameter (in our example, the parameter is the real number ϵ). We should emphasize that the parameter of the transformation does not depend on the spacetime point, for which reason we call these transformations *global*. Noether's theorem predicts that there is a conserved charge associated with this invariance, and indeed one can derive from it the conservation of the electric charge (though not complicated, we shall omit this derivation).

What happens if we make the parameter ϵ a function of coordinate, $\epsilon(x)$, that is we *localize* the transformations? Such transformations are called *local* or *gauge*. An argument for such a requirement could be as follows: the phase ϵ is immaterial—it can be chosen by someone at a certain place and time differently than by someone else, at another place and time, so that, in general, the phase can be a function of the spacetime point x. It is very easy to see that, in this case, the Lagrangian (7.52) is no more invariant. If we consider $\epsilon(x) = ef(x)$ and perform the transformations (7.53) with this parameter, we obtain that

$$\mathcal{L}' = \mathcal{L} - e\Psi\gamma^{\mu}\Psi\,\partial_{\mu}f. \tag{7.54}$$

To make the Lagrangian invariant, we need to add a new field, such that the term depending on $\partial_{\mu}f(x)$ be compensated on the right-hand side of (7.54). The new field is the electromagnetic vector potential $A_{\mu}(x)$, transforming according to the formula (7.51), and included in the Lagrangian density through the interaction term

$$\mathcal{L}_{int} = e\bar{\Psi}\gamma^{\mu}A_{\mu}\Psi,$$

which is precisely the form of the interaction term of quantum electrodynamics (7.35). The electromagnetic field is called a *gauge field*. The total Lagrangian density of electrodynamics, invariant under gauge transformations, contains the terms describing the free Dirac field, the free electromagnetic field, and their interaction:

$$\mathcal{L} = \bar{\Psi}(i\gamma^{\mu}\partial_{\mu} - m)\Psi - \frac{1}{4}F^{\mu\nu}F_{\mu\nu} + e\bar{\Psi}\gamma^{\mu}A_{\mu}\Psi. \tag{7.55}$$

We conclude this paragraph by pointing out that there exists an intimate relationship between the following four properties:

1. The free photon has only two modes of (transverse) polarization.
2. The photon mass is zero.
3. The Lagrangian density describing electromagnetic interaction is gauge invariant.
4. The electromagnetic interactions have long range (the force decreases as the inverse square of the distance or the potential decreases as the inverse of the distance).

The line of reasoning that we presented above is the prototype for constructing any modern theory of particle physics, which is required to satisfy the principle of gauge invariance (though with other groups of internal transformations, as we shall see in Chap. 11). The remarkable property of gauge field theories, like QED, is that the ultraviolet divergencies appearing in higher orders in perturbation can be removed, or renormalized, in a consistent way – such theories are called *renormalizable*.

The Aharonov–Bohm Effect. The reformulation of electrodynamics in terms of the vector potential appears so far like a device to make the theory manifestly relativistically invariant. The vector potential itself is unobservable, since it is not gauge invariant, unlike the fields **E** and **B**. However, using the concept of electromagnetic vector potential, a quantum mechanical effect was predicted, whose existence would never have been conceived, had electromagnetism been formulated only in terms of the fields **E** and **B**. This is the Aharonov–Bohm effect, predicted in 1959 by Yakir Aharonov (b. 1932) and David Bohm (1917–1992) (an earlier report had been made by Werner Ehrenberg and Raymond E. Siday in 1949). The Aharonov–Bohm effect clearly indicates that in quantum theory the electromagnetic potentials are truly fundamental, and not the electromagnetic field strength components.

When Schrödinger's equation is written in the presence of a magnetic field, the momentum operator becomes

$$\hat{\mathbf{p}} = -i\hbar\nabla + \frac{e}{c}\mathbf{A}, \tag{7.56}$$

where **A** is the vector potential. Let us assume now that we perform the double-slit experiment described at the beginning of Chap. 6, but with a very long solenoid placed behind the pierced screen, perpendicular to the straight line that joins the slits (Fig. 7.13). The magnetic field **B** is confined *inside* the solenoid, while in the outer region it vanishes. Around the solenoid, however, there is a nonvanishing vector potential **A**, whose curl is zero.

Before placing the solenoid, the electrons are described by wave functions of the type $\psi(x, t) = Ce^{i\varphi}$, where in the case of plane waves, the phase is $\varphi = (\mathbf{p} \cdot \mathbf{r} - Et)/\hbar$. For the electrons passing through the slits, the expression for the phase is more complicated, but for our illustrative purpose we may as well use the plane waves.

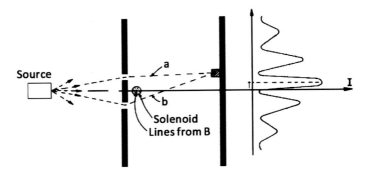

Fig. 7.13 Influence of the magnetic field inside the solenoid on the motion of electrons, according to the Aharonov–Bohm effect. The interference pattern is shifted compared to the situation when the solenoid is removed.

The interference of the amplitudes on the screen depends on the difference of these phases: if we assume the beams of electrons described by the wave functions $\psi_1 = C_1 e^{i\varphi_1}$ and $\psi_2 = C_2 e^{i\varphi_2}$, the probability density of the electrons reaching some point on the screen is given by the function

$$|\psi_1 + \psi_2|^2 = C_1^2 + C_2^2 + 2C_1 C_2 \cos(\varphi_1 - \varphi_2).$$

The last term determines the pattern of interference on the screen, through the difference of phases, $\delta = \varphi_1 - \varphi_2$.

When the solenoid is introduced, the phase acquires a new element. When in φ we substitute the momentum operator by $-i\hbar\nabla + \frac{e}{c}\mathbf{A}$, the difference of phases becomes:

$$\delta' = \delta + \frac{e}{\hbar c} \int_{(a)} \mathbf{A} \cdot d\mathbf{l} - \frac{e}{\hbar c} \int_{(b)} \mathbf{A} \cdot d\mathbf{l} \tag{7.57}$$

$$= \delta + \frac{e}{\hbar c} \oint_{(ab)} \mathbf{A} \cdot d\mathbf{l}, \tag{7.58}$$

where (a) and (b) are the trajectories of the two beams of electrons and (ab) is the closed path composed by the two trajectories of the electrons, as seen in the figure. This integral is invariant under gauge transformations; in fact, the integral represents the flux of the magnetic field $\mathbf{B}$ through the area bounded by the closed curve (ab). The interesting aspect of this effect is that although the magnetic field is zero outside the solenoid, it influences the motion of the electrons, manifested by the integral of the vector potential $\mathbf{A}$, which is not zero outside the solenoid, taken along the trajectory of the electrons. The Aharonov–Bohm effect is directly connected with the magnetic flux quantization. Observe that, using Stokes' theorem, the phase difference can be put in the form

$$\delta' - \delta = \frac{e}{\hbar c} \oint_{(ab)} \mathbf{A} \cdot d\mathbf{l} = \frac{e}{\hbar c} \int_S \mathbf{B} \cdot d\mathbf{S},$$

where S is the area subtended by (ab). If the flux is quantized in units of $\hbar c/2e$ (as is the case in superconductors), then $\delta' - \delta$ can be interpreted as a measure of twice the number of flux quanta across S, that is, across the solenoid (Fig. 7.13).

The first experimental observation of the Aharonov–Bohm effect was made as soon as 1960, by R.G. Chambers. Several other groups also attempted it. However, since the experiment requires an ideal solenoid, which does not leak any magnetic field outside, for many years the reported observations were argued to be inconclusive. The final confirmation came in 1986 from the Japanese group led by Akira Tonomura (1942–2012), a pioneer of holographic electron microscopy. In the experiment, a toroidal ferromagnet was covered with a superconductor layer to confine the field, and further with a copper layer for complete shielding from the electron wave. Tonomura performed also other fundamental experiments in quantum physics, like the demonstration of the electron two-slit interference in 1989, with the pattern developing gradually by the detection of individual electrons, or the creation of vortices in electron beams, in 2010.

7.7 *CPT* Symmetry

The charge conjugation or *C*-transformation changes a particle to its antiparticle. *C*-invariance means that every process has a symmetric process, in which electrons and positrons (or any particle and its antiparticle, even if they have no electric charge, but are different in any other way) are interchanged. Quantum electrodynamics is *C*-invariant. For instance, the Compton scattering has a symmetric process under charge conjugation: the scattering of a photon by a positron. The probabilities for the two processes are identical, i.e. they are *C*-invariant.

The *P*-invariance or parity symmetry means invariance under the inversion of space, for instance, the exchange of right and left. The mirror has the property of doing such an inversion. If we are right-handed, our mirror image is left-handed and vice versa. If we walk toward the mirror, the image also approaches us, that is, it moves in the opposite direction to us. However, the directions parallel to the mirror do not change, and if we move in these directions, the image follows us in the same direction. The *P*-invariance of quantum electrodynamics means that given a process, there exists an equivalent process in which the space has been inverted; that is, the mirror image of the phenomenon.

Time reversal is the transformation which reverts the direction of time. *T*-invariance means that an equivalent process inverting the initial and final states occurs with the same probability. For example, the scattering of two photons leads to the creation of an electron–positron pair. But the electron–positron pair may just as well annihilate, creating two photons. The processes are symmetric with respect to the inversion of time.

These transformations are called discrete, because they cannot be obtained by the continuous variation of some parameters, but rather by a "jump". They are all symmetries of quantum electrodynamics, but not all the fundamental interactions of particle physics preserve them—we shall see in Chap. 9 that weak interactions violate *C*-, *P*-, and the combined *CP*-symmetry.

Up to now, no *CPT*-invariance violation has been experimentally observed, that is, the physical processes of the atomic and subatomic world are invariant under the product *CPT* of the three transformations. As a consequence, in quantum electrodynamics the inversion of time, *T*, is equivalent to the product *CP* of charge conjugation and parity. According to Feynman, positrons can be interpreted formally as electrons moving backward in time. Thus, in a process in which electrons take part, if the charge conjugation *C* and the inversion of space *P* are performed, a process results involving positrons, and the operation is equivalent to the inversion of time.

The *CPT* symmetry is ensured by a theorem proved in 1954 by Wolfgang Pauli and Gerhart Lüders in certain conditions, all of which being fulfilled by the modern mainstream theories of particle physics.

7.8 Grassmann Variables

The quantization of the electron–positron field to account for Pauli's exclusion principle requires the use of anticommuting operators. While non-relativistic quantum mechanics was developed based on commutators of operators, having a correspondent in classical Hamiltonian mechanics in the Poisson brackets, for the anticommuting operators which describe fermions there is no classical analog.

As we have seen in Chap. 1, there is an example of anticommutation in classical mechanics—the vector product of two given vectors:

$$\mathbf{A} \times \mathbf{B} + \mathbf{B} \times \mathbf{A} = 0, \quad \mathbf{A} \times \mathbf{A} = 0.$$

This property can be generalized by introducing new mathematical entities: the anticommuting or Grassmann variables.

A set of n Grassmann variables θ_i, $i = 1, 2, \ldots, n$, satisfy the so-called exterior algebra, or Grassmann algebra anticommutation relations:

$$\theta_i \theta_j + \theta_j \theta_i = 0. \tag{7.59}$$

In particular, for any index i,

$$\theta_i^2 = 0. \tag{7.60}$$

The Grassmann numbers commute with any ordinary number c, i.e. $c\,\theta_i = \theta_i c$. Moreover, a product of two Grassmann numbers behaves like an ordinary number:

$$(\theta_i \theta_j)\theta_k = \theta_k (\theta_i \theta_j),$$

which can be easily seen by applying the property (7.59).

For the Grassmann variables there are obviously no order relations, that is, we cannot say that one Grassmann number is bigger or smaller than another.

Any function of one Grassmann variable θ, due to the property (7.60), can be written simply as

$$f(\theta) = a + b\theta,$$

since in the expansion of $f(\theta)$ higher powers in θ are zero.

One formally defines derivation and integration with respect to Grassmann variables, by analogy with the case of the ordinary numbers.

One must distinguish between left and right derivatives: if θ and η are two different Grassmann variables, then

$$\frac{\overrightarrow{\partial}}{\partial\theta}\eta\theta = -\eta, \qquad \eta\theta\frac{\overleftarrow{\partial}}{\partial\theta} = \eta, \qquad \frac{\overrightarrow{\partial}}{\partial\eta}\eta\theta = \theta. \qquad (7.61)$$

The rules for integration which comply with the anticommutativity of the Grassmann variables look rather peculiar:

$$\int d\theta = 0, \qquad (7.62)$$

$$\int d\theta\,\theta = 1. \qquad (7.63)$$

Let us justify them. Consider

$$\int d\theta d\eta = -\int d\eta d\theta = -\int d\theta d\eta,$$

where for the first equality we used the anticommutation property, and for the second we exchanged the notation $\theta \leftrightarrow \eta$, since the two variables are integration variables, and the result should not depend on them. Consequently, $\int d\theta d\eta = 0$, which leads unequivocally to (7.62). Regarding (7.63), we observe that $d\theta\,\theta$ has to behave like an ordinary number, therefore the integral is also expected to be an ordinary number. This number can be chosen arbitrarily, as long as it is used consistently. The natural choice, for simplicity, is 1. Remark that there are no limits of integration—the operation is purely formal.

It is interesting to note that, for Grassmann variables, the integration and the derivation of a function lead to the same result, as one can straightforwardly check.

The formulation of the path integral quantization of fermionic fields, as well as of the ghost fields in non-Abelian gauge theories, cannot be imagined in modern times without the use of Grassmann variables. They are essential also in supersymmetry. These aspects are far beyond the scope of the present book, but we hope that we have enticed the reader to delve into them.

Problems

Problem 7.1 Start with the non-relativistic time-independent Schrödinger equation for an electron in a strong magnetic field, viz., (6.61) in Chap. 6, and write it as a two component spinor wave function.

Problem 7.2 (i) Check using the definition (7.2) that the Pauli matrices satisfy the commutation property $[\sigma_i, \sigma_j] = 2i\epsilon_{ijk}\sigma_k$. The previous relation may be obtained from the more general one

$$\sigma_l \sigma_j = \delta_{lj} + i\epsilon_{ljk}\sigma_k. \tag{7.64}$$

(ii) Use (7.64) to find the anticommutator $\{\sigma_l, \sigma_j\} = \sigma_l \sigma_j + \sigma_j \sigma_l$.

Problem 7.3 The Bohm–Aharonov effect reveals the fundamental physical role of the electromagnetic potentials $\mathbf{A}$, ϕ (in addition to the fields $\mathbf{B}$ and $\mathbf{E}$) in quantum theory. As a consequence, it also leads to the quantization of the magnetic flux. Concerning the last property, consider an electron moving in a magnetic field, and impose the Bohr–Sommerfeld quantization condition $\oint \mathbf{P} \cdot d\mathbf{l} = nh$ for the electron in its orbit, where $\mathbf{P} = \mathbf{p} + e\mathbf{A}/c$ is the canonical momentum of the electron. Assume that the electron momentum $\mathbf{p}$ is expressed in terms of $\mathbf{B}$ through the Lorentz force. (i) Show that the magnetic flux is quantized. (ii) Discuss the difference between the Aharonov–Bohm problem and the present one. Hint: Use the relations between $\mathbf{p}$ and $\mathbf{B}$ from classical electrodynamics.

Literature

1. S.S. Schweber, *An Introduction to Relativistic Quantum Field Theory* (Harper and Row, New York, 1962). The topic of the Dirac equation is found in any textbook of quantum mechanics dealing with relativistic equations, but the present book contains a very clear and complete discussion of it
2. F. Mandl, G. Shaw, *Quantum Field Theory* (Wiley, London, 2010). An excellent book to start the study of quantum field theory
3. A.I. Akhiezer, V.B. Berestetskii, *Quantum Electrodynamics* (Interscience, Wiley, New York, 1965). This is a classic text of quantum electrodynamics
4. R.P. Feynman, *Quantum Electrodynamics* (W.A. Benjamin, New York, 1961). Contains a very original treatment of the topic, as usual in Feynman's books

5. S.S. Schweber, *QED and the Men Who Made It: Dyson, Feynman, Schwinger, and Tomonaga* (Princeton University Press, Princeton, 1994). A remarkable book on the history of quantum electrodynamics, including all the original derivations and references

6. I.D. Lawrie, *A Unified Grand Tour of Theoretical Physics* (IOP, Bristol, 2002). This book contains a good summary of the basic aspects of quantum electrodynamics

7. M. Bordag, G.L. Klimchitskaya, U. Mohideen, V.M. Mostepanenko, *Advances in the Casimir Effect* (Oxford University Press, Oxford, 2009). This book contains the latest developments, both theoretical and experimental, in Casimir effect. The first two chapters are written as an elementary introduction to the subject

Chapter 8
Fermi–Dirac and Bose–Einstein Statistics

8.1 Fermi–Dirac Statistics

As we have seen in Chap. 6, identical particles are indistinguishable, and if their spin is a half-integer multiple of $\hbar$, they are called fermions. They are then described by antisymmetric wave functions. In what follows we shall consider the statistical problem of the properties of a large number of electrons and its relation with some properties of solid bodies.

Suppose that there is a large number of electrons in a cubic box, and let L be the length of its side. We assume that the electrons are not interacting (as a matter of fact, this is equivalent to assuming that the interaction energy between the particles is negligible as compared with their mean kinetic energy). If p is the momentum of the electron and m its mass, the energy is simply

$$E = \frac{p^2}{2m}. \tag{8.1}$$

By solving the Schrödinger equation with some boundary conditions (for instance, the wave function vanishes at the walls) this leads to the quantization of the electron momentum inside the cubic box. The components p_x, p_y, p_z should have the form:

$$p_x = \frac{2\pi\hbar l_1}{L}, \qquad p_y = \frac{2\pi\hbar l_2}{L}, \qquad p_z = \frac{2\pi\hbar l_3}{L}, \tag{8.2}$$

where l_1, l_2, l_3 are integers. Given three integers, they determine the three components of the electron momentum and, as a consequence, its energy. To each of these sets of three numbers there corresponds one of the two possible orientations of the electron spin, so that the quantum state of the electron will be characterized by the four numbers l_1, l_2, l_3, and s, the latter describing the spin and taking two values.

Let us consider all possible values of p_x, p_y, p_z on a system of three coordinate axes. The result is a three-dimensional lattice of cubic boxes (Fig. 8.1). To each box one can ascribe a set of three numbers l_1, l_2, l_3, the coordinates of one of the vertices.

© The Author(s), under exclusive license to Springer-Verlag GmbH, DE,
part of Springer Nature 2021
M. Chaichian et al., *Basic Concepts in Physics*, Undergraduate Lecture Notes in Physics,
https://doi.org/10.1007/978-3-662-62313-8_8

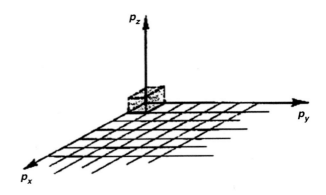

Fig. 8.1 For a gas of electrons in a cubic box, the momentum space is divided into boxes, in each of which there can be only two electrons, with antiparallel spins.

Within each box one can have only two electrons, according to the Pauli exclusion principle (Fig. 8.1).

If now N electrons are distributed in the set of boxes, the configuration of minimal energy is the one in which the boxes are filled starting from the origin of coordinates, taking care that no empty boxes remain between the filled ones, and that pairs of two electrons of opposite spins are placed in each box.

The final result is a spherical distribution of momenta (Fermi sphere), whose radius p_f corresponds to the electrons of maximum energy, called the Fermi energy, E_f:

$$p_f^2 = 2m E_f. \tag{8.3}$$

This radius p_f, called Fermi momentum, would increase if the density of electrons n (particles per unit volume) were increased. If the density or the temperature are very high, the problem must be studied from the relativistic point of view. (For instance, assume the Fermi energy $E_f \geq mc^2$. The electron kinetic energies would be of the same order as their rest energies.) Qualitatively, the relativistic problem is similar to the non-relativistic one. It is important in astrophysics, i.e., in the study of very dense stars, like white dwarfs.

For lower densities, the non-relativistic model discussed previously provides a good description of the behaviour of free electrons in a metal at low temperatures.

It is evident that, even at extremely low temperatures, the electron gas exerts some pressure, since very few electrons have momentum zero or close to zero, and a large number have values around p_f.

For zero temperature, the density of electrons n in each quantum state of energy less than or equal to E_f is 1, while for states of energy greater than E_f it is 0. This is illustrated by the curve in Fig. 8.2, where the vertical axis corresponds to the density and the horizontal axis to the energy.

If the temperature differs from zero, the thermal motion induces some electrons to occupy states of energy greater than E_f, leaving vacant some states of lower energy. Obviously, the more excitable electrons are those close to the Fermi surface, and for this reason, the mean number of electrons in each quantum state adopts the form indicated in Fig. 8.3.

Fig. 8.2 Mean density of electrons in energy quantum states at zero temperature. All the states with $E < E_f$ are occupied ($n = 1$) and all the states with $E > E_f$ are empty ($n = 0$).

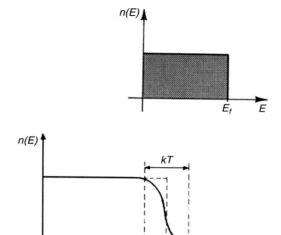

Fig. 8.3 Mean number of electrons per quantum state at a temperature $T \neq 0$. Some fraction of the electrons has been excited to states with energies greater than E_f, and the curve is smoothed around this value, over an interval of temperatures of width approximately equal to kT.

8.2   Fermi–Dirac and Bose–Einstein Distributions

To justify mathematically the previous statements, we shall calculate the average number of particles per quantum state in the ideal Bose and Fermi gases in equilibrium. We consider the problem of distributing n_i particles in g_i states of the same energy. The number g_i is in general very large. In the momentum sphere, the number of states or cells with energy between E_i and $E_{i+1} = E_i + \Delta E$ is to be understood as the number of states in the momentum interval p_i, $p_i + \Delta p$, where we take $\Delta p = 2\pi\hbar/L$, so that the spherical layer has the thickness of a single cell. This number is

$$g_i = 4V\pi p_i^2 \Delta p/(2\pi\hbar)^3$$

for one spin orientation, since the volume of the spherical layer is $4\pi p_i^2 \Delta p$ (in terms of the energy, it is $\sqrt{2}m^{3/2} E_i^{1/2} \Delta E$) and the volume in configuration space is $V = L^3$. To account for more than one spin orientation, one has to multiply this expression by the number of spin degrees of freedom which, in the case of electrons, is 2.

To give an idea about orders of magnitude, let us consider speeds between 10 and 11 cm/s in a volume $V = 1$ cm^3. A simple calculation gives $g_i \simeq 100$. For fermions, we always have $n_i \leq g_i$ due to Pauli's principle, while for bosons there is no such restriction.

Let us calculate the number of possible states for bosons. We consider the problem as analogous to the task of putting balls inside boxes. Schematically, let us mix balls and boxes and let us put them in lines, in such a way that a row begins always with a box (B), and then place balls (b) and other boxes on the right, understanding that the numbers of balls to the right of one box are contained in it. A possible configuration is

$$B\,bbb\,B\,bbbbb\,B\,b\,B\,B\,bbbbbbbbbbb \dots ,\tag{8.4}$$

that is, three balls in the first box, five in the second, one in the third, zero in the fourth, eleven in the fifth, etc. The number of possible distributions is $(n_i + (g_i - 1))!$ (the box at the left is fixed), but as the particles are indistinguishable, and the boxes also, we must divide by the number of permutations of the quantum states resulting from the distribution of the n_i boson particles in the g_i energy states E_i, i.e.,

$$\Gamma^{(b)}_{g_i,n_i} = \frac{(n_i + g_i - 1)!}{n_i!(g_i - 1)!}.\tag{8.5}$$

This is the number of microscopic states that results from the distribution of n_i bosons in the g_i one-particle quantum states with energy E_i.

For fermions, the task is even simpler because, as $n_i \leq g_i$ and as we can have at most one fermion in each state, the problem consists in calculating the combinations of order n_i of the g_i states. This gives

$$\Gamma^{(f)}_{g_i,n_i} = \frac{g_i!}{n_i!(g_i - n_i)!}.\tag{8.6}$$

Considering all possible energy levels E_i, each with an occupation number n_i, the total number of microstates is given by the product between the number of energy levels and the number of ways in which each energy level can be populated:

$$\Gamma^{(b,f)} = \Pi_{E_i} \Gamma^{(b,f)}_{g_i,n_i}.\tag{8.7}$$

We assume that in both cases the total energy, given by the formula

$$U = \sum_i n_i E_i,\tag{8.8}$$

and the total number of particles,

$$N = \sum_i n_i,\tag{8.9}$$

are conserved.

The problem is to calculate the maximum of the entropy $S = k \ln \Gamma^{(b,f)}$ subject to the conditions of conservation of energy, $dU = 0$, and conservation of the total number of particles, $dN = 0$, assuming that n_i is effectively a continuous variable. Using the Stirling formula $\ln n! \simeq n \ln n - n$, and neglecting unity in (8.5), a short calculation yields

$$dS^{(b,f)} = k d\ln\Gamma^{(b,f)} = k \sum_i dn_i \ln \frac{g_i \pm n_i}{n_i},\tag{8.10}$$

where the numbers n_i are not independent, because one has

$$dU = \sum_i dn_i \, E_i = 0, \quad dN = \sum_i dn_i = 0. \tag{8.11}$$

Imposing that $dS^{(b,f)} = 0$, one faces a problem of conditional extremum which can be treated by using Lagrange multipliers. We multiply dU and dN in (8.11) by the Lagrange multipliers α and β, and introduce the constraints in the extremum problem:

$$dS^{(b,f)} + \alpha dU + \beta dN = 0.$$

This formula is a consequence of the second law of thermodynamics written for constant volume, and thus we identify α and β, respectively, as $-1/T$ and μ/T, where μ is the chemical potential. This leads to the equation

$$dS^{(b,f)} - \frac{dU}{T} + \frac{\mu dN}{T} = k \sum_i dn_i \left(\ln \frac{g_i \pm n_i}{n_i} - \frac{E_i}{kT} + \frac{\mu}{kT} \right) = 0. \tag{8.12}$$

This equality has to be true for an arbitrary set of dn_i, therefore all the quantities in brackets have to vanish identically. For bosons, this gives the Bose–Einstein distribution:

$$n_i^{(b)} = \frac{g_i}{e^{\frac{E_i - \mu}{kT}} - 1}, \tag{8.13}$$

and for fermions, the Fermi–Dirac distribution:

$$n_i^{(f)} = \frac{g_i}{e^{\frac{E_i - \mu}{kT}} + 1}. \tag{8.14}$$

The expressions for total energy (8.8) and total particle number (8.9) become for bosons:

$$U = \sum_i E_i \frac{g_i}{e^{\frac{E_i - \mu}{kT}} - 1}, \quad N = \sum_i \frac{g_i}{e^{\frac{E_i - \mu}{kT}} - 1}, \tag{8.15}$$

and for fermions:

$$U = \sum_i E_i \frac{g_i}{e^{\frac{E_i - \mu}{kT}} + 1}, \quad N = \sum_i \frac{g_i}{e^{\frac{E_i - \mu}{kT}} + 1}. \tag{8.16}$$

8.3   The Ideal Electron Gas

For electrons, the average density of particles in a quantum state of given energy, or *occupation number*, is

$$n(E_i) = \frac{n_i^{(f)}}{g_i} = \frac{1}{e^{(E_i - E_f)/kT} + 1}, \tag{8.17}$$

where we replaced the chemical potential μ by the Fermi energy E_f. This replacement can be made for temperatures T such that $kT \ll E_f$, and we assume this is the case for the discussion below. In general, the chemical potential depends on temperature, $\mu = \mu(T)$, but $\mu(0) = E_f$. This can be seen by examining formula (8.17) at $T = 0$: if $E_i > E_f = \mu(0)$ then the exponential is infinite, and $n(E_i) = 0$. This means that at $T = 0$, all states with energy $E_i < \mu(0)$ are occupied, while all those with energy $E_i > \mu(0)$ are vacant. Thus, $\mu(0)$ coincides with E_f, by the definition of the latter. For $T \neq 0$, the occupation number $n_i(E)$ given by (8.17) is represented by the curve in Fig. 8.3. Indeed, when $T \neq 0$, for $E_i < E_f$, the exponential $e^{(E_i - E_f)/kT}$ is significantly different from zero only when $E_f - E_i \approx kT$. For such a region of values, $n(E_i) \leq 1$. For $E_i - E_f = 0$, one has $e^0 = 1$ and $n(E_i) = 1/2$. For $E_i > E_f$, the term $e^{(E_i - E_f)/kT}$ grows as E_i increases and $n(E_i) \to 0$.

For the calculation of the total energy of the electron gas it is necessary to know the number of quantum states with energies in the shell between E and $E + dE$. This is equivalent to calculating how many boxes are contained in that shell. Since the sphere of energy less or equal to E has in momentum space the radius $R = \sqrt{p_x^2 + p_y^2 + p_z^2} = \sqrt{2mE}$, we find that the number of boxes included in the sphere is equal to the volume $\frac{4\pi}{3}(2mE)^{3/2}$ divided by the volume of each elementary box, $(2\pi\hbar)^3/V$, where $V = L^3$ is the volume of the cube containing the electron gas. Since two opposite-spin states correspond to each box, we have

$$G(E) = \frac{8\pi(2mE)^{3/2}V}{3(2\pi\hbar)^3}, \tag{8.18}$$

for the total number of quantum states with energy less than or equal to E.

For $T = 0$, since each state of energy smaller than E_f is occupied by an electron and all the states of greater energy than E_f are vacant, the total number of electrons N is equal to $G(E_f)$. In order to calculate the energy of the electron gas, we first need to find the number of states in the interval of energy between E and $E + dE$. Let us call this number $g(E)dE$. We have

$$g(E)dE = G(E + dE) - G(E) = [2^{7/2}\pi m^{3/2}V/(2\pi\hbar)^3]E^{1/2}dE. \tag{8.19}$$

For $T = 0$, the energy of all the electrons between the shells E and $E + dE$ is

$$dU = E\,g(E)dE = 2^{7/2}\pi m^{3/2}V E^{3/2}dE/(2\pi\hbar)^3. \tag{8.20}$$

The total energy is obtained by integrating (8.20) between 0 and E_f, with the result

$$U = \frac{2^{9/2}\pi m^{3/2}}{5(2\pi\hbar)^3}V E_f^{5/2}. \tag{8.21}$$

Thus, the total energy of the electron gas included in a volume V is proportional to the volume V and the Fermi energy to the power $5/2$.

If the temperature is different from zero, the above formulas must be corrected. The reader interested in this point may consult any of the books on statistical physics mentioned in the list at the end of the chapter.

8.4 Heat Capacity of Metals

The model already discussed provides a good description of the electron gas in metals, and it has been used to characterize some of their properties.

If we assume a metal starting out at a temperature close to absolute zero, and heat it up to a temperature T, the thermal energy ΔE which may be absorbed by the electrons is on the average of order kT. Obviously, if $kT < E_f$, the electrons whose energy is far from the Fermi level cannot absorb this energy to be thermally excited, since they would jump to electronic states that are already occupied. The excitable electrons are those located near the Fermi surface, since if they increase their energy by ΔE, they can pass to vacant quantum states. It can be estimated that the relative fraction of excitable electrons is of the order

$$\frac{\Delta E}{E_f} = \frac{kT}{E_f} = \frac{T}{T_f},$$

(8.22)

where the Fermi temperature T_f is defined as E_f/k.

If there are N electrons per unit volume, the excitable fraction ΔN is

$$\Delta N = \frac{NT}{T_f},$$

(8.23)

and the energy absorbed by them is

$$U \approx \Delta NkT \approx \frac{NT}{T_f}kT.$$

(8.24)

The heat capacity at constant volume is defined as the derivative of the internal energy with respect to the temperature:

$$C_V = \left(\frac{\partial U}{\partial T}\right)_V \approx \frac{NkT}{T_f}.$$

(8.25)

According to the classical theory, it is expected that $C_V = \frac{3}{2}Nk$. Due to quantum effects, this number is very much reduced for the electron gas, since for metals T_f is of the order of 10^4 K, and if T is the room temperature (300 K), $T/T_f \approx 0.01$.

From this we conclude that the heat capacity of the electron gas in metals at room temperature is a hundredth of the value predicted by classical theory.

Regarding the thermal conductivity K, if Q is the flow of thermal energy (energy transmitted through unit area and per unit time) and dT/dx is the temperature gradient (variation of the temperature in the x direction), one can define K in the x direction by

$$K = \frac{Q}{(dT/dx)}. \tag{8.26}$$

In the kinetic theory, one can demonstrate that, if v is the average velocity of the particles, l the mean free path (mean distance between two collisions), and C_V the heat capacity, then approximately

$$K = \frac{1}{3} C_V v l. \tag{8.27}$$

Now, by taking $v = v_f = p_f/m$, and $l = v_f \tau$, where τ is the mean time interval between two collisions, since the Fermi temperature would be given by

$$T_f = m v_f^2 / 2k,$$

by taking for the Fermi gas $C_V = \frac{1}{2} \pi^2 N k \frac{T}{T_F}$, the final result is

$$K = \frac{\pi^2 k^2 \tau}{3m} N T, \tag{8.28}$$

that is, the thermal conductivity is proportional to the total number of electrons per unit volume. For this reason, metals are good conductors of heat, since at room temperature N is larger for metals than for solid dielectrics.

Another consequence of the Fermi–Dirac statistics for the electron gas of metals is the large electrical conductivity. If $\mathbf{j}$ is the current density (current per unit area) and $\mathbf{E}$ is the electric field, the electrical conductivity σ is given in terms of these quantities by Ohm's law:

$$\mathbf{j} = \sigma \mathbf{E}. \tag{8.29}$$

This current could also be written as $\mathbf{j} = Ne\mathbf{v}$. It is then easy to demonstrate that, in a metal, when an electric field is applied in the x direction, the Fermi sphere is displaced in that direction (Fig. 8.4), that is, all the electrons gain a momentum

$$\Delta \mathbf{p} = e\mathbf{E}\tau \tag{8.30}$$

in that direction. Here τ is the average time between collisions. As a consequence of (8.30), one can deduce an approximate expression for σ, viz.,

Fig. 8.4 An electric field displaces the Fermi sphere in the direction of the electric field.

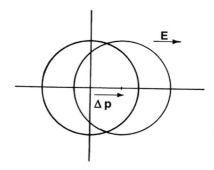

$$\sigma = \frac{Ne^2\tau}{m}, \tag{8.31}$$

that is, the conductivity is proportional to the density of the electron gas and to the average time τ between collisions. Since τ increases when the temperature decreases (as a consequence, the mean free path also increases), the conductivity increases at low temperatures.

As we have pointed out at the end of Chap. 3, the conductivity falls abruptly at very low temperatures and there appears the phenomenon of superconductivity. We shall return to this phenomenon at the end of the present chapter.

The resistivity ρ is defined as the reciprocal of the conductivity:

$$\rho = \frac{m}{Ne^2\tau}. \tag{8.32}$$

This grows with temperature T, since τ decreases with increasing T.

If we compare (8.28) and (8.31), we conclude that the quotient of the thermal and electric conductivities is proportional to the absolute temperature:

$$\frac{K}{\sigma} = \frac{\pi^2}{3} \left(\frac{k}{e}\right)^2 T. \tag{8.33}$$

This relation is known as the Wiedemann–Franz law, due to Gustav Wiedemann (1826–1899) and Rudolph Franz (1826–1902). They formulated it in 1853, and it holds true only approximately, since the value τ for electrical and thermal conductivities is not exactly the same.

Up to now we have discussed the application of Fermi–Dirac statistics to metals, where there is an electron gas which can be treated approximately as a gas of free particles. Another interesting example is the electron gas in stars, e.g., in white dwarf stars. We shall discuss this issue in Chap. 9.

8.5   Metals, Semiconductors, and Insulators

In our discussion of metals, we referred to the way they differ from semiconductors and insulators from a quantum mechanical perspective. This is a topic usually studied in solid state physics, and we shall discuss it only briefly.

A solid is characterized by order (crystal) or disorder (amorphous solid) among an enormous number of atoms, molecules, or ions. In a crystalline solid, the basic structures repeat themselves periodically in the form of a lattice of ordered cells of atoms, molecules, or ions. The vibrations of atoms around their equilibrium positions in this lattice produces elastic waves, which correspond in the quantum version to *quasi-particles* called *phonons*. These are to elastic waves what photons are to electromagnetic waves. The phonons are considered as quasi-particles, which means that they cannot exist as independent particles outside the solid.

The crystal structure of the solid creates a situation where the energy states of the electrons in the last shell of the atoms are very closely spaced, forming continuous bands of energy separated by forbidden bands. Some of these electrons remain attached to the atom, and they are called valence electrons. Some others may belong to delocalized states, being named conduction electrons, since they contribute to the electric conduction when an external electric field is applied to the sample. The periodicity of the crystalline lattice implies a periodicity of the interaction potential in which the valence electrons of the atoms or molecules can move. If the characteristic length of the lattice is denoted by a, then the relevant values of the wave vector k of a conduction or valence electron in the crystal are between 0 and π/a. The band structure is represented as the dependence of the energy of the electrons on the values of their wave vector. In Fig. 8.5 is illustrated the band structure for some prototypical crystals:

- Case (a) corresponds to an insulator. The valence band is completely filled, whereas the conduction band is empty.

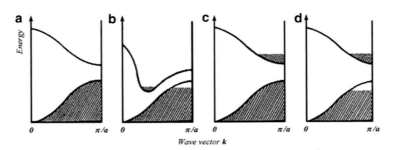

Fig. 8.5 Energy bands in **a** an insulator, **b** a semimetal, **c** a metal, and **d** a semiconductor. The lower line delimits the valence band and the upper line delimits the conduction band. The *band gap* is the energy difference between the top of the valence band and the bottom of the conduction band, which appears in certain cases.

- Case (b) represents a semimetal. The valence band is almost filled. The conduction band, with the same Fermi energy, since there is some overlap, is almost empty.
- Case (c) illustrates the situation in a metal. The valence band is filled, while the upper contains a significant number of conduction electrons.
- Case (d) is typical of a semiconductor. At absolute zero, the semiconductor would have the valence band filled and the conduction band empty. At room temperature, some electrons are excited and pass into the conduction band due to the fact that the band gap is narrow. The valence band now contains a certain number of vacant states, which are holes.

8.6 Electrons and Holes

In the previous example of a semiconductor, if an electron in the valence band is excited to the conduction band, an electron state is left vacant in the first band. This state behaves as a positive charge and is called a hole, with opposite momentum and spin to the excited electron.

This description was inspired by the Dirac model of the electron vacuum, suggested to explain the negative energy solutions of the relativistic Dirac equation (see Chap. 7). The holes play an important role as charge carriers in semiconductors. They are quasi-particles and they also obey Fermi–Dirac statistics.

The dynamic properties of electrons and holes depend on the shape of the band. In general, they behave in their motion as though they had an effective mass which differs from the free electron mass, being greater or smaller, and even negative.

8.7 Applications of the Fermi–Dirac Statistics

8.7.1 Quantum Hall Effect

If a magnetic field is applied transverse to an electric current in a plane sample, a voltage (an electric field) appears perpendicular to both the original current and the magnetic field, inducing a new current. The effect was discovered by Edwin Hall (1855–1938) in 1879, and is named Hall effect. Classically, this effect is easily understood if we start from the Lorentz force introduced in Chap. 3. Let us assume that the system of electrons lies in the plane xy, and the electrons move with velocity $\mathbf{v}$, due to an applied electric field $\mathbf{E}$ in the y-direction, i.e. $\mathbf{E} = (0, E_y, 0)$ (see Fig. 8.6).

In addition to the force term due to the electrostatic interaction, $\mathbf{F}_e = e\,\mathbf{E}$, an electric charge placed in a magnetic field $\mathbf{B}$ experiences a Lorentz force perpendicular to the magnetic field and to the initial electron velocity, $\mathbf{F}_L = e\mathbf{v}/c \times \mathbf{B}$. If the current is stationary, i.e. $\dot{\mathbf{v}} = 0$, the two forces cancel each other, which means:

$$\mathbf{E} = -\frac{\mathbf{v}}{c} \times \mathbf{B}.$$

Fig. 8.6 Schematic
representation of the Hall
experiment. The Hall
resistance is $R_H = V_H/I$.

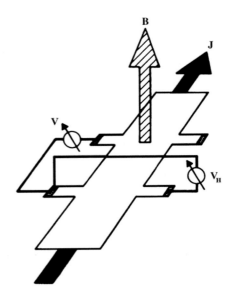

As a result, the flow of current will be only in the direction perpendicular to the
applied electric field. The current density is $\mathbf{J} = e\rho_0\mathbf{v}$, with ρ_0 being the surface
density of the electrons. Extracting $\mathbf{v}$ from the stationarity condition, we see that
the only nonvanishing component of $\mathbf{J}$ is in the x-direction, being given by the
expression:

$$J_x = \frac{ec\rho_0}{B}E_y.$$

Now we can calculate the resistivities, by the formula $R = E/J$. It turns out that a
Hall resistance R_H develops, denoted sometimes also by R_{xy}, because it is the ratio
of the electric field in the y-direction and the current in the x-direction:

$$R_H = R_{xy} = \frac{E_y}{J_x} = \frac{B}{ec\rho_0}. \tag{8.34}$$

The diagonal component of the resistivity tensor, called longitudinal resistivity R_{xx},
vanishes, since $E_x = 0$. Thus, classically, the Hall resistance is a linear function of
the transverse magnetic field, for fixed density of electrons (see Fig. 8.7).

It took one century to discover that all had not been said about the Hall effect.
In two-dimensional electron systems subjected to low temperatures and strong magnetic fields, the quantum-mechanical version of the Hall effect is observed, namely
the quantum Hall effect. In this case, the Hall resistance R_H takes on quantized values
given by the expression

$$R_H = \frac{1}{\nu}\frac{h}{e^2}, \tag{8.35}$$

Fig. 8.7 Integer quantum Hall effect schematically illustrated. In the classical theory of the electron gas, the Hall resistance R_H is proportional to the magnetic field B applied perpendicular to the sample, as indicated by line labeled *classical theory*. In an actual sample, subjected to very low temperature and very high magnetic field, the Hall resistance appears like a staircase, with the plateaux crossing the classical line at integer values of the filling factor, $\nu = 1, 2, 3, \ldots$. While the Hall resistance is on a plateau, the longitudinal resistivity R_{xx} vanishes.

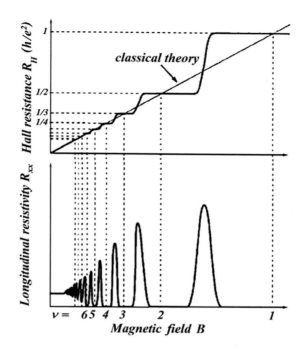

where e is the electric charge, h is Planck's constant and ν is the *filling factor*. This factor can have either integer values ($\nu = 1, 2, 3, \ldots$) for the so-called integer quantum Hall effect, or rational fraction values ($\nu = 1/3, 2/5, 3/7, 2/3, 3/5, 1/5, 2/9, 3/13, 5/2, 12/5 \ldots$) in the fractional quantum Hall effect. The profile of the Hall resistance curve when the magnetic field is increased is shown in Fig. 8.7: it appears like a staircase, with the plateaux crossing the classical line at $\nu = 1, 2, 3, \ldots$. At the same points, the longitudinal resistivity R_{xx} vanishes.

The *integer quantum Hall effect* was discovered in 1980 by Klaus von Klitzing (b. 1943) and collaborators, and the *fractional quantum Hall effect* was discovered in 1982 by Daniel Tsui (b. 1939), Horst Störmer (b. 1949) and collaborators. The Nobel Prize in 1985 was awarded to Klaus von Klitzing, and in 1998 to Robert Laughlin, Horst Störmer and Daniel Tsui, for the discovery and explanation of the fractional Hall effect.

The mysterious filling factor can be easily obtained by matching the expressions (8.34) and (8.35), with the result

$$\nu = \frac{hc\rho_0}{eB}. \tag{8.36}$$

This formula is derived exactly by performing the detailed quantum mechanical analysis, which leads also to the physical interpretation of the filling factor.

The integer quantum Hall effect is explained in terms of single-particle states of electrons in a plane system (two-dimensional electron gas) under the action of an orthogonal magnetic field. Classically, the electrons are accelerated in a cyclotron motion when placed in a magnetic field which has a component orthogonal to their direction of motion. Quantum mechanically, the cyclotron orbits are quantized, such that the planar electrons can occupy discrete energy levels, given by the same expression as in the case of a one-dimensional quantum mechanical oscillator, but with the frequency depending on the magnetic field:

$$E_n = \hbar \omega_c (n + 1/2), \quad n = 0, 1, 2, \ldots \tag{8.37}$$

The cyclotron frequency is

$$\omega_c = \frac{eB}{mc}, \tag{8.38}$$

where e is the charge of the electron, B is the value of the magnetic field orthogonal to the plane of the system, m is the mass of the electron and c is the speed of light.

The energy levels indexed by $n \geq 0$ are called *Landau levels*, in honour of Lev Landau (1908–1968). One can imagine the planar electrons as making a cyclotron motion of a radius corresponding to their energy level, such that each electron occupies an area πr_n^2, avoiding any other electron. The calculations show that the squares of the various radii are expressed as $r_n^2 = (2n + 1)l_B^2$. The quantity $l_B = \sqrt{\frac{\hbar c}{eB}}$ is called magnetic length and it gives the scale of the quantum Hall effect. Note that this length depends on the applied magnetic field and becomes smaller with the increase of the latter. For a typical magnetic field used in the experiments, of about 10^4 gauss, the magnetic length is about $100\,\text{Å}$.

The system of electrons being planar, each electron has two degrees of freedom. When the problem is reduced to a one-dimensional harmonic oscillator, one degree of freedom only is taken care of in the quantization. This means that there is degeneracy of the energy levels in the second degree of freedom, which does not appear in the expression of the energy. The second degree of freedom in this case is the coordinate of the centre of the orbit. A simple calculation using the Fermi–Dirac statistics (similar to the one in Sect. 8.3) shows that a two-dimensional system of free electrons exhibits a degeneracy of each energy level, such that the density of states is independent of the energy, $g(E) = A \frac{m}{\pi \hbar^2}$, where A is the area of the planar system. When the magnetic field is applied, the energy levels are discrete, with a constant spacing of $\hbar \omega_c$. The number of zero-field states in an interval $\hbar \omega_c$ is

$$g(E)\hbar \omega_c = A \frac{m}{\pi \hbar^2} \hbar \frac{eB}{mc} = 2 \frac{eB}{hc} A.$$

One has to include also the Zeeman splitting, for spin up and spin down electrons, which will reduce the density of states to half of the above value. This is the degeneracy of each Landau level:

Fig. 8.8 **a** Landau levels in a two-dimensional electron gas. **b** Broadening of the density of states in a strong magnetic field. The delocalized states are at the centre of the peaks, while the states localized by impurities are found in the tails.

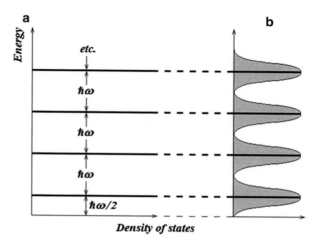

$$\text{Number of states per Landau level} = \frac{eB}{hc}A. \qquad (8.39)$$

The distribution of states is not continuous, like in the case of zero magnetic field, but rather the states are grouped at the discrete values of the energy (see Fig. 8.8a). However, the total number of states below a certain energy, much larger than the Landau level spacing, is unchanged by the magnetic field.

The filling factor is the ratio between the total number of electrons and the number of states in one Landau level:

$$\nu = \frac{\text{Number of electrons}}{\text{Number of states per Landau level}} = \frac{\rho_0 A}{eBA/hc} = \frac{hc\rho_0}{eB}. \qquad (8.40)$$

Assuming that the thermal energy is much smaller than the Landau level spacing, i.e. $kT \ll \hbar\omega_c$ (very low temperature and high magnetic fields), and keeping the number of electrons constant by adjusting the longitudinal voltage, experimentally one observes the quantization of the Hall resistance, according to the formula (8.35). The profile of the Hall resistance curve when the magnetic field is varied shows the peculiar formation of plateaux at the quantized values of R_H and sudden jumps at those values of the magnetic field where the filling factor takes integer values (see Fig. 8.7).

The formation of the plateaux is one of the most interesting features of the quantum Hall effect, and it is due to the presence of impurities in the sample. In a pure sample, the distribution of states would be exactly like in Fig. 8.8a. The impurities make the Landau energy levels spread out—in other words, they lift to some extent the degeneracy. For example, if an impurity atom has an excess of positive charge, an electron is more likely to be found near that atom, in which case its energy

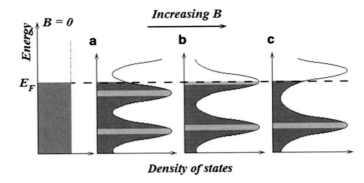

Fig. 8.9 Plateaux formation in a sample with constant density of electrons, when the applied magnetic field is varied. **a** While the Fermi level is in the sub-band of high-energy localized states, the Hall resistance does not change, since the number of charge carriers is not affected. The current is carried only by the electrons in the delocalized states. **b** When the Fermi level enters the sub-band of delocalized states, the current drops, because current-carrying electrons become trapped in localized states. The filling factor (number of completely filled Landau levels) decreases by one unit and the Hall resistance increases abruptly, until the Fermi level enters the sub-band of low-energy localized states. **c** A new plateau starts to be formed at the new value of the Hall resistance.

would be slightly lower (the electron would be more stable) than its corresponding Landau level. As a result, bands of energy form around each Landau level, as in Fig. 8.8b. These bands are still separated by energy gaps. The electrons in each band are divided into two classes: (1) those electrons which are close to the centre of the band (having the Landau energy) are spread over a large region of space (*delocalized*), and contribute to conduction; (2) the electrons which are at the margins of the energy bands, are *localized* in space due to the impurities, and do not carry the current.

Let us now explain the formation of the plateaux in the Hall resistance (see Fig. 8.9). Keeping the density of electrons constant in the sample (which means that the Fermi energy does not change over the whole process), one increases the magnetic field and measures R_H. Let us assume that we start with a value of B at which the Fermi energy is in a gap between two Landau levels, for example above the nth of them. This means that all Landau levels below the Fermi energy are completely filled, while all the Landau levels lying above are completely empty. In this case, the conduction is ensured by the *delocalized* electrons in the n completely filled Landau levels. Now suppose that the strength of the magnetic field is gradually increased and that at the same time, the current is continuously adjusted in such a way that the Hall voltage between the two edges of the sample remains constant. The Fermi level starts to enter the region of high-energy localized states (Fig. 8.9a), and the corresponding electrons vacate those states. As long as the Fermi level remains in the sub-band of high-energy localized states, all the extended states within the Landau band remain fully occupied.

While increasing the magnetic field, the degeneracy of the Landau levels increases proportionally according to the formula (8.39). The fraction of localized states changes, but this has no effect on the conduction properties of the sample. The amount of current flowing in the sample therefore remains constant as long as the sub-band of extended states is completely filled: although the increased magnetic field slows the forward motion of any current carrying electrons, this effect is precisely cancelled by the increase, due to the newly created extended states, in the number of electrons available to carry current. Since the Hall voltage is being held constant, the fact that the current does not change as the magnetic field is varied implies that the Hall resistance also remains constant and a plateau is formed.

Increasing further the magnetic field, the Fermi energy enters the sub-band of delocalized states (Fig. 8.9b). As a result, some of these states become vacant and the number of current-carrying electrons drops down. The current decreases suddenly until all the delocalized states of the nth Landau level become empty, and the Fermi energy enters the region of the low-energy localized states (Fig. 8.9c). At that moment, the gradual increase in the magnetic field ceases to have any effect on the Hall resistance, as long as the delocalized states of the $(n - 1)$th Landau level are completely filled.

In summary, the Hall resistance makes a jump between two plateaux when the delocalized states of one Landau level become empty due to the increase in the magnetic field. When localized states become empty, this has no effect on the number of current-carrying electrons, and the Hall resistance remains on a plateau.

A few words are in order about the longitudinal resistivity, whose behaviour is depicted also in Fig 8.7. According to the classical formula, $R_{xx} = 0$. This is the case also in the quantum version of the Hall effect, as long as the Hall resistance is on a plateau. As we have just explained, a plateau means that the delocalized sub-bands are completely filled. The conduction electrons cannot jump from one energy level to another, since there are no available energy levels for them. As a result, the scattering of conduction electrons, with loss of energy, cannot happen. A state similar to superconductivity in the longitudinal direction is thus attained. It should be emphasized that the vanishing of the longitudinal resistivity does not require the absence of scattering centres, but the absence of possibilities for electrons to scatter. However, when the magnetic field is such that the Fermi energy is in a sub-band of delocalized states and these are no more completely filled, the possibility of scattering suddenly reappears, and the longitudinal resistivity becomes finite. The clean jump of the Hall resistance at the plateau transition and the peak in the longitudinal resistivity have, therefore, the same origin.

The Hall resistance quantization is universal in nature, depending only on universal constants and integer numbers. The form of the localization potential and the distribution of scatterers are of no importance for quantization.

The *fractional quantum Hall effect* is more complicated but it is also understood as a form of integer quantum Hall effect, in which the electrons form bound states with an even number of magnetic flux quanta. The resulting particles are called composite fermions.

The quantum Hall effect has important applications even in metrology. Based on it, a resistance standard has been introduced and named the von Klitzing constant, $R_K = h/e^2 = 25812.807557(18)\ \Omega$. Moreover, the fine structure constant, α, is related to the von Klitzing constant by

$$\alpha = \frac{\mu_0 c}{2} \frac{e^2}{h} = \frac{\mu_0 c}{2} R_K^{-1}.$$

Since the speed of light is known with great accuracy and the magnetic permeability of the vacuum by definition equals $4\pi \times 10^{-7}$ H/m, this relation permits an independent high-precision determination of α, one of the fundamental constants of Nature, characterizing the strength of electromagnetic interactions.

8.7.2 Graphene

Some chemical elements exist in two or more different forms, known as allotropes, which are different structural modifications of an element, in the sense that the atoms of the element are bound in a different manner. Carbon's allotropes include diamond—in which carbon atoms are bound in a tetrahedral lattice, graphite—in which carbon atoms are bound in sheets of hexagonal form, as a honeycomb, graphene—which represents single sheets of graphite, and fullerene, whose carbon atoms are bound for instance in a spherical lattice (see Fig. 8.10).

Graphene is at present a very important material, of special interest in the area of nanoscience and nanotechnology. The Nobel Prize in Physics in 2010 was awarded to Andre Geim (b. 1958) and Konstantin Novoselov (b. 1974) "for groundbreaking experiments regarding the two-dimensional material graphene."

Graphene is a unique two-dimensional material (0.335 nm = 3.35 Å thickness). It appears to be one of the strongest materials ever tested, having a breaking strength near 200 times greater than steel. Its high mobility and conductivity suggest unbelievable microelectronic possibilities, from transistors and scroll screens, to photovoltaic cells, bio-devices, and many other applications.

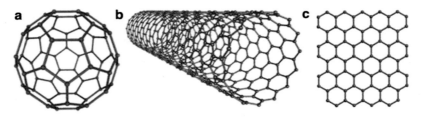

Fig. 8.10 In the figure are shown three allotropes of carbon: **a** Buckminsterfullerene—a polyhedron form of fullerene, consisting of 60 atoms, **b** carbon nanotube, and **c** graphene, a two-dimensional arrangement. Graphite, not shown, is a three-dimensional structure composed of graphene layers.

Fig. 8.11 A cat on a
graphene hammock.

A customary intuitive picture of graphene is of a 1 m² hammock tied between two
trees, and holding up to 4 kg without breaking. To make the picture more attractive,
one can imagine the hammock holding a cat (Fig. 8.11), while the weight of the
hammock itself is less than 1 mg, corresponding to the weight of one of the cat's
wiskers!

Carbon is an element with six electrons, populating two atomic levels. The four
electrons on the last layer are distributed as follows: two on the $2s$ orbital and two
on the $2p$ orbitals. When chemical bonds are formed, the $2s$ and $2p$ orbitals mix in
a superposition of the two states, named hybridized state. In one of the hybridized
states, named sp^2, the $2s$ orbital mixes with two of the three $2p$ orbitals, forming
three degenerate sp^2 orbitals placed in the xy-plane, with equal angles of $120°$
among themselves. The remaining $2p$ orbital is oriented orthogonal to this plane. It
is energetically favourable to have one electron on each of these four valence orbitals.
When neighbouring atoms come close, strong covalent bonds are formed, leading to
a hexagonal lattice. The $2p_z$ orbitals of adjacent atoms overlap, forming a delocalized
state. The electrons in this state have the ability to move relatively freely above and
below the plane of the nuclei, forming two bands with zero gap. Thus, graphene
differs from most conventional three-dimensional materials and is a semi-metal or
zero-gap semiconductor (having massless quasi-particles).

In the presence of a magnetic field, graphene displays an anomalous quantum
Hall effect with the sequence of steps shifted by 1/2 with respect to the standard
sequence, and with an additional factor of 4:

$$R_H = \pm \frac{h}{e^2} \frac{1}{4(m + 1/2)}, \qquad m \text{ integer.} \qquad (8.41)$$

Thus, the filling factor is $\nu^G = \pm 4(m + 1/2)$, which is the fingerprint of the
relativistic Landau quantization. The steps in units of 4 reflect the so-called
"spin-valley degeneracy", which is characteristic for graphene. The Hall resistance
shows the same universality as in the non-relativistic case, being determined by the

universal constant $R_K = h/e^2$ and integer numbers. The impurities have the same essential role in the formation of the plateaux in the Hall resistance.

From the side of pure science, the links between condensed matter physics and high energy physics established by quantum Hall effect and graphene is profound: for instance, the measurement, with high accuracy of quantities like the resistivity quantum h/e^2 and the fine-structure constant $\alpha = e^2/\hbar c$. This makes possible the study of experimental problems related to high energy physics (usually demanding large installations) in the frame of a "table condensed matter laboratory".

8.8   Bose–Einstein Statistics

The identical particles obeying Bose–Einstein statistics, commonly called bosons, have integer spin and, like fermions, are indistinguishable. The essential difference from fermions is that there can be an arbitrary number of bosons in a quantum state. In other words, they do not obey the Pauli exclusion principle.

The expression (8.13) gives the Bose–Einstein distribution. Thus, the average density of particles per quantum state for an ideal Bose–Einstein gas is:

$$n(E_i) = \frac{1}{e^{(E_i - \mu)/kT} - 1}. \tag{8.42}$$

If we compare (8.42) with (8.17), we see that the difference is in the sign in front of unity in the denominator. It is now subtracted from the exponential. This is enough to guarantee that, whenever $E_i - \mu > 0$, n_i could take any positive value, where the chemical potential μ plays a role analogous to the Fermi energy in the case of fermions. If the minimum value of the one-particle energy is 0, then $\mu \leq 0$. Photons obey Bose–Einstein statistics with $\mu = 0$. The same is true for phonons in a solid.

For a gas of photons, the average energy $\langle \varepsilon \rangle$ in a mode with energy $\varepsilon = h\nu$ is equal to the average number of photons in that state multiplied by $h\nu$:

$$\langle \varepsilon \rangle = \frac{h\nu}{e^{h\nu/kT} - 1}. \tag{8.43}$$

The number of states per unit volume contained in a shell of energy $(h\nu, h\nu + h\,d\nu)$ is $4\pi p^2 dp/h^3$, with $p = h\nu/c$. For photons we have to take into account the two orthogonal polarizations, resulting in the doubling of the above number of states:

$$n(\nu)d\nu = 2\frac{4\pi\, p^2\, dp}{h^3} = \frac{8\pi}{c^3}\nu^2 d\nu. \tag{8.44}$$

If we multiply (8.43) by (8.44), the energy density for electromagnetic radiation in thermal equilibrium is obtained as a function of the frequency:

$$U(\nu, T)d\nu = \frac{8\pi h}{c^3} \frac{\nu^3 \, d\nu}{e^{h\nu/kT} - 1}. \tag{8.45}$$

This is Planck's law (see Chap. 4, (4.51)). If U is graphed as a function of ν at several temperatures, the typical curves of the Planck distribution are obtained (Fig. 4.25). If (8.45) is integrated with respect to ν, one obtains the total energy density

$$U_t = \int_0^\infty U(\nu, T)d\nu = \frac{8\pi (kT)^4}{(hc)^3} \int_0^\infty \frac{x^3 dx}{e^x - 1} = 4\sigma \frac{T^4}{c}. \tag{8.46}$$

This is the Stefan–Boltzmann law: the total energy density of the black body radiation is proportional to the fourth power of the absolute temperature, where the coefficient of proportionality in the CGS system of units is $(4/c)\sigma$, where $\sigma = 2\pi^5 k^4/15h^3c^2 \simeq 5.67 \times 10^{-5}$ erg cm^{-2} s^{-1} K^{-4} and is called the Stefan–Boltzmann constant.

Another system obeying Bose–Einstein statistics is a gas of normal helium, ^{4}He, whose nucleus contains two protons and two neutrons. This guarantees that the spin of the nucleus is an integer (the isotope ^{3}He has half-integer spin and obeys Fermi–Dirac statistics).

8.9 Einstein–Debye Theory of Heat Capacity

We have seen the role of Fermi statistics in calculating the specific heat (heat capacity per unit mass) of metals. For insulators, the thermal properties are determined by the lattice oscillations or elastic (sound) waves. Their quanta, named *phonons*, obey Bose–Einstein statistics. The calculation of the specific heat is another good illustration of the application of Bose–Einstein statistics in condensed matter.

Einstein was the first who calculated in 1907 the specific heat due to elastic waves, of special interest for non-metallic solids (for metals, in the limit $T \to 0$, the Fermi–Dirac specific heat behaviour linear in T is usually the dominant term). In Einstein's approach, each atom is treated as an independent quantum harmonic oscillator (i.e., without interaction), and the frequency is the same for all atoms. By assuming N oscillators per unit mass, in one dimension, with an energy per oscillator $E = \hbar\omega(n + \frac{1}{2})$, where $n = (e^{\hbar\omega/kT} - 1)^{-1}$, and defining the Einstein temperature as $\Theta_E = \hbar\omega/k$, one finds for the energy per unit of mass

$$U = 3N\hbar\omega \left(\frac{1}{e^{\Theta_E/T} - 1} + \frac{1}{2} \right), \tag{8.47}$$

where we have introduced a factor 3, to account for the fact that each atom has actually three degrees of freedom. From here it follows that

$$C_V = \left(\frac{\partial U}{\partial T}\right)_V = 3Nk \left(\frac{\Theta_E}{T}\right)^2 \frac{e^{\Theta_E/T}}{(e^{\Theta_E/T} - 1)^2}. \tag{8.48}$$

The high temperature limit leads to $C_V = 3Nk$, which is the empirical Dulong–Petit law, according to which the heat capacity of solids does not depend on the temperature. At low temperatures, the energy (8.47) tends to a constant, or zero point energy, and $C_V \to 0$. Thus, the entropy also tends to zero, $S \to 0$, according the third law of thermodynamics. Einstein's work provided a first theoretical basis to understand the departure of the specific heat of solids from the Dulong–Petit law at low temperatures.

Peter Debye (1884–1966) refined the model in 1912, including the interaction of the atoms in the lattice, which are viewed as coupled oscillators, resulting in a collective phenomenon of lattice oscillation. Quantum mechanically, the oscillation of the lattice can be given an interpretation in terms of fictitious particles, *phonons*, thus named in 1932 by Igor Tamm (1895–1971). The phonons have a continuous spectrum of frequencies, from zero up to a cut–off frequency ω_D, which is bound by the atomic lattice, thus ensuring a finite number of degrees of freedom, $3N$. Denoting by $g(\omega)$ the density of frequencies, we have

$$\int_0^{\omega_D} g(\omega)d\omega = 3N. \tag{8.49}$$

Calling respectively c_L and c_T the longitudinal and transverse velocities for modes propagating in the solid, if we do the counting of states similarly as we did earlier (for instance, for the black body modes) and take into account the fact that a phonon has two transverse and one longitudinal degrees of freedom, one can write

$$\int_0^{\omega_D} V\left(\frac{\omega^2 d\omega}{2\pi^2 c_L^3} + \frac{\omega^2 d\omega}{\pi^2 c_T^3}\right) = 3N. \tag{8.50}$$

Defining an effective sonic velocity c_s by the formula $c_s^{-3} = (1/3c_L^{-3} + 2/3c_T^{-3})$, and comparing equations (8.49) and (8.50), we obtain

$$g(\omega) = 3V\frac{\omega^2}{2\pi^2 c_s^3}. \tag{8.51}$$

Introducing (8.51) into (8.49), the expression of ω_D which bounds the possible frequencies of the phonons is found to be:

$$\omega_D = c_s \left(6\pi^2 \frac{N}{V}\right)^{\frac{1}{3}}.$$ (8.52)

One can now calculate the total energy U of the phonons:

$$U = \int_0^{\omega_D} d\omega \, g(\omega) n(\omega) \hbar\omega = 3NkT \frac{3}{x_D^3} \int_0^{x_D} dx \frac{x^3}{e^x - 1} = 3NkT D_3(x_D),$$ (8.53)

where we made the notations $x = \hbar\omega/kT$ and $x_D = \hbar\omega_D/kT = \Theta_D/T$ (the temperature Θ_D is called Debye temperature). The function $D_3(x_D)$ is the third Debye function. The heat capacity has the expression:

$$C_V = \left(\frac{\partial U}{\partial T}\right)_V = 9Nk \frac{1}{x_D^3} \int_0^{x_D} \frac{x^4 e^x dx}{(e^x - 1)^2}.$$ (8.54)

Debye's theory leads to satisfactory theoretical results both for $T \to 0$ and $T \to \infty$, although the simplifications inherent to the model make the curve of C_V as a function of T to depart from the experimental observation for intermediate temperatures.

For $x_D \ll 1$, which means $T \gg \Theta_D$, by expanding the exponential as $e^x \approx 1 + x$ in (8.54), we obtain

$$C_V \approx 9Nk \frac{1}{x_D^3} \int_0^{x_D} \frac{x^4}{x^2} dx = 3Nk,$$

which is again the classical Dulong–Petit behaviour.

In the opposite limit, $T \ll \Theta_D$, x_D is a large number, and the upper limit of integration in (8.53) may be taken to be infinity. In that case, the integral can be calculated exactly, with the result $\pi^4/5$. Taking the derivative with respect to T, we obtain:

$$C_V \simeq \frac{12\pi^4}{5} Nk \left(\frac{T}{\Theta_D}\right)^3,$$ (8.55)

i.e. a T^3-dependence, which—unlike Einstein's theory—reproduces correctly the observed heat capacity behaviour also in the low-temperature limit. The entropy, evaluated from $S = \int_0^T C_V(t) dt/t$, leads to $S = C_V(T)/3$, and thus the entropy also vanishes as T^3 for $T \to 0$.

Debye's model is in good agreement with the empirical data. From the values of N/V and the elastic properties (expressed by the velocities c_L and c_T), the Debye temperature Θ_D can be determined. The pure T^3-behaviour is satisfied reasonably well below $\Theta_D/50$, but the curve $C_V(T)$ is in a wider range of temperature values in agreement with experimental results.

8.10 Bose–Einstein Condensation

As we have already seen, for bosons, $-\mu > 0$. In general, μ decreases with the temperature. One can write the number of particles N in a volume V for an atomic or molecular Bose gas as

$$N = \frac{4V\pi}{(2\pi\hbar)^3} \int_0^\infty \frac{p^2 dp}{e^{(E(p)-\mu)/kT} - 1},$$ (8.56)

where $E(p) = p^2/2m$. If μ decreases with temperature, there should be a temperature T_c for which $\mu = 0$, keeping the same number of particles N:

$$N = \frac{4V\pi}{(2\pi\hbar)^3} \int_0^\infty \frac{p^2 dp}{e^{p^2/2mkT_c} - 1}.$$ (8.57)

Equation (8.57) for a given density N/V gives a critical temperature T_c at which condensation starts. If we take the formula (8.57) with an arbitrary temperature T on the right-hand side and perform the change of variable $p^2/2mkT = y$, the integral becomes

$$N = \frac{2V(2\pi mkT)^{3/2}}{\pi^{1/2}(2\pi\hbar)^3} \int_0^\infty \frac{y^{1/2} dy}{e^y - 1}.$$ (8.58)

If the temperature decreases the expression on the right in (8.58) gives a number $N' < N$. What has happened to the remaining molecules $N - N' = N_0$? A macroscopic number of them occupies the ground state $E = 0$. This phenomenon is called Bose–Einstein condensation. It was discovered by Einstein in 1925 when studying the Bose distribution.

In a system in which condensation occurs, two phases appear: a normal phase and a condensed phase. As T decreases below T_c, the number of molecules in the condensed phase increases, and at $T = 0$, all the molecules will be in the condensed phase. One can write approximately $N_0 = N[1 - (T/T_c)^{3/2}]$. If the critical temperature is calculated for a gas with the density of ^{4}He, it is found to be 3.2 K.

In 1937, Pyotr Kapitsa in Moscow, and independently John Allen and Don Misener in Cambridge discovered that, at 2.26 K, liquid helium ^{4}He exhibits the phenomenon of superfluidity, i.e., loss of viscosity, manifested by the property of being able to flow through thin capillaries. Under such conditions the liquid helium appears as a mixture of normal fluid and superfluid. Although the gas model cannot strictly be applied to liquid helium, it is generally accepted that the phenomenon of superfluidity corresponds qualitatively to a sort of Bose–Einstein condensation. Pyotr Kapitsa (1894–1984) was awarded the Nobel Prize in Physics in 1978, for his work in low-temperature physics. In 1938, Fritz London and Lazlo Tisza elaborated models to explain superfluidity by Bose–Einstein condensation. In 1941, Lev Landau proposed an essentially phenomenological model of superfluidity as a quantized theory of hydrodynamics, for which he was awarded the Nobel Prize in 1962.

The numerical value of the integral in (8.58) is $2.612\pi^{1/2}/2$, and finally one can write, for $T = T_c$,

$$\frac{N}{V} = d^{-3} = \frac{2.612}{\lambda_c^3}, \tag{8.59}$$

where λ_c^3 is the de Broglie wavelength corresponding to the critical temperature, and the critical conditions can be written as

$$\lambda_c/d = 2.612^{1/3} = 1.38. \tag{8.60}$$

Thus, condensation sets in when the thermal wavelength is longer by a factor near 1.4 of the average distance between particles. This is the *quantum degeneracy* condition for the Bose–Einstein gas.

Superconductivity and ferromagnetism (which we have encountered in Chap. 3) can also be understood qualitatively as forms of a Bose–Einstein condensation. In the phenomenon of superconductivity, an important role is played by systems of paired electrons, of opposite spin and momenta, called Cooper pairs. The pair of electrons behaves as a composite particle of spin 0, and as a consequence, it can be treated as a bosonic system, in which condensation can take place. For ferromagnetism, there is an exchange interaction between electrons such that, under certain conditions, an enormous number of them go into a single quantum state (when all the elementary dipoles align). In fact, this has interest essentially from the qualitative and conceptual points of view. The problems of superfluidity, like those of superconductivity and ferromagnetism, require more advanced models.

Bose–Einstein condensation was observed directly for the first time in 1995 in an experiment led by Carl E. Wieman and Eric A. Cornell, with their collaborators M. H. Anderson, J. R. Ensher, and M. R. Mattews. This remarkable experiment involved rubidium-87 atoms (which bear a nonzero magnetic moment), slowed down by laser cooling and caught in a magnetic trap.

To understand the essence of the experiment, let us see first how magnetic trapping works. To start with, recall that an atom in a magnetic field has the energy $E = -\boldsymbol{\mu} \cdot \mathbf{B}$, where $\boldsymbol{\mu}$ is its magnetic moment. The magnetic moments aligned parallel to the field have lower energies than those antiparallel to it. In a nonuniform magnetic field, the parallel magnetic moments seek for higher field intensity, while the antiparallel ones tend to seek for lower field intensity, in order to achieve minimum energy of the whole system. As local minimum fields can be produced in the laboratory, they serve to trap low-energy atoms, that is, atoms whose kinetic energies correspond to temperatures of a fraction of a kelvin.

The mechanism for slowing down atoms was laser cooling. This phenomenon is rather counterintuitive, since one would expect radiation to warm up and not cool down a system. The great ingenuity consisted in tuning the laser slightly below the frequency of a certain transition that can take place in the atoms which are supposed to collide with the radiation. Upon head-on scattering, the atoms absorb one photon and correspondingly lose a part of their momentum equal to the momentum of the photon that they had scattered. If the photon's momentum is in the same direction with the momentum of the atom that absorbs it, the effect is quite the opposite, that is, the atom is speeded up. Tuning the frequency of the laser slightly below the absorbtion frequency of the atom ensures that the head-on collisions dominate, since the atom which is approaching the photon, by Doppler effect, "sees" the photon as having the resonant frequency to be absorbed. The slowed down, but excited atom emits radiation in a random direction, thus returning to the ground state and the cycle can start again. In this way, the atoms are drastically slowed down by the interaction with the laser radiation. The laser cooling technique was invented by Steven Chu (b. 1948), Claude Cohen-Tannoudji (b. 1933), and William Daniel Phillips (b. 1948), who were awarded the Nobel Prize in 1997. This mechanism is used in the experiment in such a way that an atom in the magnetic trap becomes more likely to get a photon kick toward the centre of the trap, and decrease its energy. The atom cloud is kept inside the trap, and it tends to thermal equilibrium, decreasing its temperature down to a lower limit. Achieving Bose–Einstein condensation requires an additional step of cooling the atoms beyond the limits of laser cooling, by means of the so-called forced evaporative cooling, which consists in rotating the magnetic trap, to selectively remove high-energy atoms from the previously laser-cooled atom cloud until the remaining cloud is cooled below the critical condensation temperature. The experimental setup acts as a Maxwell demon, the atoms with nonzero speed being removed, and leaving a sample of nearly pure condensate, at a temperature of 1.7×10^{-7} K. The condensate in the original experiment contained about 2,000 atoms, corresponding to a density of 2.5×10^{12} cm^{-3}, and lasted for 15 s.

Examining Fig. 8.12, it must be remarked that the peak is not infinitesimaly narrow because of the Heisenberg uncertainty principle: since the atoms are trapped in a particular region of space, their velocity distribution necessarily possesses a certain minimum width, determined by the size of the magnetic trapping potential. More confined space leads to larger widths in the velocity distribution, since in a direction of diameter L, $\Delta p \sim 2\pi\hbar/L$, from which $\Delta v = \Delta p/m$.

About 4 months after the Wieman–Cornell experiment, independently, Wolfang Ketterle with collaborators at MIT created a condensate of sodium-23 atoms. Later on, in 1997, his group reported the observation of quantum mechanical interference between two different condensates, as well as several other important results. E. A. Cornell (b. 1961), C. E. Wieman (b. 1951), and W. Ketterle (b. 1957) were awarded the 2001 Nobel Prize in Physics for their achievements.

Bose–Einstein condensation for photons was observed in 2010 by J. Klaers, J. Schmitt, F. Vewinger, and M. Weitz. Free photons have mass zero and in the case of the black body radiation, the number of particles does not conserve when temperature is varied, which in principle makes condensation impossible. The experimental setup, using a curved-mirror optical resonator filled with a dye solution, acts as "white-walls" which do not absorb the low-energy photons, thus ensuring the conservation of their number during thermalization and making the system formally equivalent to a two-dimensional gas of trapped, massive bosons.

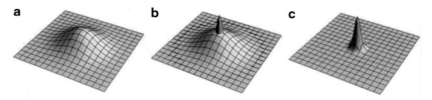

Fig. 8.12 Bose–Einstein condensation experiment. **a** For frequencies of the rotating magnetic trap above 4.23 MHz, the atoms show a single, smooth, Gaussian-like distribution. **b** At 4.23 MHz, a sharp central peak in the distribution begins to appear, and at lower frequencies two distinct components to the cloud are visible, the smooth broad curve and a narrow central peak, identified as the noncondensate and condensate fractions, respectively. **c** As the cooling progresses the noncondensate fraction is reduced until, at a frequency of 4.1 MHz, there remains an almost pure condensate.

Experiments include the observation of interference fringes between condensates, as a consequence of wave–particle duality. Other experiments produced manifestations of superfluidity and quantized vortices, and several other phenomena.

8.11 Quantum Coherence

In the ideal Bose gas theory, as the temperature is decreased, the number of particles in the condensate, that is, in the ground state, increases, and at 0 K, all the particles of the gas would be in the ground state. When the whole system is in a single quantum state, it is said to be in a *pure* state. The system can be described by a wave function. It is also said to be in a coherent state. Coherent states are quantum states that are closest to the classical states. They were introduced in 1960 by John Klauder (b. 1932). The technique of coherent multiphoton states, which are essential for the description of lasers, was developed in 1963 by Roy Glauber (1925–2018), who was awarded the Nobel Prize in 2005 for his contribution to the quantum theory of optical coherence.

When a system cannot be described by a single wave function, because our (limited) information about the system is compatible with several quantum states, and it is not possible to determine precisely in which of these states it is, the system is in a *mixed* state, and instead of the wave function one must use a function which is an average of the wave functions compatible with the macroscopic information one has about the system. Then a *density matrix* ρ is used to describe the system. Actually, all the quantum statistical systems discussed in Chap. 2 and in the present chapter are described by density matrices.

If we start from a condensate at 0 K, that is to say, from a pure state, and we increase the temperature slightly, the system acquires the possibility of being in a very large number of possible states: the system passes from a pure to a mixed state and there is a loss of *quantum coherence*.

It is not necessary for a system to be at zero temperature in order to be in a coherent state. Laser radiation provides an example of this. Because of the analogy between the coherent state of thousands or millions of atoms in the condensate and the coherent state of photons in the laser, one concludes that it is possible to make a laser of atoms, instead of photons.

8.12   Nonrelativistic Quantum Gases

We may express the general non-relativistic quantum gas density in terms of the particle mass and the temperature by means of an integral of a function of the particle energy E. Assume there is spin degeneracy with a factor g, and the volume of the system is V. The quantum gas is assumed to be composed of non-interacting particles of energy $E = p^2/2m$. Note that quantum properties become manifest in the ideal gas when $\lambda/d \gtrsim 1$, where λ is the thermal de Broglie wavelength, and d the average interparticle separation. Thus, the quantum properties become relevant, not only due to the increase in the term λ when the temperature decreases, but also due to the decrease in d when the density grows enough. Later, we shall discuss the lowest order temperature correction to the Fermi degenerate gas.

We start from the density of particles in momentum space for quantum gases $n(p) = 1/(e^{(E-\mu)/kT} \pm 1)$. The density of states is $gV4\pi p^2 dp/(2\pi)^3\hbar^3$. For $T \neq 0$, (8.15), (8.16) imply

$$N = \frac{4\pi V g}{(2\pi)^3\hbar^3} \int_0^\infty \frac{p^2 dp}{e^{(E-\mu)/kT} \pm 1},$$
(8.61)

where the $+$ sign corresponds to fermions and the $-$ sign to bosons. We now express the integral in terms of the energy E, noting that $p\,dp/m = dE$ and $p = \sqrt{2mE}$. We have

$$N = \frac{gVm^{3/2}}{2^{1/2}\pi^2\hbar^3} \int_0^\infty \frac{\sqrt{E}\,dE}{e^{(E-\mu)/kT} \pm 1}.$$
(8.62)

We shall obtain an expression for the total energy of the system, and its relation to the thermodynamic potential Ω. One can get an integral expression for the energy $U = \int_0^\infty E\,dN(E)$ as

$$U = \frac{gVm^{3/2}}{2^{1/2}\pi^2\hbar^3} \int_0^\infty \frac{E^{3/2}\,dE}{e^{(E-\mu)/kT} \pm 1},$$
(8.63)

and rewrite (8.62) in the form

$$N = \frac{gVm^{3/2}}{2^{1/2}\pi^2\hbar^3} \int_0^\infty \frac{e^{-(E-\mu)/kT}\sqrt{E}\,dE}{1 \pm e^{-(E-\mu)/kT}} = -\frac{\partial}{\partial\mu}\Omega,$$
(8.64)

where

$$\Omega = \mp \frac{gVm^{3/2}kT}{2^{1/2}\pi^2\hbar^3} \int_0^\infty \sqrt{E} \ln(1 \pm e^{-(E-\mu)/kT}) dE.$$

Then, integrating by parts, one gets

$$\Omega = -\frac{2}{3} \frac{gVm^{3/2}}{2^{1/2}\pi^2\hbar^3} \int_0^\infty \frac{E^{3/2}dE}{e^{(E-\mu)/kT} \pm 1}. \tag{8.65}$$

As $\Omega = -pV$, we have

$$pV = \frac{2}{3}U. \tag{8.66}$$

The Boltzmann approximation is valid when $e^{(E-\mu)/kT} \gg 1$. This holds in the limit $e^{\mu/kT} \ll 1$, which implies that $\mu < 0$. We shall use this to express the first order quantum corrections to thermodynamic quantities.

Let us consider the term

$$\frac{1}{e^{(E-\mu)/kT} \pm 1} = \frac{e^{-(E-\mu)/kT}}{1 \pm e^{-(E-\mu)/kT}}.$$

It can be written approximately as

$$\frac{1}{e^{(E-\mu)/kT} \pm 1} = \frac{e^{-(E-\mu)/kT}}{(1 \mp e^{-(E-\mu)/kT})} \approx e^{-(E-\mu)/kT} \left[1 \mp e^{-(E-\mu)/kT}\right]. \tag{8.67}$$

The term in square brackets on the right-hand side is obtained by going up to the term linear in $\epsilon (= e^{-(E-\mu)/kT})$, in the series expansion $1/(1 \pm \epsilon) = 1 \mp \epsilon + \epsilon^2 \cdots$. We obtain the Boltzmann approximation by taking $\epsilon/(1 + \epsilon) \sim \epsilon$, that is, by neglecting ϵ in the denominator:

$$N_B = \frac{g4\pi V}{(2\pi)^3\hbar^3} \int_0^\infty e^{-(E-\mu)/kT} p^2 dp. \tag{8.68}$$

Introducing the integration variable $x = p/2mkT$, this gives

$$N_B = \frac{gV}{2\pi^2} \frac{(2mkT)^{3/2}}{\hbar^3} e^{\mu/kT} \int_0^\infty e^{-x^2} x^2 dx.$$

Replacing the integral by $\sqrt{\pi}/4$ and taking into account the de Broglie thermal wavelength defined by $\lambda = h/\sqrt{2\pi mkT}$ (see Chap. 2), we have

$$N_B = \frac{gV}{8\pi^{3/2}} \frac{(2mkT)^{3/2}}{\hbar^3} e^{\mu/kT} = \frac{gV}{\lambda^3} e^{\mu/kT}. \tag{8.69}$$

The quantity $p(T) = \sqrt{2\pi mkT}$ is called the *thermal momentum*.

From (8.69), we have

$$\mu = kT \ln \frac{N_B \lambda^3}{gV} = kT \ln \frac{\lambda^3}{gd^3},$$

where $d^3 = V/N$ is the average volume per molecule. As g is of order unity, for μ to be significantly negative, we must have $d \gg \lambda$. This corresponds to the classical or Boltzmann limit. As pointed out earlier, quantum effects become significant when $\lambda \gtrsim d$.

By integrating N_B with respect to μ, and multiplying by -1 we get,

$$\Omega_B = -\frac{gV}{8\pi^{3/2}} \frac{(2mkT)^{3/2}}{\hbar^3} kT e^{\mu/kT} = -\frac{gV}{\lambda^3} kT e^{\mu/kT}.$$

Since $\Omega = -\frac{2}{3}U$, we get

$$U_B = -\frac{3}{2}\Omega_B = \frac{3gV}{2\lambda^3} kT e^{\mu/kT} = \frac{3}{2}kT N_B, \tag{8.70}$$

as expected in the Boltzmann limit of a quantum gas.

If we calculate the first quantum correction (see the literature), we obtain

$$\Omega = -pV = -kT \left(N_B \pm \frac{\lambda^3}{d^3} \frac{1}{2^{5/2}g} \right). \tag{8.71}$$

This is negligibly small if the average distance between particles $d \gg \lambda$. For constant N and V, since λ varies with the temperature as $T^{-1/2}$, when the temperature is low enough, λ may become $\sim d$, and the quantum properties will become manifest. Similarly, for very high densities, d decreases and quantum effects may become relevant even at very high temperatures, leading to macroscopic quantum properties, as happens in astrophysical objects like white dwarf stars. We observe that, for fermions, the pressure increases, and we may say that exchange effects lead to an effective repulsion among the particles, while for bosons, the pressure decreases, and quantum corrections lead to an effective attraction among the particles, when compared to the Boltzmann gas.

For the fermion case, we now consider the previous expression for the number of particles and calculate it in the zero temperature limit for electrons (degenerate electron gas). We shall also calculate the energy in terms of p_f. In this case $1/(e^{(E-\mu)/kT} + 1) \to \theta(\mu - E)$, that is, the distribution becomes a step function, such that for $\mu \geq E$, $\theta(\mu - E) = 1$ and for $\mu < E$, $\theta(\mu - E) = 0$, as discussed in Sect. 8.3. We take the degeneracy factor $g = 2$. Instead of (8.19), it is simpler to write the density of states in terms of the momentum as $g(p)dp = 8\pi V p^2 dp/h^3$. Call the maximum momentum p_f and recall that it must satisfy $p_f^2/2m = E_f = \mu$. The expression (8.62) becomes

$$N = \frac{8\pi V}{h^3} \int_0^{p_f} p^2 dp = \frac{8\pi V}{3h^3} p_f^3, \tag{8.72}$$

from which

$$p_f = (3/8\pi)^{1/3} h/d, \qquad E_f = (3/8\pi)^{2/3} h^2/2md^2. \tag{8.73}$$

The energy is given by

$$U_0 = \frac{8\pi V}{2mh^3} \int_0^{p_f} p^4 dp = \frac{8\pi V}{10mh^3} p_f^5 = -\frac{3}{2}\Omega. \tag{8.74}$$

Substituting in $p_f = (2mE_f)^{1/2}$, we obtain (8.21).

The first temperature dependent term is given by the expression (see the bibliography)

$$U = U_0 + \frac{\pi^2}{4} \frac{N(kT)^2}{E_f} = U_0 + \frac{\pi^2}{4} \frac{NkT^2}{T_f}, \tag{8.75}$$

where $T_f = E_f/k$ is the Fermi temperature. By dividing by V, one can write the specific heat as

$$C_V = \frac{\pi^2}{2} \frac{nkT}{T_f}, \tag{8.76}$$

where $n = N/V$ is the electron density. For metals, the Fermi temperature is usually of order 10^4 K and the electron density n is of order 10^{22} cm^{-3}. Equation (8.76) is a more exact version of (8.25). When applying this to practical problems in condensed matter physics, it should be remembered that the effective electron mass is different from the physical mass, and also that the Debye term contributes (nonlinearly) to the total specific heat.

Problems

Problem 8.1 Consider a photon gas in equilibrium at temperature T enclosed in a volume V. The photon is a massless particle, so its energy is $E = pc$ and its chemical potential $\mu = 0$. The average number of photons in the volume V depends on $(kT/\hbar c)^3$. Find the photon density in the early Universe, when $T = 10^{15}$ K.

Problem 8.2 When we integrate (4.52) over ν, we obtain the so-called Stefan–Boltzmann law (8.46), which gives the total energy density of the black body radiation at temperature T as $u = aT^4$, where $a = 4\sigma/c$ is called the radiation constant. The emitted power is the energy emitted per unit area and per unit time, namely, $E_b = \sigma T^4$, while the luminosity expresses the total radiated power, defined for a black body by the product

$$L = \sigma A T^4,$$

where A is the surface area of the body. Treating the Sun as a black body, calculate:
(i) the solar luminosity, assuming the temperature of the photosphere to be $T = 5700$

K and its surface area to be $A = 6.09 \times 10^{12}$ km^2; (ii) the mass equivalent of the energy radiated throughout its life, estimating the Sun's age as 4.6×10^9 years and assuming that it has maintained the present luminosity on the average right through its existence.

Problem 8.3 From the thermodynamic potential Ω, find the pressure of the photon gas in terms of the internal energy.

Problem 8.4 Starting from $U = aT^4$ and $p = U/3V$, according to the Big Bang theory, the radiation energy of the Universe was initially confined to a small region and expanded adiabatically in a spherically symmetric manner. The radiation would have cooled down as it expanded. Find the relation between the temperature T and the radius R of the spherical volume of radiation, on the basis of thermodynamic considerations.

Problem 8.5 Find the total entropy of a photon gas as a function of its temperature T, volume V, and the constants σ, a. Recall that $dS = \frac{dU}{T} + \frac{p}{T}dV$.

Literature

1. R.K. Pathria, *Statistical Mechanics*, 2nd edn. (Elsevier, Oxford, 2006). This book deals with quantum statistics at an advanced level
2. K. Huang, *Statistical Mechanics* (Wiley, New York, 1963). Quantum statistics is very clearly discussed in this book
3. C. Kittel, *Introduction to Solid State Physics*, 7th edn. (Wiley, New York, 1996). A basic and well-known treatise on solid state physics
4. J.K. Jain, *Composite Fermions* (Cambridge University Press, Cambridge, 2007). Excellent introduction to the novel topic of composite fermions
5. L.D. Landau, E.M. Lifshitz, *Statistical Physics* (Pergamon, Oxford, 1975). The discussion of the lowest order quantum corrections to the ideal gas follows the treatment of the problem in this excellent textbook

Chapter 9
Four Fundamental Forces

At present we know of four types of forces as basic interactions in Nature. The strongest of them is the *nuclear force*, attracting protons and neutrons inside the atomic nucleus, although its range is limited to distances of the order of the diameter of the atomic nucleus, i.e., 10^{-13} cm. After this, the next strongest is the *electromagnetic force*, which is exerted between electrically charged particles, and in particular which attracts protons and electrons to form the atom. Then follows the so-called *weak force*, mediating the beta decay of nuclei. This is also a short-range force. As a consequence of beta decay, electrons and neutrinos are produced. Finally, the weakest is the *gravitational force*. Like the electromagnetic force, this has long range.

All other interactions observed in Nature can be reduced to these four forces. For instance, molecular forces are a consequence of electromagnetic interactions. It must be pointed out, however, that purely quantum effects, such as Pauli's principle, lead to effects close to the idea of forces. The exchange interaction is a good example for the case of fermions. For bosons, Bose–Einstein condensation is a representative example.

Atmospheric and oceanic pressure are both determined by gravity. If one presses down on the table with the hand, the forces intervening in the process, at the level of atoms and molecules, are electromagnetic forces combined with quantum effects.

In what follows, we shall examine some characteristics of these four forces in more detail.

9.1 Gravity and Electromagnetism

We have already dealt earlier with two fundamental forces: the gravitational and the electromagnetic forces. They have some similar properties, for instance, the force between two electric charges depends on the inverse square of the distance, and gravity obeys a similar law.

On the other hand, electromagnetic interactions can propagate to large distances, the intensity of the force decreasing with the square of the distance. Gravitational

M. Chaichian et al., *Basic Concepts in Physics*, Undergraduate Lecture Notes in Physics, https://doi.org/10.1007/978-3-662-62313-8_9

interactions behave similarly from this point of view. For these reasons, gravity and electromagnetism are said to be long-range forces. From the quantum field theoretical perspective, this is reflected in the fact that the quanta of the electromagnetic field (photons) and the gravitational field (gravitons) are massless particles, and hence propagate at the velocity of light. (Although gravitons have not yet been observed, it is believed that they do exist.) From the mathematical point of view, gravity and electromagnetism are both described by theories with the property of gauge invariance (see Chaps. 5 and 7).

On the other hand, there exist essential differences between electromagnetism and gravity. The existence of electric charges of opposite signs leads to the screening of the electrostatic force. An electric charge always tends to be screened by attracting opposite charges. This occurs in atoms, in molecules, and in any macroscopic body. However, the gravitational *charges* cannot be screened since the force is only attractive, whence the gravitational interaction plays a dominant role at cosmic scale. It determines the motion of the planets around the Sun. The Sun with its planetary system, like other stars in our galaxy, are coupled by gravitation in a complex rotational motion around a common centre. And galaxies, in turn, interact with one another through gravity over enormous distances, forming clusters of galaxies.

Although electrostatic interactions are screened at the cosmic scale, another form of the long range electromagnetic field is manifested in planets and stars, namely, magnetic forces. Magnetic fields are believed to be produced mostly by rotational motions of electric charges inside those bodies. These magnetic fields may become very strong in objects like pulsars, which are neutron stars rotating at high frequency. It is estimated that fields of the order of 10^{14} gauss and even higher are generated in them.

9.2 Atomic Nuclei and Nuclear Phenomena

Heavy nuclei may be treated as classical systems, since they contain hundreds of nucleons. For instance, in the liquid-drop model, the nucleus energy is considered as arising both from surface tension and from electrical repulsion of the protons. The liquid-drop model can reproduce many features of nuclei, including the binding energy and the phenomenon of nuclear fission.

The nuclear shell model was proposed first by Dmitry Ivanenko (1904–1994) in 1932, and was developed in 1949 mainly by Eugene Wigner, Maria Goeppert Mayer and J. Hans D. Jensen (1907–1973), who were awarded the 1963 Nobel Prize in Physics. It includes quantum mechanical effects to some extent analogous to the electron shells in the atom structure. One interesting consequence is that nuclei with certain numbers of neutrons and protons (the magic numbers (2, 8, 20, 28, 50, 82, 126,...) are particularly stable, because their shells are filled.

Nuclear Decay and Radioactivity. Among the known elements, eighty have at least one stable isotope never observed to decay, giving a total of about 254 known stable isotopes. But thousands of isotopes have been characterized as unstable. Radioisotopes decay over diverse time scales, ranging from fractions of a second to billions of years. The phenomenon of radioactivity was discovered in 1896 by the French physicist Henri Becquerel (1852–1908), while working with phosphorescent materials. Fundamental contributions were also made by Marie Skłodowska Curie (1987–1934), a Polish and naturalized-French physicist and chemist who conducted pioneering research on radioactivity. She was the first woman to win a Nobel Prize (in Physics, shared with Pierre Curie and Henri Becquerel in 1903), and the only person to win a Nobel Prize in two different sciences (in 1911 she was awarded the Nobel Prize for Chemistry, among other reasons, for her discovery of two new elements, radium and polonium).

Nuclei are more stable when there is a balance in the numbers of neutrons and protons. Too few or too many neutrons may cause a nucleus to decay. For instance, a nitrogen-16 atom, whose nucleus contains 7 protons and 9 neutrons, will decay in a few seconds to an oxygen-16 atom, with 8 protons and 8 neutrons. This is due to a weak interaction process, namely beta decay of a neutron inside the nitrogen nucleus, producing a proton, an electron, and an antineutrino.

Alpha decay is characterized by the emission of a helium nucleus, which is called an α particle (containing 2 protons and 2 neutrons), to give another element. In many cases this process continues in several steps, including other types of decays, until a stable element is formed.

In gamma decay, the nucleus decays from an excited state into a lower energy state by emitting a gamma ray, and no nuclear transmutation is involved.

Nuclear fusion. Nuclear fusion is produced when two low mass nuclei come into very close contact with each other, whereupon the strong force acts and fuses them together. A large amount of energy is required to overcome the repulsion between the nuclei, so that the strong force can produce this effect. For this reason, nuclear fusion takes place at very high temperatures or high pressures. If the fusion process succeeds, a very large amount of energy is released and the combined nucleus assumes a lower energy level. The binding energy per nucleon increases with mass number A up to nickel-62. The power of stars like the Sun comes from a nuclear reaction consisting in the fusion of four protons into a helium nucleus, two positrons, and two neutrinos. Natural nuclear fusion is the origin of the light and energy produced by the core of all stars including the Sun. Various laboratories around the world are working on the development of an economically viable method for using energy produced in a controlled fusion reaction.

Nuclear fission. The reverse process of nuclear fusion is nuclear fission. The binding energy per nucleon decreases with the mass number for nuclei beyond nickel-62. It is possible to release energy if a nucleus breaks apart into two lighter ones. The process of α decay is actually a sort of spontaneous nuclear fission, which is highly asymmetrical because the four particles composing the α particle are tightly bound to each other.

For some heavy nuclei which produce neutrons upon undergoing fission, and which also easily absorb neutrons to initiate fission, a self-igniting type of neutron-initiated fission can be obtained. This is known as a chain reaction. The fission or nuclear chain reaction, using fission-produced neutrons, is the source of energy for nuclear power plants. For a neutron-initiated chain reaction to occur, there must be a critical mass of the element present in a certain volume and under certain conditions.

9.3 Strong Interactions

The strongest known force in Nature is exerted between the constituent particles of the atomic nucleus: protons and neutrons. In order to get a comparative idea, it is enough to point out that the attraction between two protons due to that force is 100 to 1000 times stronger than their electrostatic repulsion, while the latter is greater than the gravitational attraction by a factor of order of 10^{37}. But the nuclear force has the feature of being short range: its action is limited to distances of the same order as the dimensions of the atomic nucleus, that is 10^{-13} cm.

In 1935, the Japanese physicist Hideki Yukawa (1907–1981) predicted the existence of mesons as the particles mediating the strong interactions, by analogy with photons mediating in the electromagnetic interactions. Since the strong interactions are short range, the mesons should have nonzero mass (differing in this respect from the photon), and this mass should be 200–300 times the electron mass. The counterpart to the electrostatic potential is then the Yukawa potential:

$$\Phi(r) = \frac{g}{r} e^{-r/r_0}, \tag{9.1}$$

where g is a constant characteristic of the strong interactions. The expression (9.1) is significant for r smaller or near r_0, where r_0 is of the order 10^{-13} cm (this is a unit called fermi or femtometer, denoted by fm). One can also write (9.1) in the form

$$\Phi(r) = \frac{g}{r} e^{-mr}, \tag{9.2}$$

where $m \sim 1/r_0$ is proportional to the mass of the π mesons, in natural units. The Fourier transform of (9.2) is

$$\tilde{\Phi}(|\mathbf{k}|) = \frac{g}{\mathbf{k}^2 + m^2}. \tag{9.3}$$

The interesting thing about (9.3) is that, if it is compared with (4.44) in Chap. 4, it can be seen that what makes the interaction short range is the mass of the mediator, i.e., if the bosons mediating in the interactions have nonzero mass m, the resulting forces are short range, and the damping factor of the interaction is proportional to e^{-mr}. As the photon has zero rest mass, the Coulomb force is long range.

The Yukawa mesons were discovered in 1947 in cosmic rays experiments and called π mesons, or pions. There are three types of pions: positive (π^+), negative (π^-), and neutral (π^0). Their masses are approximately

$$m_{\pi^\pm} = 273 \ m_e,$$
$$m_{\pi^0} = 264 \ m_e, \tag{9.4}$$

where m_e is the electron mass.

The π^0 pions are not stable particles and decay in different ways, e.g., into two photons:

$$\pi^0 \rightarrow 2\gamma, \tag{9.5}$$

in a time of order 10^{-16} s. The $\pi^\pm$ pions have a longer lifetime, of order 10^{-8} s and decay in several ways.

Let e^- and e^+ be the electron and positron, respectively, let ν_e and $\bar{\nu}_e$ be the neutrino and antineutrino associated with the electron, let μ^+ and μ^- be the particles known as muons (or heavy electrons, first called μ mesons), with mass approximately 207 times the electron mass, and let ν_μ and $\bar{\nu}_\mu$ be the neutrino and antineutrino, respectively, associated with the muon. Then the pions $\pi^\pm$ decay weakly, with the highest probability in the following ways:

$$\pi^+ \rightarrow \mu^+ + \nu_\mu,$$
$$\pi^- \rightarrow \mu^- + \bar{\nu}_\mu,$$

but also to

$$\pi^+ \rightarrow e^+ + \nu_e, \tag{9.6}$$
$$\pi^- \rightarrow e^- + \bar{\nu}_e,$$

or, with extremely small probability, to

$$\pi^+ \rightarrow \pi^0 + e^+ + \nu_e,$$
$$\pi^- \rightarrow \pi^0 + e^- + \bar{\nu}_e.$$

Other decay modes have also been observed. From the theoretical point of view the pions are described by pseudoscalar fields. They behave as scalars under all Lorentz transformations, except the inversion of the space coordinates (parity transformation), under which they change sign. In this respect, they differ radically from the photons, whose field operator A_μ is a four-vector. The photons have spin unity, and the π mesons have spin 0. Since they have integer spin, they obey the Bose–Einstein statistics.

The mechanism of attraction between protons and neutrons inside the nuclei was explained by an exchange of pions, i.e., of π^0 between nucleons of identical charge, and of π^0 and $\pi^\pm$ for interactions between nucleons of different charge. Neutrons and protons can be understood as two *isotopic spin* states of the same particle, called *nucleon*. The neutron absorbs a π^+ or emits a π^- and becomes a proton. The proton absorbs a π^- or emits a π^+ and becomes a neutron. (To be precise, these processes happen when the exchanged particle is virtual, otherwise energy and momentum would not be conserved.) As mentioned earlier, for some time it was believed that the pions were the quanta of the strong interactions, and the existence of three types with charges $(\pm, 0)$ is necessary in order to achieve the different possible interactions between protons and neutrons.

Neutrons and protons, together with other heavy particles called hyperons form a set of particles known as baryons. The hyperons are characterized by having a higher mass than the proton, and a lifetime of the order 10^{-10} s, specific of decay by weak interactions.

Strong forces are known at present to be determined by the interaction between quarks, elementary particles of fractional charge $\pm 2e/3$, $\pm e/3$, from which the hadrons are composed. Quarks interact through the *gluon field*, which is the carrier of the fundamental strong force, and all previous models are modified (this is discussed in detail in Chap. 11). The nuclear force is actually a residual force, whose relationship with the gluon field should be understood as similar to the relation between molecular forces and the interatomic Coulomb force.

A neutron outside the nucleus is unstable, and its mean lifetime is approximately 12 minutes. Protons are assumed to be stable, although some Grand Unification Theories, for instance, the one based on the $SU(5)$ gauge symmetry group, predict a lifetime of the order 10^{33} years for the proton. This hypothesis remains unconfirmed, but if it were true it would violate the principle of baryon number conservation, which is assumed to be valid in any interaction process of fundamental particles.

9.4   Weak Interactions

Some atomic nuclei decay by emitting an electron and an antineutrino, causing the nucleus to increase its atomic number Z by one unit. Since the electrons emerging from these decays were originally called beta particles, this process emitting an electron and an antineutrino is called beta decay. The beta decay of carbon 14 gives a nitrogen nucleus, an electron and an antineutrino:

$$\;^{14}_{6}\text{C} \rightarrow \;^{14}_{7}\text{N} + e^- + \bar{\nu}_e. \tag{9.7}$$

A free neutron decays similarly into a proton, an electron, and an antineutrino:

$$n \rightarrow p + e^- + \bar{\nu}_e. \tag{9.8}$$

An antineutron $\bar{n}$ decays into an antiproton $\bar{p}$, a positron e^+, and a neutrino ν_e:

$$\bar{n} \rightarrow \bar{p} + e^+ + \nu_e. \tag{9.9}$$

The weak interactions are responsible for the decay of many other particles, such as the pions, as can be seen in (9.6), and other mesons, as well as all the hyperons.

Weak interactions are short range. The characteristic length is of the order 10^{-15}–10^{-16} cm. An important consequence is that the probability of interaction of neutrinos with other particles is very small at the usual energies encountered on Earth. For this reason, we call them weak interactions. The weak force is stronger than gravity, but the force of gravity acts more strongly at great distances because it is long-range and proportional to the masses of the interacting bodies.

The weak interactions have the fundamental property of not conserving parity. Let us discuss this now.

9.5 Parity Non-Conservation in Beta Decay

Until 1956, it was believed that parity (P) conservation was a fundamental property of the atomic world. If a process exists, its mirror image should also exist. In that year, the Chinese–American physicists Chen Ning Yang (b.1922) and Tsung-Dao Lee (b.1926) realized that this property was not in fact valid for weak interactions. Their hypothesis was based on a comparative study of the decay of K^+ mesons in two final states of different parity. In 1957, the non-conservation of parity was confirmed by Chien-Shiung Wu (1912–1997) with collaborators, in an experimental study of the beta decay of spin-polarized nuclei of cobalt 60. Figure 9.1 illustrates the process of beta decay of such a nucleus. The nucleus emits an electron L (where L means left-handed) with negative helicity (or left chirality), i.e., the direction of spin rotation and the direction of its momentum are as in a left-hand screw. The electron L is *always* emitted in the direction opposite to the spin polarization of the nucleus. The nucleus also emits an antineutrino $\bar{\nu}$ with positive helicity, i.e., the direction of the spin rotation and the momentum combines as in a right-hand screw. The antineutrino is *always* emitted in the direction of the spin of the decaying nucleus. The fact that in the experiment of C.S. Wu the electrons coming out of the decay of the polarized cobalt 60 nucleus fly only in one direction and never in the opposite direction was suggested as parity violation by Lee and Yang. At the very same time, the parity violation was checked also by Leon Lederman, together with Richard Garwin and Marcel Weinrich, in an experiment of pion and muon decay. In 1957, the Nobel Prize in Physics was awarded to Lee and Yang for the hypothesis of parity violation in weak interactions.

Thus, the mirror image of beta decay does not occur in Nature. However, if besides taking the mirror image, the charge is conjugated, that is, the nucleus is replaced by an antinucleus, the electron by a positron, and the antineutrino by a neutrino, the resulting process does occur in Nature. That is, parity P alone is not conserved, but the symmetry is recovered if parity P and charge conjugation C are combined into a CP conjugation operation. Thus individually, P and C fail to be symmetries, but the combined symmetry CP remains valid. Comparing (9.8) and (9.9), it can be seen that the decay of a neutron and of an antineutron are related by a CP transformation (see also Fig. 9.1).

Neutrinos, as massless particles, violate parity maximally. Strictly speaking (see Chap. 11), recent results suggest that neutrinos actually have a very tiny mass, so the

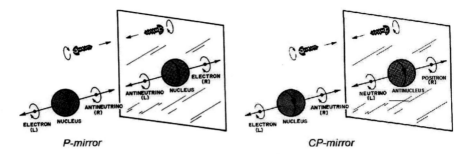

P-mirror   **CP-mirror**

Fig. 9.1 Parity non-conservation. A nucleus emits an electron L of negative helicity (as in a *left-handed screw*) and an antineutrino R of positive helicity (as in a *right-handed screw*), so beta decay distinguishes right from left. The mirror image (under P inversion) would give an electron R and an antineutrino L. This process has not been observed in Nature. In a hypothetical CP mirror, in which the images of particles are antiparticles, the antinucleus emits a positron *R* and a neutrino L, and this process would be observable (the mirror image of a screw is an antiscrew, made by antimatter).

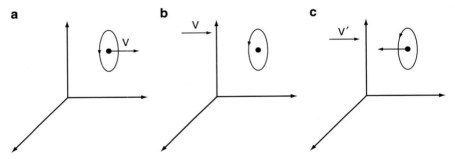

Fig. 9.2 (**a**) Electron with velocity V and helicity R in some reference system. (**b**) Electron at rest in a system moving with velocity V toward the right with respect to the previous one. (**c**) Electron with negative helicity in a system moving with velocity $V' > V$ toward the right.

following discussion, based on a massless neutrino is to be taken as approximate. In the zero mass limit, one can be sure that the antineutrino has positive helicity R in any frame of reference (R means right-handed), and that the neutrino has negative helicity L.

For a massive particle like the electron, the notion of helicity is not relativistically invariant, since if the electron moves with the velocity V and has positive helicity (R), by choosing a system of reference moving at the velocity V in the same direction as the electron, the particle will be seen to be at rest. If now the velocity of the system of reference is increased to a value $V' > V$ in the same direction, the relative velocity of the electron will be seen as negative, and its helicity also negative (L) (Fig. 9.2). However, this cannot be done with a massless fermion, which would move at the speed of light, whence there would be no reference system in which it were at rest, and as a consequence, no reference system in which its momentum and helicity were inverted. A new kind of symmetry would then arise, viz., *chiral* invariance, associated with such massless fermions, manifested by their always having the same helicity. If neutrinos (and antineutrinos) have a very small mass, they will not travel at the velocity of light, but at velocities very close to it, and the conservation of helicity will only hold approximately. It should be emphasized, however, that all weak interaction processes are parity violating, whether they involve or not neutrinos. We shall return to this issue in Chap. 11.

9.6 Violation of *CP* and *T* Invariance

Charge conjugation transformation *C* and space inversion, or parity, transformation *P* are combined in the *CP* transformation. By applying the *CP* operator to a particle with a certain "handedness," we obtain the antiparticle with the opposite handedness (for example, a left-handed electron by *CP* transformation goes into a right-handed positron). Symmetry under *CP* transformation means, actually, symmetry between matter and antimatter. In 1980, the Nobel Prize in Physics was awarded to Val Fitch (1923–2015) and James W. Cronin (1931–2016), for their outstanding discovery made in 1964: the violation of *CP* invariance in certain decays.

With regard to the weak interaction, the neutral *K* mesons have some special features. Both the electric charge and the baryonic charge of K^0 and $\bar{K}^0$ are zero. They only differ in strangeness S ($S = +1$ for K^0 and $S = -1$ for $\bar{K}^0$). Since the strangeness is not conserved in weak interactions (see Chap. 11 for details), K^0 and $\bar{K}^0$ are identical with respect to them. In particular, they can convert into each other.

All particles can be classified into two groups. One group comprises particles (e.g., proton, electron, hyperons, etc.) which differ from their antiparticles by some strictly conserved quantum numbers (e.g., electric charge, baryonic charge, etc.). The second group comprises particles identical with their antiparticles or truly neutral particles (e.g., photon or π^0 meson). Neutral K^0 mesons are on the border between these two groups: K^0 and $\bar{K}^0$ differ in strangeness, but this difference is relevant for strong interaction and irrelevant for weak interaction. While K^0 and $\bar{K}^0$ have a definite strangeness, they do not have a definite *CP* parity – under *CP* transformations, K^0 transforms into $\bar{K}^0$ and vice versa:

$$CP|K^0\rangle = |\bar{K}^0\rangle, \quad CP|\bar{K}^0\rangle = |K^0\rangle. \tag{9.10}$$

A definite *CP* parity can be attributed to the quantum mechanical linear combination of K^0 and $\bar{K}^0$:

$$|K_1^0\rangle = \frac{1}{\sqrt{2}}(|K^0\rangle + |\bar{K}^0\rangle), \quad |K_2^0\rangle = \frac{1}{\sqrt{2}}(|K^0\rangle - |\bar{K}^0\rangle). \tag{9.11}$$

Under *CP* transformation, K^0 converts into $\bar{K}^0$ and $\bar{K}^0$ converts into K^0, so that the state $|K_1^0\rangle$ will transform into itself, while $|K_2^0\rangle$ will reverse its sign:

$$CP|K_1^0\rangle = |K_1^0\rangle, \quad CP|K_2^0\rangle = -|K_2^0\rangle. \tag{9.12}$$

Hence, K_1^0 has positive and K_2^0 has negative CP parity, but neither K_1^0 nor K_2^0 possess a definite value of strangeness.

If the CP invariance holds, the K_1^0 meson may only decay into two π mesons, whose total CP parity is positive:

$$K_1^0 \begin{smallmatrix} \nearrow \pi^+ + \pi^- \\ \searrow \pi^0 + \pi^0 \end{smallmatrix} ,$$

while K_2^0 may only decay into three π mesons, their total CP parity being negative, as the π mesons are pseudoscalars:

$$K_2^0 \begin{smallmatrix} \nearrow \pi^+ + \pi^- + \pi^0 \\ \searrow \pi^0 + \pi^0 + \pi^0 \end{smallmatrix} .$$

Both kinds of decay have been observed. The lifetime of the K_1^0 meson (0.86×10^{-10} s) is shorter than that of the K_2^0 meson (5×10^{-8} s). For this reason, K_1^0 and K_2^0 are also referred to as the *short-lived* K_S^0 meson and the *long-lived* K_L^0 meson, respectively.

In the absence of CP invariance, the K_L^0 meson can also decay into two π mesons and the K_S^0 meson into three π mesons:

$$K_L^0 \begin{smallmatrix} \nearrow \pi^+ + \pi^- \\ \searrow \pi^0 + \pi^0 \end{smallmatrix} , \quad K_S^0 \begin{smallmatrix} \nearrow \pi^+ + \pi^- + \pi^0 \\ \searrow \pi^0 + \pi^0 + \pi^0 \end{smallmatrix} . \tag{9.13}$$

The experiment of Cronin and Fitch consisted of injecting a pure K^0 beam into a vacuum tube of 15 m length. K_1^0 should have decayed within 6 cm of length, such that at the end of the tube only the long-lived K_2^0 should have been detected, by its decay to three pions. However, surprisingly, also decays to two pions were observed at the end of the tube, signaling the presence of K_1^0. Thus, the mass eigenstates $|K_S^0\rangle$ and $|K_L^0\rangle$, with definite lifetime, cannot be identified with the CP eigenstates $|K_1^0\rangle$ and $|K_2^0\rangle$, respectively, but they represent rather a superposition of the two:

$$|K_S^0\rangle = \frac{|K_1^0\rangle + \epsilon|K_2^0\rangle}{\sqrt{1 + \epsilon^2}}, \tag{9.14}$$

$$|K_L^0\rangle = \frac{|K_2^0\rangle - \epsilon|K_2^0\rangle}{\sqrt{1 + \epsilon^2}}. \tag{9.15}$$

Here, ϵ is a complex parameter which encodes the magnitude of the CP violation.

The effect of CP violation is very small: for example, only 0.20% of all K_L^0 mesons do decay into $\pi^+ + \pi^-$; the ratio of the decay probabilities of $K_L^0 \to \pi^+ + \pi^-$ and $K_S^0 \to \pi^+ + \pi^-$ amounts to only $(3.69 \pm 0.15) \times 10^{-6}$.

In case of P invariance, a particle can be figuratively represented as a "nail"; in case of CP invariance the particle can be imagined as a "screw" with a certain direction of screw thread, whose length is the same for the particle and the antiparticle, due to the fact that only a relative difference exists between them under *CP* invariance. If *CP* is violated, different "screws" have to be attributed to a particle and its antiparticle (both the sense of the screw thread and its length being different), to represent an absolute difference between the particles and the antiparticles. As a consequence, the probabilities of, e.g., the following lepton decays of K^0_L mesons with the production of particles or antiparticles will be different:

$$K^0_L \begin{smallmatrix} \nearrow \pi^+ + e^- + \bar{\nu}_e \\ \searrow \pi^- + e^+ + \nu_e \end{smallmatrix}, \quad K^0_L \begin{smallmatrix} \nearrow \pi^+ + \mu^- + \bar{\nu}_\mu \\ \searrow \pi^- + \mu^+ + \nu_\mu \end{smallmatrix}. \tag{9.16}$$

That is to say, in the case of *CP* non-invariance the probabilities of creating particles with opposite charge (e.g., e^- or e^+, μ^- or μ^+) will be different (charge asymmetry). Charge asymmetry in the mentioned leptonic decays of K^0_L mesons has been experimentally observed. In 1967, Andrei Sakharov (1921–1989) used the *CP* violation argument to justify the matter–antimatter asymmetry in the Universe (we shall return to this point in Chap. 11).

   While *CP* symmetry is violated in these processes, it turns out that the time-reversal symmetry, *T*, is also violated, such that the combined *CPT* symmetry is preserved. Actually, *CPT* is a symmetry of every known process. All the experimental searches for *CPT* violation have failed to date, but the quest continues. The *CPT* symmetry was proved theoretically in 1954 to be intrinsic to relativistic quantum field theories by Gerhart Lüders (1920–1995) and, independently, by Wolfgang Pauli; the proof of the *CPT* theorem in axiomatic quantum field theory was given in 1957 by Res Jost (1918–1990).

9.7   Some Significant Numbers

The Compton wavelength of a particle of mass m is $\lambda_c = \frac{h}{mc}$. Relativistic quantum behaviour, described by the theory of quantized fields as creation and annihilation of particles and antiparticles, manifests itself at distances of this order. Mass scales in quantum physics are customarily represented by the reduced Compton wavelength, defined as $\lambdabar_c = \frac{\hbar}{mc}$. The reason is simple, if we look at it in natural units ($\hbar = c = 1$): $\lambdabar_c = 1/m$. For an electron the reduced Compton wavelength is:

$$\lambdabar^e_c = \frac{\hbar}{m_e c} = 3.86 \times 10^{-11} \text{ cm.} \tag{9.17}$$

This is a number 100 times smaller than the radius of the smallest atom, which is of the order 10^{-9} cm.

For the proton, we have

$$\chi_c^p = \frac{\hbar}{m_p c} = 0.21 \times 10^{-13} \text{ cm},\tag{9.18}$$

which is of the order of the nuclear size ~ 1 fm. For pions we have

$$\chi_c^\pi = \frac{\hbar}{m_\pi c} = 1.41 \times 10^{-13} \text{ cm},\tag{9.19}$$

a number giving the range of the nuclear force.

The weak interactions have a smaller range, roughly of the order 10^{-15} cm. Let us estimate the masses of the particles mediating that interaction:

$$m_W = \frac{\hbar}{c \times 10^{-15} \text{ cm}} \approx 10^{-22} \text{ g},\tag{9.20}$$

which is about 100 times larger than the proton mass. The subscript W stands for the W vector bosons, which mediate the beta decay. We conclude that the smaller the range of an interaction, the greater the mass of the particles mediating it. Conversely, the greater the range, the smaller the mass of the particles mediating the interaction. Since the electromagnetic and gravitational forces have long (actually, infinite) range, it can be deduced from the previous relations that the mass of the mediating particles (the photon and the graviton) should be zero.

The corresponding energies associated with these masses are frequently given in electron volts (eV) or their multiples ($\text{MeV} = 10^6$ eV and $\text{GeV} = 10^9$ eV). We have

$$1 \text{ eV} = 1.6 \times 10^{-12} \text{ erg}.\tag{9.21}$$

The mass of the proton in natural units (previously multiplied by c^2) would be

$$m_p = 9.38 \times 10^2 \text{ MeV} \approx 1\text{GeV},\tag{9.22}$$

and for the W bosons mediating the weak interactions:

$$m_W \approx 80\text{GeV}.\tag{9.23}$$

The so-called classical radius of the electron corresponds to the distance at which the electrostatic energy is of the same order as the rest energy of the electron:

$$mc^2 = e^2/r,\tag{9.24}$$

leading to

$$r = e^2/mc^2 \approx 2.82 \times 10^{-13} \text{cm}. \tag{9.25}$$

Once again we obtain distances characterizing nuclear dimensions. This suggests that new phenomena arise at distances of that order.

9.8 Death of Stars

To conclude this chapter we discuss a cosmic phenomenon in which the four fundamental forces of Nature and Pauli's principle (through the so-called *fermion degeneracy pressure*) come together, namely, the death of stars. It should be mentioned, however, that after undergoing the final stages of stellar evolution, the remnants of stars are still very active objects.

Inside stars, matter is not organized in the form of atoms and molecules: the atoms are completely ionized, and the nuclei and electrons move independently, forming a plasma at temperatures of order $10^7 - 10^9$ K. At such temperatures, thermonuclear reactions can occur. The strong, weak, and electromagnetic interactions make the nucleosynthesis of helium (and some heavier elements) possible from hydrogen.

At present the generally accepted hypothesis is that the mass of a star plays an important role in its evolution. When a star with mass less than 1.44 times the solar mass has transformed all its hydrogen reserves to helium, it becomes a white dwarf, according to the well-known appellation. At a certain moment, a contraction process will begin in a white dwarf, due to the gravitational force. This contraction is balanced only by the pressure exerted by the electron gas, which is a purely quantum effect, due to Pauli's principle. Owing to its high temperature (near 10^7 K), the gas also has relativistic behaviour, with some electrons moving at velocities near c. Although there is a contribution to the pressure from the nuclei, the main contribution comes from the extremely dense electron gas (near 10^{30} cm^{-3}). The brightness of such stars is due to the release of gravitational energy in the slow contraction process.

Only stars of mass smaller than 1.44 solar masses can become white dwarfs. Higher mass stars collapse. Let us outline a demonstration based on elementary considerations. Assume that we have N electrons in a star of radius R. As a consequence, its density is $n \sim N/R^3$, and the volume per fermion is $1/n$. The uncertainty principle implies for the electron $p \sim \hbar n^{1/3}$.

The repulsive Fermi energy for relativistic electrons can be then immediately written as:

$$E_F = \frac{\hbar c N^{1/3}}{R}. \tag{9.26}$$

The total attractive gravitational energy is given practically by the interaction of the baryons (since they are much heavier than the electrons)

$$E_G = -\frac{GM^2}{R}, \tag{9.27}$$

with $M = Nm_B$, where m_B is the baryon mass, the baryons being protons and neutrons, and it is assumed that the number of nucleons is approximately equal to the number of electrons. The condition of zero energy is $E = NE_F + E_G = 0$, and from this we obtain

$$N_{max} = \left(\frac{\hbar c}{Gm_B^2}\right)^{3/2} \sim 2 \times 10^{57}. \tag{9.28}$$

Equation (9.28) (which can be written as $N_{max} \sim (m_P/m_B)^3$, where $m_P = \sqrt{\hbar c/G} \sim 10^{-5}$ g is the Planck mass, see Chap. 10) implies $M_{max} \sim 1.85 M_\odot$, where $M_\odot \sim 1.98 \times 10^{33}$g is the solar mass. We observe that, except for some numerical constants dependent on the star composition, N_{max} and M_{max} depend on fundamental physical constants. This value for M_{max} is a rough estimate of the quantity $1.44 M_\odot$, known as the Chandrasekhar limit, in honour of Subrahmanyan Chandrasekhar (1910–1995) who found it first and was awarded the Nobel Prize in Physics in 1983 for "for his theoretical studies of the physical processes of importance to the structure and evolution of the stars".

This hand-waving argument can be made more precise as follows. Let us denote by P_0 the pressure of the degenerate electron gas. The work done in order to change the volume by a small amount dV is $dW_0 = -P_0 dV$ (where $dV = 4\pi R^2 dR$). At equilibrium, this must be equal to the work done by the gravitational field due to the small change dR of the star radius, $dW_0 = \eta(GM^2/R^2)dR$, where η, of order unity, depends on the density inside the star. By equating the two expressions for dW_0, one obtains an equation from which two important limits can be calculated. By defining the relative Fermi momentum as $x = p_F/mc$, these limits are the non-relativistic one, for $x \ll 1$ and the ultra-relativistic, for $x \gg 1$. Recalling that $\lambdabar_c = \hbar/mc$ is the reduced Compton wavelength of the electron, in the non-relativistic limit one obtains the mass–radius equation

$$RM^{1/3} \simeq \frac{3(9\pi)^{2/3}}{40\eta} \frac{\lambdabar_c m_P^2}{m_B^{2/3}}, \tag{9.29}$$

i.e., the white dwarf mass decreases with its radius as $M = \text{const.} \times R^{-3}$. In the ultra-relativistic case, we obtain another mass–radius equation for the star equilibrium:

$$R \simeq \frac{(9\pi)^{1/3}}{2}\left(\frac{M}{m_B}\right)^{1/3}\lambdabar_c\left[1 - \left(\frac{M}{M_0}\right)^{2/3}\right]^{1/2}, \quad \text{where} \quad M_0 = \frac{9}{64}\left(\frac{3\pi}{\eta^3}\right)\frac{m_P^3}{m_B^2}. \tag{9.30}$$

We see that in these two expressions the electron Compton wavelength, the baryon mass, and the Planck mass play a fundamental role. From (9.30), it is seen that a white dwarf star in equilibrium must have a mass $M < M_0$, where $M_0 \sim 10^{33}$ g, a result fully upheld by observation. Thus, from (9.29) and (9.30) we conclude that in both cases the white dwarf radius decreases with its mass, but in the latter, for $M \geq M_0$, R would vanish or become an imaginary number: this implies that the star is not stable and would explode.

A white dwarf becomes a hot carbon ball, and it may increase its mass by accretion, for instance of hydrogen, from a neighbouring star. The fusion process continues and the star may explode in a type Ia supernova. These supernovas have brightness greater than the whole galaxy, and they are so similar in their characteristics that they have been taken as 'standard candles' to estimate intergalactic distances. Recently, it has been argued that the accretion mechanism may be able to produce 5% of the observed type Ia supernovas in some galaxies. It has also been suggested that collisions may be a significant mechanism for producing supernovas. Since colliding white dwarfs could have a range of masses, this in turn would weaken arguments for using exploding white dwarfs as standard candles for determining the nature of the Universe.

A neutron star may be the final result after the gravitational collapse of a massive star suffering a type II, type Ib, or type Ic supernova event.

A typical neutron star has a mass between 1.35 and about 2.1 solar masses. In general, compact stars of less than 1.44 solar masses are white dwarfs. Above 2 to 3 solar masses, it is believed that a quark star might be created, but this is uncertain. Neutron stars having masses less than 2–3 solar masses are stabilized by the quantum degeneracy pressure of neutrons, which opposes the gravitational collapse. The mass limit for neutron stars is called the Tolman–Oppenheimer–Volkoff limit, and it was found in 1939 by Robert Oppenheimer (1904–1967) and George Volkoff (1914–2000). For star masses $M > 4M_\odot$, gravitational collapse will always occur, with the inevitable creation of a black hole.

9.9 Neutron Stars and Pulsars

In 1967, Jocelyn Bell (b. 1943) and Antony Hewish (b. 1924) in Cambridge discovered cosmic radio waves consisting of short pulses, received at regular intervals. Later, similar sources of pulsed radiation were discovered. The objects emitting these pulses, called pulsars, seem to be bodies of small dimensions (diameters around 10 km), rotating around their axis with periods of the order of the duration of the pulses they emit. The period of the pulses varies from some tens of milliseconds to a few seconds. It is believed that such objects could originate precisely in the compression of the nucleus of a star when it explodes in a supernova. The resulting superdense body comprises mainly neutrons, since under such conditions it is thermodynamically more favourable for electrons, protons, and neutrinos to form neutrons by means of the weak interaction. A neutron star has a density comparable to the atomic nucleus, i.e., of the order 10^{11} kg/cm^3. Antony Hewish was awarded the Nobel Prize in Physics in 1974 "for his decisive role in the discovery of pulsars".

Associated with a neutron star there is usually a strong magnetic field, maybe billions of times stronger than the Earth's, which would accelerate streams of electrons emerging from the star. The accelerated electrons emit radiation at various frequencies (radio waves, visible light, X rays), but after going through several stages of supernova evolution, the relic radiation consists mainly of radio waves.

The enormous angular velocity of neutron stars, and probably also the extremely strong magnetic field, are due to the fact that, when compressed, the star conserves its angular momentum, and thus increases its angular velocity. This is similar to a skater who rotates with arms outstretched, then folds them in to increase his or her angular velocity. The magnetic flux is also conserved, and the field intensity is thereby greatly increased.

The rotation energy of a neutron star with a period of 10 milliseconds and a mass of the same order as the Sun, is comparable to the energy radiated by a star throughout its whole life.

In 1054, there appeared a supernova in the constellation of Taurus, recorded by Arab, Chinese and Japanese astronomers, and now in its place we observe a bright mass of expanding gas. This is known as the Crab Nebula. In its central part, a pulsar has been discovered with a period of rotation of almost 30 milliseconds and a diameter of about 30 km.

This discovery and other observations justify the hypothesis that pulsars are neutron stars with an enormous spinning angular velocity. The rotation period of the pulsar in the Crab Nebula decreases by one part in 2,400 per year. The corresponding decrease in rotational energy is enough to account for the energy radiated by the entire Nebula.

The rotation energy of one solar mass $M_\odot$, concentrated in a radius $r \simeq 10\,\mathrm{km}$ ($= 10^6\,\mathrm{cm}$), and having an angular frequency of the order of $\omega = 200\,\mathrm{rad/s}$, is $E_r \simeq mr^2\omega^2 \simeq 10^{50}\,\mathrm{erg}$. If this is compared with the rest energy $E_T = M_\odot c^2 \simeq 10^{54}\,\mathrm{erg}$, we see that $E_r \simeq 10^{-4}E_T$. This is indeed a significant fraction of the rest energy. In recent years pulsars with accompanying planets have been discovered. An example is PSR B1257+12, with two bodies much more massive than the Earth orbiting around it, and a third body much smaller, like the Moon. These planets should have a chemical composition very different from the planets of our Solar System, subjected as they are to the radiation wind issuing from the pulsar.

Problems

Problem 9.1 Explain why the following decays are forbidden

(a) $n \rightarrow p + e^+ + \nu^e$
(b) $n \rightarrow p + e^- + \nu^e$
(c) $p \rightarrow \pi^+ + \pi^0$

Problem 9.2 Is the following process allowed

$$p + p \rightarrow p + n + e^+ + \nu \,? \tag{9.31}$$

Problem 9.3 The mass of the Sun is estimated as 1.98×10^{33} g. (i) Assuming that it is composed mainly of baryons (protons and neutrons) of average mass 1.67×10^{-24}g, estimate the number of baryons contained in the Sun. (ii) Is this number constant?

Literature

1. T.D. Lee, *Particle Physics and Introduction to Field Theory* (Harwood Academic Publishers, New York, 1991). A unified presentation of particle physics at an advanced level
2. S.L. Shapiro, S.A. Teukolsky, *Black Holes, White Dwarfs, and Neutron Stars: The Physics of Compact Objects* (Wiley, New York, 1983). A comprehensive treatise on modern astrophysics
3. L.D. Landau, E.M. Lifshitz, *Statistical Physics*, 3rd edn. (Pergamon, London, 1981). This book contains an illuminating presentation of the thermodynamics of dense matter
4. R.K. Pathria, *Statistical Mechanics*, 2nd edn. (Elsevier, Oxford, 2006). The statistical equilibrium of white dwarf stars is very clearly discussed in this book

Chapter 10
General Relativity and Cosmology

The general theory of relativity is considered to be Albert Einstein's masterpiece in theoretical physics. In contrast with special relativity, where scientists like Hendrik Lorentz and Henri Poincaré worked in parallel, motivated by the unsolved physical problems existing at the beginning of the twentieth century (for instance, motion with respect to the æther and the negative result of the Michelson–Morley experiment), there was no such motivation for general relativity. With the exception of an anomaly in the precession of Mercury's orbit, the Newtonian theory of gravitation did not manifest symptoms of obsolescence.

The general theory of relativity was constructed by Einstein in a purely deductive form, using as basic postulates the principles of covariance and equivalence. A suitable mathematical tool had just been invented, thanks to the works of the Italian mathematicians Gregorio Ricci-Curbastro (1853–1925) and Tullio Levi-Civita (1873–1941), who had developed the so-called absolute differential calculus. Einstein was introduced to the formal aspects of non-Euclidean geometry by his friend, the mathematician Marcel Grossmann (1878–1936).

In the summer of 1915, Einstein was invited by David Hilbert (1862–1943), an outstanding mathematician, to visit Göttingen in order to lecture on his work on the theory of gravitation. In November 1915, independently, Einstein and Hilbert presented the equations of the gravitational field, which Hilbert had derived by variational principle. Therefore, the gravitational field action is customarily called Einstein–Hilbert action. However, the scheme of general relativity was developed by Einstein, therefore the new theory of gravity is Einstein's general relativity.

The final version of the theory was published by Einstein in 1916. The most spectacular confirmation was obtained in 1919, when Arthur Eddington (1882–1944) together with a team observed the bending of light from a distant star as it passed close by the Sun during a solar eclipse. This and other predictions of general relativity were subsequently confirmed in several experiments, making it an essential tool in cosmological research.

© The Author(s), under exclusive license to Springer-Verlag GmbH, DE, part of Springer Nature 2021
M. Chaichian et al., *Basic Concepts in Physics*, Undergraduate Lecture Notes in Physics, https://doi.org/10.1007/978-3-662-62313-8_10

10.1   Principle of Equivalence and General Relativity

It is customary to distinguish between two forms of the principle of equivalence, referred to as weak and strong. The weak principle of equivalence establishes the equality of the inertial and gravitational masses. The inertial mass m_i of a body is the coefficient of the acceleration **a** in Newton's second law:

$$\mathbf{F} = m_i \mathbf{a}. \tag{10.1}$$

The gravitational mass of the same body, for example, in its interaction with the Earth, is the one which appears in the expression for the force of gravitational attraction, i.e.,

$$\mathbf{F} = -\frac{GMm_g}{r^2} \mathbf{r}_0, \tag{10.2}$$

between, say, the Earth, of mass M, and the body of interest, of mass m_g. The unit vector $\mathbf{r}_0$ is along the line joining the body with the Earth's centre. We have the equivalence between these two masses expressed by means of the equality $m_i = m_g$. As a consequence, the acceleration due to gravity is the same for all bodies, if air resistance is neglected.

Imagine an elevator falling freely. An observer inside it would feel weightless. If the observer has a ball and lets go of it, without pushing it in any way, it will hang in the air, falling together with the system. When falling freely under the action of gravity, everything happens for the observer as if gravity were zero inside the elevator. For an observer inside an artificial satellite, this produces the effect of feeling weightless.

Returning to the elevator, if we accelerated it, for instance, by doubling the acceleration produced by the Earth attraction, our observer would feel weight in the opposite direction, that is, he would feel attracted toward the ceiling, as though there were a gravitational field in that direction. We see in this way that an accelerated system and a gravitational field produce similar effects, or in other words, motion in accelerated systems is equivalent to motion produced by a gravitational field (Fig. 10.1).

If the elevator were to ascend with some acceleration g', however, the observer of mass m would experience an increase in weight by an amount mg'. That is, it would seem as though the Earth's gravitational field had increased, and the observer's weight would now be $m(g + g')$, instead of mg. In conclusion, a local equivalence exists (that is, in a small region of space) between an accelerated reference frame and a gravitational field.

The *strong principle of equivalence* establishes that in every gravitational field, an elevator falling freely turns *locally* into a system in which the laws of physics are the same as in special relativity, that is, in an inertial system. The case is the same for an artificial satellite, in which the weightlessness effect is produced as a consequence of the satellite falling continuously toward the Earth as it moves around its orbit (as we pointed out in Chap. 1, the closed orbits result from the combination of this free-fall effect with a large enough tangential velocity).

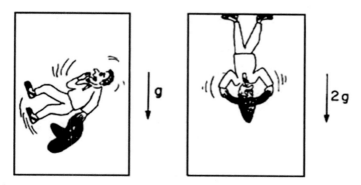

Fig. 10.1 For an observer inside an elevator falling freely under the action of gravity, the gravitational field force acting on him is canceled, and he feels as though he is floating or weightless. If the elevator is accelerated with twice the acceleration due to gravity, the observer inside feels a force equal to the Earth's gravity acting on him, but directed toward the ceiling of the elevator.

Fig. 10.2 For a very large elevator falling freely, the forces F_1, F_2, F_3 are not parallel, whence the elevator tends to adopt the form of an arch.

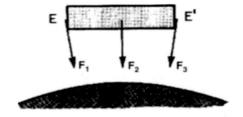

So both the falling elevator and the satellite can be treated as inertial systems, if their dimensions are small (strictly, pointlike). In the same way the Earth could be considered as an inertial system with respect to the Sun if its dimensions were negligibly small. In the case of the Earth, the fact of not being pointlike causes the tidal forces due to the solar attraction (there are also tidal forces due to the Moon). The atmospheric and oceanic masses are more sensitive to the tidal forces.

It is easy to understand the origin of tidal forces if we consider an extremely large elevator (Fig. 10.2). Its centre of mass M moves with the acceleration due to gravity, and falls freely. But the forces exerted on the ends E and E' are not parallel to the one which acts on M, since, due to the curvature of the Earth's surface, F_1, F_2, F_3 are directed toward the Earth's centre, whence the elevator tends to adopt the form of an arch.

Similarly, the trajectory followed by the Earth in its motion around the Sun (without considering the effect of the Moon) corresponds to a pointlike mass located at the Earth's centre of mass. The centre of mass behaves like a freely falling body during its motion. But because of the Earth's extension, the points distant from the centre of mass do not rigorously follow the *free-falling* motion. The result is that a small residual force is exerted on them by the Sun, producing tides. It must be emphasized, however, that the most notable tides are produced by the Moon, and have a similar

origin. For artificial satellites, this tidal effect is very small, and it can be neglected in
the first approximation. One can thus consider that the satellite satisfies the condition
of the principle of equivalence: for observers inside it, there the Earth's gravitational
field vanishes.

10.2 Gravitational Field and Geometry

The potential of the gravitational field near the surface of the Earth is

$$V(r) = -\frac{GM}{r}, \tag{10.3}$$

where r is the Earth's radius, G is the constant of gravitation, and M is the Earth's
mass.

Imagine now the following *Gedanken experiment*: suppose that at some height l
with respect to some reference system on the Earth's surface we have an electron and
a positron at rest. The mass of each is m. The potential energy of the two particles at
that height, putting $l \ll r$, is

$$E = -\frac{2mGM}{r+l} = -\frac{2mGM}{r}\frac{1}{1+\frac{l}{r}} \approx 2mV\left(1 - \frac{l}{r}\right) = 2m(V + \Delta V), \tag{10.4}$$

where $\Delta V = GMl/r^2$. If the two particles now fall to the Earth's surface, their
potential energy decreases to $2mV$, and their kinetic energy will be equal to $2m\Delta V$. If
now the electron and positron annihilate to produce two photons of angular frequency
ω, the following equation will be satisfied:

$$2\hbar\omega = 2mc^2 + 2m\Delta V. \tag{10.5}$$

That is, the energy of the two photons will be equal to the sum of the rest energy of
the electron and positron, plus their kinetic energy. We assume that the velocity of
these particles is not very large, so that we can use the approximation

$$\frac{mc^2}{\sqrt{1 - v^2/c^2}} \approx mc^2 + \frac{1}{2}mv^2,$$

where $\frac{1}{2}mv^2 = m\Delta V$. Now, by means of a suitable mirror, let the two photons be
reflected back up to the initial level of height l. At this height l, let the two photons
create the electron–positron pair again. The pair will be at rest, since otherwise
there would be a gain of energy in the cyclic process, implying the possibility of
constructing a perpetual motor of the second kind.

The frequency ω' of the two photons at the height l is different from the frequency
ω at the level of the Earth's surface, and should satisfy

$$2\hbar\omega' = 2mc^2. \tag{10.6}$$

Comparing (10.5) and (10.6), we deduce that

$$\frac{\omega - \omega'}{\omega'} = \frac{\Delta V}{c^2}. \tag{10.7}$$

So the frequency of the radiation varies in a gravitational field. Since ΔV is positive in our case, (10.7) implies that radiation emitted away from the surface of the Earth has frequency diminished by an amount

$$\Delta\omega = \omega - \omega' = \frac{\Delta V}{c^2}\omega, \tag{10.8}$$

where in writing the second equality we have assumed that ω and ω' are much larger than their difference.

Assume that a source on the Earth emits radiation at some frequency. The observer at some height will measure a lower frequency, i.e., shifted toward the red. This effect was measured for the first time by the American physicists Robert Pound and Glen Rebka in 1960, using a source of γ rays and the Mössbauer effect. These and other experiments reached an accuracy of 7×10^{-5}. In 2010, a much more exact measurement of the gravitational red shift based on quantum interference of matter waves within an accuracy of 7×10^{-9} was reported by H. Müller, A. Peters, and S. Chu.

Let us now examine the phenomenon from the wave point of view. If the frequency varies in a gravitational field, this should be caused by a time dilation. Actually, if a train of waves is sent from the Earth's surface, containing n complete oscillations during the time T_1, the relation between the angular frequency and the interval T_1 is

$$T_1 = 2\pi n/\omega. \tag{10.9}$$

The angular frequency of the same train of waves at the height l can be measured by dividing n by the duration of the train. The number obtained, ω', is different from ω, and this means that the interval T_2 that corresponds to n oscillations is

$$T_2 = 2\pi n/\omega'. \tag{10.10}$$

From (10.8) to (10.10), it follows that

$$\frac{T_2 - T_1}{T_1} = \frac{\Delta V}{c^2}. \tag{10.11}$$

That is, a clock at a height l measures for the duration of the wave train an interval of time longer than a clock located at the Earth's surface, and $T_2 = (1 + \Delta V/c^2)T_1$.

Fig. 10.3 Spacetime
diagram of the propagation
of a wave train in the
gravitational field of the
Earth from its surface to
some height l. The duration
of the wave train is different
for observers located at the
two points.

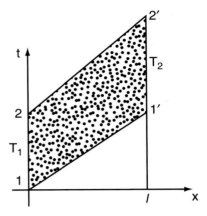

A clock on the Earth goes more slowly than another one placed at some height above the Earth's surface. In general, a clock located in a gravitational field goes more slowly than another clock located where the field is zero.

Let us draw a picture in which we mark on the horizontal axis x the height above the Earth and on the vertical axis the time t (Fig. 10.3). The event marked as 1 corresponds to the origin of the wave train as measured by the observer on the Earth. The event $1'$ corresponds to the origin of the train as measured by a second observer located at the height l. Similarly, the point 2 marks the end of the wave train as measured by the terrestrial observer, and $2'$ the same event as measured by the second observer. The lines $11'$ and $22'$ are the graphs of the propagation of the origin and the end of the wave train in spacetime. But as we have seen, the duration of the train, considered as the segments $12 = T_1$, $1'2' = T_2$, are different for the two observers:

$$T_2 > T_1. \tag{10.12}$$

On the other hand, the lines $11'$ and $22'$ should be parallel, since they correspond to the same phenomenon (the propagation of the signal) in a static gravitational field (it does not vary in time), and they differ only in that they have been measured by two different observers.

But the figure $11'2'2$ is not a parallelogram. The only solution to this paradox is that, in the presence of a gravitational field, the spacetime is curved. Hence, instead of taking the axes x, t on a plane, they must be taken on a curved surface. Then, by redefining the condition of parallelism on the surface, the lines $11'$ and $22'$ can be made to satisfy it on this surface.

A fundamental consequence of the general theory of relativity is that the effect of a gravitational field is described by spacetime curvature. Let us compare a plane and a curved surface like the surface of a sphere. Mark two points in the plane. Geometry demonstrates that the geodesic or shortest distance between those two points is the straight line segment joining them. In the geometry of the plane, the geodesics are straight lines extending across the whole plane toward infinity. Three points that are not aligned determine a triangle, the sum of whose internal angles is 180°. In other words, we can say that the plane is a two-dimensional Euclidean space.

Considering the same problem on the surface of the sphere leads to the conclusion that the geodesics are arcs of great circles (a great circle on the spherical surface is one whose centre coincides with the centre of the sphere). On the sphere, geodesics are finite in extent, and so is the total area of the sphere. Furthermore, a triangle on the spherical surface has the property that the sum of its internal angles is greater than 180°. The spherical surface is an example of a two-dimensional non-Euclidean space.

If α, β, and γ are the internal angles of a spherical triangle, A the area of this triangle, and R the radius of the sphere, we have the relation

$$\frac{A}{R^2} = \alpha + \beta + \gamma - \pi. \tag{10.13}$$

If A is kept constant and R tends to infinity, (10.13) gives the planar limit

$$\alpha + \beta + \gamma = \pi. \tag{10.14}$$

On the other hand, from (10.13), one can define the reciprocal of the square of the radius of the sphere, $K = 1/R^2$, by

$$K = \frac{\alpha + \beta + \gamma - \pi}{A}. \tag{10.15}$$

If the area A tends to zero in the expression (10.15), the resulting expression allows us to define the *curvature* in the neighbourhood of any point on the surface as

$$K = \lim_{A \to 0} \frac{\alpha + \beta + \gamma - \pi}{A}, \tag{10.16}$$

i.e., the excess over π of the sum of the internal angles of a triangle divided by the area of such triangle, in the limit of the area going to zero.

At a given point the curvature can be positive, zero, or negative. For example, K is positive everywhere in the case of a sphere, zero in the case of a plane, and negative on a saddle-shaped surface (Fig. 10.4).

Our intuition suggests that the three-dimensional physical space has the geometric properties resulting from generalizing the plane by adding one more dimension to obtain a three-dimensional Euclidean space. In this case, if we start from a point and move along a geodesic, that is, in a straight line, we move away from our starting point toward infinity.

In contrast, if our three-dimensional physical space had the geometrical properties which result from the generalization of the spherical surface to three dimensions, the geodesics would be closed curves. In contrast, if the geometry of space were of saddle

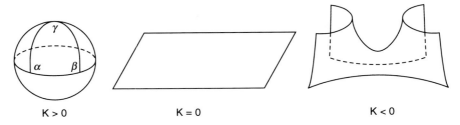

Fig. 10.4 A sphere has positive curvature, the plane has zero curvature, and a saddle-shaped surface has negative curvature.

type (negative curvature), the geodesics would not be closed, but open curves, and they would extend toward infinity.

According to general relativity, the planets, in their orbital motion around the Sun, describe geodesics in a four-dimensional curved spacetime, which is deformed by the mass of the Sun. In addition, when the rays of light emitted by a distant star pass close to the Sun, they follow a geodesic curve and hence deviate from the straight line trajectory. By defining $b = cL/E$, where L is the angular momentum of the beam and E its energy, the shifted angle is given approximately by

$$\delta\phi = 4GM_\odot/bc^2,$$

where $M_\odot$ is the mass of the Sun. Notice that b has dimension of length. The effect is $1.75''$ for light coming from distant stars and grazing the Sun's limb.

Actually, according to classical Newtonian mechanics and special relativity, some deviation of the light rays would be expected near the large solar mass, and it is not difficult to calculate this effect, which has been mentioned also in Sect. 1.5.3. But general relativity predicts a result twice as large, and this was confirmed by the observations made later by Eddington and other observers. The doubling of the deviation can be explained only in the framework of the general relativity, as a consequence of the curvature of space. It is found from the solution of the equation of motion for a light ray (the so-called eikonal equation) in a centrally symmetric gravitational field.

The geodesic curves described by the planets according to general relativity are not ellipses (as predicted by Newtonian mechanics), but more complicated curves in the form of almost-ellipses whose major axes precess around their focus (Fig. 10.5). An anomalous effect of this sort had been known since the nineteenth century in the orbit of Mercury. The observed precession of Mercury's perihelion (the point of closest approach to the Sun on the orbit) is 574" (arc-seconds) per century. The gravitational tugs of other planets, calculated by Newton's theory, could explain a precession of about 531". The origin of such a difference was not known. The calculations performed by Einstein in 1915 in the framework of general relativity provided the extra amount of + 43", in perfect agreement with the observed data. This was the first observational fact explained by the theory of general relativity. The

Fig. 10.5 The true trajectories of the planets around the Sun are precessing ellipses, resulting in curves in the shape of rosettes. This precession effect is very small, and is more perceptible in the case of Mercury owing to its proximity to the Sun.

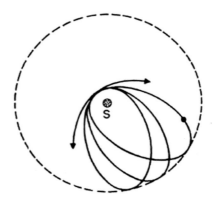

effect is more perceptible in Mercury's orbit because of its high eccentricity and its proximity to the Sun, but other planets display it in smaller amounts. In particular, for the Earth, this effect is about 3.84″ per century.

It is interesting to note that, according to general relativity, a body moving with some velocity **V** in a gravitational field is under the action of two forces: one corresponding to the usual gravitational attraction of Newtonian mechanics, and another one perpendicular to its velocity. This has a close analogy with the electromagnetic case, in which a charged particle in motion suffers the action of the Lorentz force, with two components: the electric force, independent of the velocity of the particle, and the magnetic force, perpendicular to its velocity. The additional force exerted by the gravitational field on a particle in motion in that field is the analog of the magnetic force. This second gravitational force is not very significant for low velocities since, as in the magnetic case, the term describing it contains the factor V/c.

According to the principle of equivalence, this second force of gravity corresponds more properly to the Coriolis force, appearing in a rotating (non-inertial) system of reference as a force perpendicular to the velocity of a particle moving in such a system.

General relativity also predicts that massive rotating bodies "drag" spacetime in their vicinity. This effect was first derived from general relativity by Josef Lense (1890–1985) and Hans Thirring (1888–1976) in 1918, and is also known as the Lense–Thirring effect.

Lensing effect. The lensing effect is due to the deflection of light coming from a distant object by a massive body. For small angles, it can be expressed as $\theta = 4GM_\odot/bc^2$, where $b = cL/E$ (see Sect. 10.2). Since the light is made up of photons, for which $p = E/c$, we have $L = Er/c$, which implies $b = r$, where r is the shortest distance from the photon beam to the body's centre. Thus, one can write

$$\theta = 2r_g/r. \tag{10.17}$$

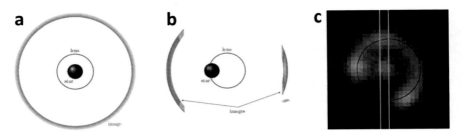

Fig. 10.6 a For perfect alignment of the observer, lens, and star we get a complete Einstein ring. **b** If this is not the case, only part of a ring will be observed. **c** An interesting example of the latter case is the Canarias ring, discovered by Margherita Bettinelli et al. in the constellation of the Sculptor in 2016. The maximal intensities are indicated by A, B, and C. The lens is a massive galaxy with redshift $z = 0.581$, and the source is also a galaxy, with $z = 1.165$. The ring covers around $300°$.

Consider a massive object, which may be a star or a galaxy, and call it the lens. This is located between the observer and a still more distant star (galaxy). When it passes near the lens, light coming from this last object can be bent round toward the eye of the observer. This gravitational lensing phenomenon was first mentioned in 1924 by the physicist Orest Chwolson (1852–1934) in Saint Petersburg, and treated quantitatively by Albert Einstein in 1936. If the object, the lens, and the observer are perfectly aligned, the image of the body will be a circular ring, known as an Einstein ring, centred on the lens (Figs. 10.6).

GPS time correction due to general relativistic effects. GPS (Global Positioning System) satellites form a global navigation system. Each carries a very accurate atomic clock that provides geolocation and time information to a GPS receiver. GPS satellites are located at a height of approximately 26600 km from the centre of the Earth, and describe two full orbits every sidereal day. For a position accuracy of $\Delta x = 15$ m, the time aboard GPS satellites must be known to an accuracy of $\Delta t = \Delta x / c = 50 \times 10^{-9}$ s. The time measured by the satellite clocks must therefore be corrected due to effects from special and general relativity.

   To calculate the relativistic effects on the time measured by a clock aboard a GPS satellite, which runs faster than a clock on Earth surface, we must compare the proper times measured on the satellite and on the Earth's surface. We take the Earth's mass to be 5.924×10^{27} g and its radius to be $R = 6378$ km. We need the velocities of the satellite v and the Earth's surface V, as well as r_g, to write the proper time at the satellite in the form $d\tau_S = ds/c$, where ds is obtained from the Schwarzschild metric (10.18). We should set r constant as well as $\theta = \pi/2$. Then, as $v = r d\phi / dt$, we have $-r^2 d\phi^2 / c^2 = -(v^2/c^2) dt^2$ and

$$d\tau_S = \sqrt{1 - \frac{r_g}{r} - \frac{v^2}{c^2}} \, dt, \qquad (10.18)$$

where the denominator of the third term in (10.28) has been approximated by one, i.e., $dr^2/(1 - \frac{r_g}{r}) \approx dr^2$, since it would only contribute to a small second order term.

(Note that, if r_g is neglected in (10.18), one is left with the special relativistic proper time relation in terms of v/c.) For the proper time at the Earth's surface, we have

$$d\tau_E = \sqrt{1 - \frac{r_g}{R} - \frac{V^2}{c^2}} \, dt. \tag{10.19}$$

Setting $\epsilon = r_g/r + v^2/c^2$ and $\eta = r_g/R + V^2/c^2$, we obtain an expression of the form

$$\frac{d\tau_S}{d\tau_E} = \frac{\sqrt{1 - \epsilon}}{\sqrt{1 - \eta}},$$

for the ratio of the proper times on the satellite and on Earth, where ϵ and η are very small quantities compared to 1.

Approximating by $\frac{d\tau_S}{d\tau_E} = (1 - \epsilon/2)(1 + \eta/2)$, and neglecting square terms, we get $d\tau_S - d\tau_E = (-\epsilon/2 + \eta/2)d\tau_E$. From this, we can integrate $d\tau$ over the interval of time to be corrected in one day. It is calculated as a problem below, leading to a value of 38×10^{-6} s.

10.3 Affine Connection and Metric Tensor

We saw in Chap. 5 how the concept of interval is used to characterize the distance between two events in spacetime. This concept remains valid in general relativity, and in fact the whole mathematical formulation of this theory starts from the expression for the infinitesimal interval between two events. In special relativity, if two events A and B have the spacetime coordinates $A = (x, y, z, ct)$ and $B = (x + dx, y + dy, z + dz, ct + cdt)$, the interval would have the form

$$ds^2_{AB} = c^2 dt^2 - dx^2 - dy^2 - dz^2. \tag{10.20}$$

Observe that the coefficients of the squares of the differentials of the coordinates are the constant numbers $(1, -1, -1, -1)$. It is customary to refer to (10.20) as the expression for the interval in the flat spacetime, and to call the set of four numbers $(1, -1, -1, -1)$ the Minkowski metric. With the notation introduced in Sect. 5.8, in the case of general relativity, the interval between events A and B would have the general form

$$ds^2_{AB} = g_{00}dx^2_0 + g_{11}dx^2_1 + g_{22}dx^2_2 + g_{33}dx^2_3 + 2g_{12}dx_1dx_2 + 2g_{23}dx_2dx_3 + \cdots, \tag{10.21}$$

with ten general, spacetime-dependent coefficients $g_{\mu\nu} = g_{\mu\nu}(x)$, where $\mu, \nu = 0, 1, 2, 3$, and by x we denote the spacetime four-vector x^μ. In weak gravitational fields, $g_{\mu\nu}$ approach their special relativity values, i.e., the Minkowski metric. The quantities $g_{\mu\nu} = g_{\mu\nu}(x)$ form a mathematical entity, the *metric tensor of spacetime*. Recall that a tensor is an object which transforms as the product of vectors. The metric tensor is symmetric, i.e., $g_{\mu\nu} = g_{\nu\mu}$.

As a consequence of the curvature of spacetime, systems of curvilinear coordinates are more convenient. Recall also the *contravariant* quantities, transforming like the coordinate differentials $dx^\mu = (dx^0, dx^1, dx^2, dx^3)$, and covariant quantities, transforming like the partial derivatives $\frac{\partial}{\partial x^\mu} = (\frac{\partial}{\partial x^0}, \frac{\partial}{\partial x^1}, \frac{\partial}{\partial x^2}, \frac{\partial}{\partial x^3})$, where by x^μ we denote the generalized coordinate. As examples of curvilinear coordinates, we have cylindrical coordinates $x^\mu = ct, \rho, \varphi, z$, and spherical coordinates $x^\mu = ct, r, \theta, \varphi$. A typical case of a covariant vector is the vector formed by the derivative of a scalar function f with respect to the (contravariant) coordinates:

$$\frac{\partial f(x)}{\partial x^\mu}.$$

Another example of a covariant quantity is the metric tensor $g_{\mu\nu}$. Given a contravariant vector, (A^0, A^1, A^2, A^3), we can transform it to a covariant one by multiplying it by the matrix formed by the metric tensor. We write $A_\mu = \sum_\nu g_{\mu\nu} A^\nu$, but from now on we drop the summation symbol, understanding that when repeated indices appear, like ν in the previous expression, we sum over them. This is Einstein's summation convention, introduced by Albert Einstein in his general relativity paper of 1916. We define δ^ν_μ to be the unit four-dimensional tensor, or Kronecker symbol, with all components equal to zero but with units down the main diagonal. Then the contravariant metric tensor $g^{\mu\lambda}$ satisfies the property

$$g^{\mu\lambda} g_{\lambda\nu} = \delta^\mu_\nu.$$

The task of defining the derivative of a vector with respect to the coordinates is more complicated. We must bear in mind that the variation of each of the components of a vector depends also on the other components. That is, this derivative which we will represent by ∇_λ, and is called the covariant derivative, or affine connection (an affine transformation has the general form $y = ax + b$), has two terms:

$$\nabla_\lambda A^\mu = \frac{\partial A^\mu}{\partial x^\lambda} + \Gamma^\mu_{\eta\lambda} A^\eta, \tag{10.22}$$

where $\Gamma^\mu_{\eta\lambda} = g^{\mu\xi} \Gamma_{\xi\eta\lambda}$. Note that $\Gamma^\mu_{\eta\lambda}$ and $\Gamma_{\xi\eta\lambda}$ are not tensors. They are called Christoffel symbols, and are defined in terms of $g_{\mu\nu}$ by the relation

$$\Gamma_{\xi\eta\lambda} = \frac{1}{2} \left(\frac{\partial g_{\xi\eta}}{\partial x^\lambda} + \frac{\partial g_{\xi\lambda}}{\partial x^\eta} - \frac{\partial g_{\lambda\eta}}{\partial x^\xi} \right). \tag{10.23}$$

We would like to point out the analogy between (10.22) and (1.27). The latter equation can be written as $dx_i'/dt = dx_i/dt - \epsilon_{ijk}\omega_j x_k$, and expresses the transformation of the velocity of a body from an inertial to a rotating (non-inertial) frame in Newtonian mechanics. Actually, (10.22) contains a generalization of (1.27), as a covariant derivative since, due to the principle of equivalence, the rotating system is equivalent (locally) to a gravitational field. Incidentally, covariant derivatives related to gauge transformations are defined also in the theory of Yang–Mills fields (see Chap. 11).

The metric tensor $g_{\mu\nu}(x)$ describes the gravitational field in the general theory of relativity. If a falling elevator is used as a system of reference, in such a system the interval between two very close events will take the form (10.20). That is, by making a transformation of coordinates to such a system, the expression (10.21) takes the form (10.20), and the Einstein metric becomes locally Minkowskian. We say 'locally' since this transformation is only valid in an infinitesimally small region. The point is that a gravitational field can only be made to vanish in the neighbourhood of a given point. As pointed out by Einstein:

> In the immediate vicinity of an observer that falls freely in a gravitational field, the gravitational field does not exist.

This establishes an essential difference between a real gravitational field and a fictitious one (created by a non-inertial system). The fictitious gravitational field can simply be eliminated *at all spacetime points* by making an appropriate transformation of coordinates. A real gravitational field cannot be eliminated in this way.

Starting from the metric tensor $g_{\mu\nu}$ (and its contravariant associated tensor $g^{\lambda\eta}$), it is possible to build other mathematical entities, such as the Riemann–Christoffel tensor $R_{\mu\nu\lambda\eta}$, the Ricci tensor $R_{\mu\nu}$, which describes the curvature of spacetime, and the scalar curvature $R = g^{\mu\nu}R_{\mu\nu}$. The tensor $R_{\mu\nu}$ is defined by

$$R_{\mu\nu} = \frac{\partial \Gamma^{\eta}_{\mu\nu}}{\partial x^{\eta}} - \frac{\partial \Gamma^{\eta}_{\mu\eta}}{\partial x^{\nu}} + \Gamma^{\eta}_{\mu\nu}\Gamma^{\lambda}_{\eta\lambda} - \Gamma^{\lambda}_{\mu\eta}\Gamma^{\eta}_{\nu\lambda}, \tag{10.24}$$

and the scalar curvature is $R = g^{\mu\nu}R_{\mu\nu}$. Remark that in general relativity the tensors are covariant under general coordinate transformations. A non-vanishing Riemann tensor is the covariant criterion to define a curved spacetime, as this tensor is identically zero for the flat Minkowski spacetime.

A distribution of matter or radiation is described in general relativity by means of another mathematical entity: the energy–momentum tensor $T_{\mu\nu}$. For a relativistic fluid in thermal equilibrium having pressure p, energy density ϵ, and velocity four vector u_{μ}, one finds:

$$T_{\mu\nu} = (p + \epsilon)\frac{u_{\mu}u_{\nu}}{c^2} - pg_{\mu\nu}. \tag{10.25}$$

10.4 Gravitational Field Equations

The gravitational field equations in general relativity, named Einstein's equations, establish a relation between the geometrical properties of the spacetime, expressed by the metric tensor $g_{\mu\nu}$, the Ricci tensor $R_{\mu\nu}$, and the spacetime curvature R on the one hand, and the distribution of mass and energy, represented by the energy–momentum tensor of matter, $T_{\mu\nu}$, on the other hand:

$$R_{\mu\nu} - \frac{1}{2}g_{\mu\nu}R = \frac{8\pi G}{c^4}T_{\mu\nu}, \qquad (10.26)$$

where G is the gravitational constant. Einstein's equations for the gravitational field are analogous to Maxwell's equations in classical electrodynamics. There are, however, three important differences:

1. Maxwell's equations apply to inertial systems. The equations of the gravitational field apply to arbitrarily moving systems;
2. Maxwell's equations do not contain the equations of motion of the charges which produce the electromagnetic field. However, the gravitational field equations provide the equations of motion for the particles producing the field;
3. Maxwell's equations are linear differential equations in the electromagnetic potential $A_\mu(x)$, while the gravitational field equations are highly non-linear in $g_{\mu\nu}(x)$, whose components represent the generalized gravitational potential.

In particular, from the latter feature, in the quantum version of the theory, we would expect the gravitons or quanta of the gravitational field (the gravitational analog of photons) to be able to split and generate other gravitons. Photons, on the other hand, do not split into pairs of photons (in vacuum), in standard quantum electrodynamics. Moreover, there is an analogy between the Lorentz force in electromagnetism and the gravitational force on a moving mass, as pointed out previously. If we denote $h = -g_{00}$ and if we define the three-dimensional vector $\mathbf{g}$ with components $g_i = g_{0i}/g_{00}$, where $i = 1, 2, 3$, for a constant gravitational field (the components of the metric tensor do not depend on time), one can write this force as

$$\mathbf{F} = \frac{mc^2}{\sqrt{1 - V^2/c^2}}\left\{-\nabla \ln \sqrt{h} + \sqrt{h}\frac{\mathbf{v}}{c} \times (\nabla \times \mathbf{g})\right\}. \qquad (10.27)$$

For small velocities, the first term corresponds to the well-known force of gravity, and it is the analog of the electrostatic attraction, while the second term depends on the velocity, as does the magnetic force, and it is equal to the Coriolis force in a rotating system with angular velocity $\Omega = \frac{c}{2}\sqrt{h}\nabla \times \mathbf{g}$. But for the latter to become significant, e.g., in the case of the planets, they would have to move at high speed, comparable with the speed of light.

In addition, as for the electromagnetic field, there should be gravitational waves, that is, deformations of the spacetime geometry propagating at the speed of light. But even for very massive astronomical objects, the amount of gravitational energy radiated is extremely small. For example, for a system of binary stars, the radiation emitted in a year would be 10^{-12} of the total energy of the system. The so-called Hulse–Taylor binary is a pair of stars, one of which is a pulsar. They each have masses around 1.4 $M_\odot$ and the distance between them is around 2×10^6 km, of the order of the Sun's diameter. They are expected to radiate 10^{22} times the gravitational energy radiated by the Earth–Sun system. This causes the stars to gradually move closer together, in what is known as an *inspiral*, and this has an effect on the observed pulsar's signals.

Russell Hulse (b. 1950) and Joseph Taylor (b. 1941) were awarded the Nobel Prize in 1993 for their measurements which led to the discovery of the first binary pulsar, and allowed them to show that the gravitational radiation predicted by general relativity matched the results of these observations with a precision within 0.2%. This was the first indirect evidence for gravitational energy radiation, which is understood as a wave phenomenon.

Observation of Gravitational Waves

The search for direct evidence of gravitational waves lead to a great success, using mainly detectors based on laser interferometry, like LIGO on Earth ground (Laser Interferometer Gravitational Wave Observatory) in Livingstone, Louisiana, and the Hanford Site in the state of Washington. The Laser Interferometer Space Antenna (LISA) is designed to detect gravitational waves at frequencies not observable by ground based interferometry, and planned to operate in the near future. LISA is a giant interferometer, composed of three satellites forming an equilateral triangle with the sides 2.5 million km long.

As a gravitational wave passes through matter, a distortion in space-time produced by the gravitational wave leads to a tiny lengthening or contraction of objects, like the arms of an interferometer. This makes interferometry-based devices particularly useful for the detection of such waves. A modified Michelson interferometer is used to measure gravitational-wave strain through the difference in length of its orthogonal arms. LIGO is the largest interferometer ever built and the most sensitive detector, possessing a measurement sensitivity of about one part in 5×10^{22}. Each arm is formed by two mirrors, acting as test masses, separated by a distance $L_x = L_y = L = 4$ km. When a gravitational wave passes, it alters the arm lengths such that the measured difference is $\Delta L = \delta L_x - \delta L_y = h(t)L$, where h is the gravitational-wave strain amplitude projected onto the detector. This length variation produces a phase difference between the two light beams returning to the splitter, transmitting an optical signal proportional to the gravitational-wave strain to the output photodetector.

When a gravitational wave enters, one of the arms of the interferometer is lengthened. Mirrors placed near the beam splitter cause multiple reflections of the laser beam, increasing the distance traveled in each arm to 1120 km. This system of mirrors forms an optical resonator known as a Fabry–Pérot cavity. The output is the signal coming from the interference of the two beams, showing the shape of the incoming gravitational wave.

Up to 2015, evidence for black holes could only be obtained through electromagnetic signals, although evidence for the radiation of gravitational waves was provided by the Hulse–Taylor observations. However, the merging of two black holes by detection of the emitted gravitational waves was first observed on 14 September 2015. LIGO reported the observation of a signal corresponding to the wave predicted by general relativity for the merger of two black holes with masses $29M_\odot$ and $36M_\odot$ about 1.3 billion light years away. The final black hole mass was estimated to be of order $62M_\odot$. The difference of $3M_\odot$ was radiated as gravitational waves. A second set of gravitational waves was reported in December 2015. They represented the merger of two black holes about 1.4 billion light years aways, with masses of about 14.2 and 7.5 solar masses, yielding a final black hole of around 20.8 solar masses, with one solar mass radiated away as gravitational waves. In 2017, the Nobel Prize in Physics was awarded to Rainer Weiss (b. 1932), Kip Thorne (b. 1940), and Barry Barish (b. 1936) "for decisive contributions to the LIGO detector and the observation of gravitational waves".

On 17 August 2017, scientists also witnessed a process in which two neutron stars spiralled into each other and merged, producing a black hole. The event was first detected by the gravitational waves this generated. Scientists immediately knew it was due to two spiralling neutron stars, which were already emitting radiation before they merged. The radiation was detected by 70 observatories around the world, ranging from gamma ray detectors to radio telescopes. They confirmed several key astrophysical models, and revealed the birthplace of some heavy elements like gold and platinum. Above all, they were able to further test general theory of relativity.

10.5   Cosmology

If Einstein's equations (10.26) are solved for a gravitational field produced in vacuum by a body of mass M with spherical symmetry, and such that the metric does not depend on time and is asymptotically flat, the interval ds^2 is given by the expression obtained by Karl Schwarzschild (1873–1916) in 1915:

$$ds^2 = \left(1 - \frac{r_g}{r}\right)c^2 dt^2 - r^2(\sin^2\theta d\varphi^2 + d\theta^2) - \frac{dr^2}{1 - \frac{r_g}{r}}, \qquad (10.28)$$

where $r_g = 2GM/c^2$ is the Schwarzschild radius of a spherical body of mass M. For $r = r_g$, $g_{00} = 0$ and $g_{11} \to \infty$ with the formation of the so-called event horizon of a black hole. An event horizon is a boundary in spacetime beyond which events cannot affect an outside observer. Such a region of spacetime is called a *black hole*. In 2020, the British mathematician Roger Penrose was awarded the Nobel Prize in Physics "for the discovery that black hole formation is a robust prediction of the general theory of relativity".

The Russian physicist Alexander A. Friedmann (1888–1925) studied the Einstein equations as applied to the Universe, assuming a homogeneous and isotropic density, and he concluded that there are two possible solutions: the closed and the open models. The latter leads to a perpetual expansion. At the boundary between the open and the closed models, there is the flat solution. Physically, the condition for open, closed, or flat Universe is determined by the density (of matter or energy).

If the distance between two galaxies is taken as $d(t) = R(t)d_0$, their relative speed can be written as $v = [\dot{R}(t)/R(t)]d(t)$, i.e., the speed is proportional to the separation between the two galaxies, with a proportionality factor $H(t) = \dot{R}(t)/R(t)$ which is called the Hubble parameter. Its present value is usually represented by H_0 and called Hubble's constant. We call $R(t)$ the *cosmic scale factor*, and here we take it to be dimensionless, while d_0 has the dimension of length. Below we shall consider $R(t)$ frequently as containing implicitly the d_0 factor and having dimensions of length. Concerning $H(t)$, it has the dimension of inverse time.

We shall discuss the problem of the motion of a galaxy by using the Newtonian mechanics of Chap. 1, but taking into account Hubble's law. Let us consider the mass of the galaxy as m, under the gravitational attraction of the rest of the Universe, of mass M. As $M \gg m$, one has $M + m \simeq M$ and the total energy is

$$\frac{1}{2}mv^2 - \frac{GMm}{r} = E. \tag{10.29}$$

Let us write $v = \dot{R}(t) = H(t)R(t)$ and $r = R$, where $H(t)$ is the Hubble parameter and R is the radius of the Universe. For a spherical mass distribution, the total mass is $M = \frac{4}{3}\pi R^3 \rho$, where ρ is the average mass density of the Universe, and we substitute this expression into (10.29). This gives

$$\frac{\dot{R}^2(t)}{2} - \frac{4\pi\rho G R^2(t)}{3} = \frac{E}{m} = \frac{-K}{2}. \tag{10.30}$$

This is a non-relativistic way of obtaining Einstein's equation from the Friedmann model for the expansion of the homogeneous and isotropic Universe. The latter is identical to the one obtained using the relativistic formalism starting from the Robertson–Walker metric, which is a metric compatible with the conditions of homogeneity and isotropy (these conditions are sometimes called *cosmological principle*):

$$ds^2 = c^2 dt^2 - R^2(t)\left[\frac{dr^2}{1 - kr^2} + r^2(d\theta^2 + \sin^2\theta \, d\varphi^2)\right]. \tag{10.31}$$

Here $k = -1, 0, 1$ correspond to open, flat, and closed cosmologies, respectively. Observe that K in (10.30) has the dimension of the square of a velocity, while k in (10.31) is dimensionless, because $R(t)$ has the dimension of length, and r is dimensionless. Then we have $K \sim kc^2$. According to (10.30), the critical condition to bring the expansion asymptotically to a halt occurs for $k = 0$, that is to say, for the density

$$\rho_c = \frac{3H^2}{8\pi G}. \tag{10.32}$$

With the present-day value of the Hubble parameter, H_0, the value of ρ_c is of the order of 10^{-29}g cm^{-3}.

But the Robertson–Walker metric does not tell us anything about the time dependence of the scale factor $R(t)$. To obtain this information, one must solve not only the Einstein equations, that is, (10.30) and (10.34) below, but also the equation of conservation of energy and the equation of state. Let us discuss the simplest case of a flat Universe. If we expand $R(t)$ in a power series around the reference time t_0, taken as the present time, we get $R(t) = R(t_0)[1 + H_0(t - t_0) - \frac{1}{2}q(t_0)H_0^2(t - t_0)^2 + \cdots]$, where the so-called *deceleration parameter* is given by

$$q(t) = -\frac{\ddot{R}(t)R(t)}{\dot{R}^2(t)}. \tag{10.33}$$

This quantity was estimated to be of the order of -0.5 at present, indicating that the expansion of the Universe is accelerated. The value of the deceleration parameter is a major topic in the present day cosmological research.

Together with (10.30) we must consider the other Einstein equation,

$$\ddot{R}(t) = -\frac{4\pi G}{3} R(t) \left(\rho + \frac{3p}{c^2}\right). \tag{10.34}$$

For $\rho > 0$ and $p > 0$, the acceleration $\ddot{R}$ is negative, and consistent with a positive deceleration. But as will be pointed out later, dark energy may provide a negative value for the factor $(\rho + 3p/c^2)$, producing an accelerated expansion of the Universe. We postpone the discussion of this case and continue with the solutions for standard cosmology. We denote $\Omega = \rho/\rho_c$. Then we can write (10.30) in terms of the Hubble parameter as follows:

$$H^2(\Omega - 1) = KR^{-2}(t). \tag{10.35}$$

If one assumes the pressure to be negligible compared with the density, that is to say $p \simeq 0$, simple solutions of the Friedmann model are found. In the flat case ($k = 0$, $q_0 < 0.5$, $\Omega = 1$), one has

$$R(t) = [3GM/\pi]^{1/3}t^{2/3}, \qquad H = 2/3t. \tag{10.36}$$

In the closed case ($k = +1$, $q_0 > 0$, $\Omega > 1$), the Universe has a finite volume, but it is unbounded (this corresponds to the previously mentioned space which can be regarded as a generalization of the spherical surface to three dimensions). In such a case, one obtains solutions in terms of a parameter η, defined by $d\eta = R(t)dt$:

$$R(\eta) = (2GM/3\pi c^2)(1 - \cos\eta), \qquad t(\eta) = (2GM/3\pi c^3)(\eta - \sin\eta). \tag{10.37}$$

In both the open ($k = -1$, $\Omega < 1$) and the flat cases, the Universe is infinite and unbounded. In the open case, one has

$$R(\eta) = (2GM/3\pi c^2)(\cosh \eta - 1), \qquad t(\eta) = (2GM/3\pi c^3)(\sinh \eta - \eta).$$
$$(10.38)$$

In none of the three cases is the Universe static, and it should be either expanding or contracting. Expansion is interpreted as meaning that the galaxies separate with increasing speed because their mutual separation increases. But if this occurs, there should be a *redshift* in the spectra of light coming from remote galaxies. The effect was observed for the first time in 1912 by Vesto Slipher (1875–1969) at the Lowell Observatory in Flagstaff, Arizona.

If ν_E is the emitted frequency and ν_O the observed one, the redshift is measured by a quantity $z = (\nu_E/\nu_O) - 1$. If $\nu_E > \nu_O$, the light is redshifted and $z > 0$. In the opposite case, if $\nu_E < \nu_O$, then $z < 0$, and the spectrum is shifted to the blue.

Edwin Hubble (1889–1953) discovered that the distances to the far-away galaxies are roughly proportional with their redshifts, which is now known as Hubble's law. Hubble reached this conclusion by interpreting his own measurements of galaxy distances and the galactic redshift measurements of Slipher. George Lemaître (1894–1966) had been the first to report this result in 1927 and to propose the theory of the expansion of the Universe. As pointed out before, as our Universe expands, the galaxies recede from each other with increasing speed. This expansion suggests that there was necessarily an initial moment in which all the matter composing these galaxies, and all intergalactic matter, was concentrated in a small region of the Universe. A great explosion, the Big Bang, occurred at some time around 10 to 20 billion years ago. The most recent estimate by the Planck collaboration for the age of the Universe, i.e. the time since the Big Bang, is 13.79 billion years. The Big Bang theory was proposed by Lemaître in 1931, but the term Big Bang was coined later.

Over the last few decades a theory has been proposed on the hypothesis that, in the early stages of the Universe, there was an exponential expansion. This phase was called *inflation* in the 1980s. It has been suggested that this could be described by a coupling between the gravitational field and some scalar field which is displaced from its equilibrium configuration. This point will be discussed further in Chap. 11.

With regard to the distribution of galaxies, moving away from each other in all space directions, observations indicate that they are grouped into clusters or superclusters, separated by empty space, with a cellular distribution. This in turn suggests a three-dimensional structure of these clusters separated by empty space, with some regularity, on a gigantic scale of 390 million light-years, in a form similar to a honeycomb.

The temperature of the *primeval fireball* in which the matter composing our visible Universe was concentrated was extraordinarily large, of the order of 10^{32} K, but it would have decreased quickly to values between 10^{10} and 10^9 K a few seconds after the Big Bang. This stage is said to be radiation dominated, because the density of the radiation was significantly greater than the density of matter. For instance, the photon density was much higher than the baryon density. As the initial ball cooled down in the process of expansion, the atoms of the light elements would have condensed out, while heavier atoms would have formed later inside the stars.

With the expansion of the Universe, the average temperature has decreased, and the whole system has cooled down, going through a matter-dominated era, when most of the energy of the Universe was concentrated in the masses of the nuclear particles. At present, the Universe is dominated by dark energy, which drives the cosmic acceleration.

As a result of this cosmological process, one may expect some fingerprint of the radiation-dominated era during the first stages after the Big Bang. George Gamow (1904–1968) predicted the existence of a background radiation, corresponding to a black body at very low temperatures. In 1964, Arno Allan Penzias (b. 1933) and Robert Woodrow Wilson (b. 1936) discovered this fossil radiation, a discovery for which they were awarded the Nobel Prize in 1978. The background radiation comes from all directions of space and it corresponds to a black body radiation at a temperature of about 2.725 K. It is called the cosmic microwave background (CMB). This radiation has a density of 4.40×10^{-34} g/cm^3, while the density of matter is of the order of 10^{-29} g/cm^3, that is, 10^5 times greater. For a certain time this justified the claim that we live in a matter-dominated era. At the present time, this view has changed due to current hypotheses about dark matter and dark energy, which we shall come back to in Sect. 10.6.2.

10.6   Gravitational Radius and Collapse

The idea of escape velocity is well known: it is the minimum velocity one must give a body so that it can escape from the Earth's gravitational field. If one neglects air resistance, the problem reduces to solving the equation in which the total energy of the particle in the gravitational field is equal to zero, viz.,

$$\frac{1}{2}mv^2 - \frac{GMm}{r} = E = 0. \tag{10.39}$$

Taking M and r as the Earth's mass and radius, this gives

$$v = \sqrt{\frac{2GM}{r}}. \tag{10.40}$$

If the Earth's radius decreased to one quarter (but keeping the same total mass), the escape velocity is doubled. But one can also consider the opposite problem: to which radius would we have to compress the Earth to reach a given value of the escape velocity? Let us suppose $v = c$, the speed of light. Then the value obtained for the radius R is the gravitational or Schwarzschild radius mentioned above,

$$R = \frac{2GM}{c^2} \equiv r_g. \tag{10.41}$$

For M of the order of the Earth's mass (6×10^{27} g), $R \approx 0.9$ cm. So if the Earth's mass were compressed to such an incredibly small size, no object could escape from inside, and only light emitted vertically would be able to get outside.

For a radius smaller than this value of R, the Earth would be transformed into a black hole, and even light could not escape from it. A black hole would absorb all the substance and radiation in its surrounding space. The existence of black holes, which we have argued mainly from non-relativistic mechanics, is a consequence of the general theory of relativity. For every body of mass M, a corresponding gravitational radius can be calculated by dividing its mass M (multiplied by the gravitational constant G), by the square of the speed of light. We have already seen that the Earth's gravitational radius is of the order of 0.45 cm. A similar calculation carried out for the Sun would give a sphere of radius about 3 km. Assuming a spherical shape and density ρ, its mass would be $M = \frac{4}{3}\pi R^3 \rho$. Then

$$R = \frac{8\pi}{3c^2} G R^3 \rho, \tag{10.42}$$

which implies that $\rho = 3c^2/(8\pi G R^2)$, that is, the density required to achieve the gravitational radius condition decreases as the reciprocal of the square of the radius. In other words, the larger the mass, the smaller the density required to achieve the gravitational radius condition. For instance, for our galaxy, if we assume a mass 10^{44} g (that is 10^{11} times that of the Sun, whose mass is about 2×10^{33} g), the gravitational radius is

$$R \approx 10^{11} \text{ km}, \tag{10.43}$$

which is about a hundredth of a light-year (one light-year is approximately 9.4×10^{12} km). The radius of our galaxy is about 55,000 light-years, i.e., $\sim 5.2 \times 10^{17}$ km. The gravitational radius would be reached by reducing the galactic radius to one millionth of its present size.

If for the Universe we estimate a mass of 10^{80} times the proton mass, that is, about 10^{56} g, the corresponding gravitational radius would be of the order of 10^{10} light-years. This is of the same order as the estimated radius of the Universe, the distance of the most remote cosmic objects. It has thus been speculated that the whole of our visible Universe is a black hole. Such an idea is in contradiction with the current cosmology.

If a star explodes in a supernova, its nucleus may be compressed to such a density that it becomes a neutron star, with a density of about 10^{15} g/cm^3. If its mass is greater than 2.5 times the mass of the Sun, gravity dominates over any other force resisting the compression. A gravitational collapse then occurs leading to the formation of a black hole. The gravitational radius determines the so-called event horizon. All the radiation and matter surrounding it would be absorbed by the black hole, and it would disappear below the horizon (Fig. 10.7). An observer inside a black hole (if it could survive the forces generated inside) could find out about what happens outside, but could never communicate with external observers, since it would be impossible to send out a signal.

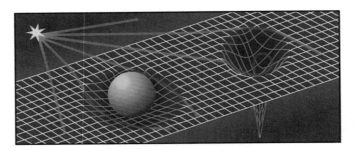

Fig. 10.7 Light from a distant star is deviated by a heavy body, which deforms spacetime around it (part of it is represented schematically by a two-dimensional mattress). A black hole captures light as well as matter incident on it.

Under such extreme conditions in which the gravitational force becomes so large, classical ideas cease to be valid, and we are in a situation similar to that of atomic theory as described by classical electrodynamics, according to which the atom would disappear in a collapse. Under such extreme conditions, quantum effects would thus enter the game in a predominant way.

Stephen Hawking (1942–2018) suggested in 1974 that black holes can evaporate in a gas of photons and other particles, by a quantum mechanism: the tunnel effect. In Chap. 7, we saw that particle and antiparticle pairs are created and annihilated spontaneously in vacuum. The process of pair formation has a characteristic time, given by the Heisenberg uncertainty principle:

$$\tau = h/E, \tag{10.44}$$

where E is the energy required for pair formation. Associated with the black hole, there is also a characteristic time τ' given by

$$\tau' = \frac{R}{c}, \tag{10.45}$$

where $R = 2GM/c^2$ is the gravitational radius of the black hole. If $\tau' < \tau$, the pair production process may be possible, at the expense of the mass (energy) of the black hole, and one particle of the pair can tunnel out of the black hole. The black hole temperature is inversely proportional to its mass and it would radiate energy proportionally to the fourth power of the temperature. Jacob Bekenstein (1947–2015) conjectured in 1972 that the area of the event horizon is proportional to the black hole entropy. This is intuitively comprehensible if one remembers that, when two black holes of masses M_1 and M_2 (and radii r_1 and r_2) join together, the area of the event horizon of the resulting black hole is always *larger* than the sum of the areas of the original black holes, because $(r_1 + r_2)^2 > r_1^2 + r_2^2$. As we see, the horizon surface area always grows with mass. In his work on black hole radiation of 1974, Hawking confirmed Bekenstein's conjecture and fixed the proportionality constant.

Fig. 10.8 MERLIN imaging of relativistic jets in the microquasar GRS1915+105, which is an X-ray binary system assumed to be composed of a rotating black hole and a normal star. The black hole has an accretion disk fed by gas from the star. It was the first known galactic source which ejects material with apparently superluminal velocities. This seems to be due to a relativistic effect known as Doppler boosting, produced by jets of particles moving with the speed of $0.9c$ (Courtesy of Fender et al., University of Manchester, Jodrell Bank Centre for Astrophysics).

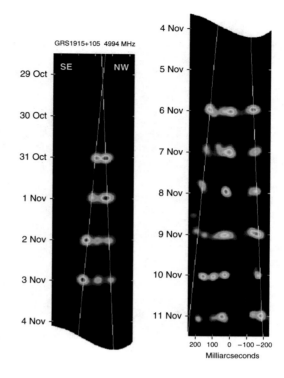

Let us assume then that the entropy of a black hole is proportional to the area of the event horizon, $S = kA$. But $A = k'R^2 = 4k'G^2M^2/c^4$, where k and k' are constants of proportionality. Then $S \sim U^2$, with U the internal energy, which is proportional to the mass M. From this, one has $1/T = \partial S/\partial U \sim U$, and evidently $T \sim 1/M$. From here we deduce that a big black hole will radiate less than a small one, according to the law $1/M^4$, since the radiation power is proportional to $T^4 \sim 1/M^4$. A black hole is a system which loses information. If the black hole is in a pure quantum state when it begins, as it radiates thermal energy, it will pass to a mixed state with consequent loss of *quantum coherence*. This hypothesis is due to Hawking.

It is believed that the first indirect observation of a black hole was the binary system GRO J1655-40, assumed to comprise a black hole and a star, like the system GRS1915+105 (Fig. 10.8). Orbiting around the black hole, there is an accretion disk made up of material fed to it by the normal star, and this disk radiates in the X-ray region. Both binary systems are galactic 'microquasars' and may provide a link between the supermassive black holes which are believed to power extragalactic quasars and more local, accreting black hole systems.

In 2008, astrophysicists found compelling evidence that a supermassive black hole, called Sagittario, of more than 4 million solar masses is located at the centre of the Milky Way. Supermassive black holes were subsequently found at the centre of all known galaxies.

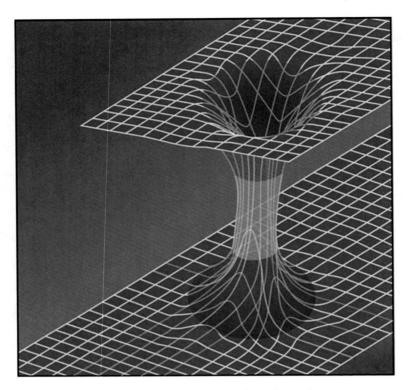

Fig. 10.9   A wormhole is a shortcut between separate regions of spacetime.

It has been suggested that instead of an infinite compression, the substance contained inside the black hole may emerge in another region of the spacetime. This would lead to a *white hole*, which would be a black hole running backward in time. In other words, gravitational collapse might cause an interconnection between two remote regions of spacetime.

10.6.1   Wormholes

The hypothetical bridges between separate regions of spacetime are called wormholes. These would be shortcuts between areas of space otherwise separated by long distances (Fig. 10.9). The wormhole would have two mouths (which are spheres in 3D space) and a throat between them. Standard theory indicates that they would not generally be stable. To be stable, or *traversable*, some exotic matter with negative energy density would be required in the throat. But this point remains open, since the assumption of extra dimensions would provide new scenarios. It should be noted that, although faster-than-light speeds remain forbidden locally, through a wormhole, even moving at speeds smaller than c, it would be possible to connect two events A, B by an interval of time $\tau < t_{AB}$, where $t_{AB} = l_{AB}/c$ and l_{AB} is the distance across standard space. This leads to an effective faster-than-light communication. Even time travel seems possible through wormholes.

10.6.2 Dark Matter, Dark Energy, and Accelerated Expansion

In an open Universe there would be perpetual expansion, whereas in a closed one the expansions and contractions would alternate in huge cycles. The time required for each of these cycles defies all imagination.

We have seen from (10.32) how to find ρ_c. The most recent value measured for the Hubble constant H_0 is 67.8 km/s Mpc^{-1}, where 1 Mpc $= 3.26 \times 10^6$ light-years. The critical density ρ_c is of order 10^{-29} g/cm^3. There is strong evidence, most recently from the Planck space mission, that the observable Universe is flat, i.e. $\rho = \rho_c$, which is consistent with inflationary models. But the flatness implies the existence of *dark matter* and *dark energy*, in significant amounts compared to the usual matter.

At present, dark matter appears to be an unavoidable hypothesis, providing some missing matter needed to explain the observed rotational velocities of galaxies, orbital velocities of galaxies in clusters, gravitational lensing of background objects by galaxy clusters, and other observable phenomena.

Most dark matter does not interact with electromagnetic radiation. It is thus transparent. However, there is not yet any satisfactory model for dark matter. For instance, it could be that some as-yet undiscovered weakly interacting particles were created during the Big Bang and today remain in significantly large amounts to account for the dark matter. The name of *weakly interacting massive particles* (WIMPs) has been suggested for some of these candidates for dark matter, assuming that it is nonbaryonic, i.e., that it contains no atoms. In addition to WIMPs, the nonbaryonic candidates for dark matter include neutrinos and hypothetical particles such as axions or supersymmetric particles (see Chap. 11).

However, certain astronomical objects may constitute the dark matter, but escape detection. For instance, brown dwarf stars with very small mass or black hole remnants of an early generation of stars would be similarly invisible. A small fraction of this hypothetical dark matter is referred to as MACHO an acronym for *massive (astrophysical) compact halo object*, made up of baryonic matter. Yet a large fraction of the dark matter has to be of a non-baryonic nature.

At present it is believed that ordinary matter constitutes only around 4.9% of the mass of the Universe, whereas dark matter would make up 26.8%, and the remaining 68.3% is thought to be due to dark energy. These percentages have varied over the last few years, since some measurements have been refined.

Since 1997, observations of supernovas of type Ia, which are excellent standard candles for measuring cosmological distances, suggest that the expansion of the Universe is actually accelerating. The Nobel Prize in Physics in 2011 was awarded to Saul Perlmutter (b. 1959), Brian Schmidt (b. 1967), and Adam Riess (b. 1969) for their discovery of the accelerated expansion of the Universe. This cannot be explained on the basis of the present gravitational interaction, and it requires an assumption of additional energies able to act as a repulsive force, for instance, whence the idea of dark energy. Such dark energy is assumed to be transparent.

The quantum vacuum was suggested in 1967 by Yakov Zel'dovich (1914–1987) as a candidate for dark energy, but present estimates give an extremely large figure for this, not compatible with what is expected by observation.

Let us consider the amount of dark energy inside a cylindrical cavity with a piston. The energy associated with a change of volume dV is $dE = -pdV$. If ρ_E is the energy density, we have $dE = \rho_E dV$. Thus, $\rho_E = -p$. The vacuum pressure is minus its energy density. In ordinary matter, we usually have $|p| \ll \rho_E$. This leads us to conclude that dark energy is essentially relativistic, able to interact in a repulsive way with ordinary matter. This would give a negative pressure term in the Einstein equations.

When Einstein wrote his equations, there was no knowledge of the expansion of the Universe. Hence, to make a static Universe from his model, Einstein introduced a cosmological constant. For years, this cosmological constant was taken as zero by cosmologists. The quantum vacuum effect is equivalent to assuming a nonvanishing cosmological constant.

Other researchers work with models based on appropriate scalar fields, called *quintessence*, able to generate similar effects. The problem is still open.

10.7 Gravitation and Quantum Effects

If we combine the constant of gravitation G, the reduced Planck constant $\hbar$, and the speed of light c, it is possible to estimate the order of magnitude at which quantum gravity phenomena are likely to manifest themselves. The combination with the dimension of length is

$$l_P = \sqrt{\frac{G\hbar}{c^3}} \approx 10^{-33} \text{ cm.} \tag{10.46}$$

This is the so-called *Planck length*. It indicates the order of distances at which quantum gravitational effects are expected to appear. Starting from this value, it is possible to derive a number with dimensions of mass, named the *Planck mass*:

$$m_P = \sqrt{\frac{c\hbar}{G}} \approx 10^{-4} \text{ g.} \tag{10.47}$$

The Planck mass can be interpreted as the mass of a body whose reduced Compton wavelength (characteristic of relativistic quantum effects) is equal to its

Schwarzschild or gravitational radius:

$$\frac{\hbar}{m_P c} = 2\frac{G m_P}{c^2},$$

which leads to the above expression for m_P. This mass has a macroscopic value, and can be used to obtain the equivalent energy

$$E_P = m_P c^2 = \sqrt{\frac{c^5 \hbar}{G}} \approx 10^{16}\text{erg} \approx 10^{19} \text{ GeV}. \tag{10.48}$$

This energy is so large that the gravitational field can give rise to the spontaneous creation of particle–antiparticle pairs. The average temperature associated with that energy is 10^{32} K which is believed to be the initial temperature of the *primeval fireball* from which the Big Bang was produced.

10.8 Cosmic Numbers

As pointed out earlier, the mass of the visible Universe is estimated as being 10^{80} times the proton mass. This is an incredibly large number. Other very large numbers (called cosmic numbers) appear in the physics of the microscopic as well as the macroscopic world. The first cosmic number is the ratio of the electromagnetic and gravitational forces exerted between an electron and a proton. Letting F_e and F_G be the moduli of these forces, one has

$$F_e = \frac{e^2}{r^2}, \quad F_G = \frac{G m_p m_e}{r^2}, \tag{10.49}$$

where e is the electron charge, G the constant of gravitation, and m_p and m_e the proton and electron masses. The first cosmic number N_1 is then

$$N_1 = \frac{F_e}{F_G} = \frac{e^2}{G m_p m_e} = 0.23 \times 10^{40}. \tag{10.50}$$

Being the ratio of two forces, it is a dimensionless number.

The second cosmic number is the quotient of the radius of the Universe L and the proton radius r_P. The number L is of order 10^{10} light-years, and one light-year is $\simeq 10^{18}$ cm, so that $L \approx 10^{28}$ cm. On the other hand, $r_P \simeq 10^{-13}$ cm. Dividing L by r_P, one obtains the second cosmic number

$$N_2 = \frac{L}{r_P} \approx 10^{40}. \tag{10.51}$$

The coincidence in the orders of magnitude of the two numbers is very striking, and Dirac suggested that there should be some relation between them. Now, as L increases with time, N_2 also increases, and if there is a relation between N_2 and N_1, the latter should vary with time. There is no evidence that e^2, m_e, or m_p vary with time, so this would leave open the possibility that G might be time-dependent. This topic remains open to speculation.

Problems

Problem 10.1 Calculate the relativistic effects on time measured by a clock on a GPS satellite, which runs faster than a clock on the Earth's surface, and show that it amounts to around 38×10^{-6} s per day.

Problem 10.2 Starting from $dl^2 = (1 - r_g/r)^{-1}dr^2 + r^2(\sin^2\theta d\varphi^2 + d\theta^2)$, which is the spatial part of ds^2 in the expression (10.28) for the metric outside a spherically symmetric gravitating body, (i) find an expression for the radial distance $(l_2 - l_1)$ between two circles of radii r_1 and r_2 concentric to the body's centre, in this geometry; and (ii) obtain the limit $r_2 > r_1 \gg r_g$. (iii) Apply to the case $r_1 = 7 \times 10^8$ m, which is the order of the average solar radius, $r_2 = 5.8 \times 10^{10}$ m, which is of the order of the average radius of Mercury's orbit, and $r_g = 3 \times 10^3$ m, which is approximately the Sun's gravitational radius.

Problem 10.3 On an intuitive quantum mechanical basis, justify the Hawking–Bekenstein expression for the black hole temperature $T = \frac{\hbar c^3}{8\pi GMk}$ (up to a constant factor).

Problem 10.4 Once a black hole has formed, its mass is extremely unevenly distributed within it. The usual concept of density applies more to the body undergoing gravitational collapse, so the density to which we refer in the present and following problems corresponds to the density of the collapsing body rather than to the resulting black hole. In any case, these problems serve to show the scales involved in black holes. As an example, calculate the size and density of the black hole at the centre of our Galaxy, called Sagittarius A^*, estimated to have a mass $M = 4 \times 10^6 M_\odot \sim 8 \times 10^{36}$ kg.

Problem 10.5 Calculate the Schwarzschild radius for a black hole with density equal to that of (a) water, (b) the estimated density of (normal) matter in the Universe, which is 10^{-29} g/cm^3.

Problem 10.6 The Hawking–Bekenstein black hole entropy formula is $S = \frac{kA}{4l^2}$, where $l = \sqrt{\frac{G\hbar}{c^3}}$ is the Planck length and $A = 4\pi R^2 = 16\pi G^2 M^2/c^4$ is the area of the event horizon, since $R = 2GM/c^2$ is the gravitational radius. Use this to calculate the black hole temperature and internal energy.

Problem 10.7 Calculate the black hole heat capacity.

Fig. 10.10 We assume a perfect alignment and hence a complete Einstein ring.

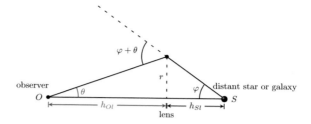

Problem 10.8 Einstein ring due to a lens star. If the distance and mass of the lens are approximately known, along with the distance to the more remote object, the radius of the Einstein ring can be calculated. Consider a lens mass similar to the Sun's mass, and assume the distance from the observer to the remote star to be $H = 60$ kpc. Assume also that the lens is located at a distance $H/3$ from the observer. Calculate the Einstein ring radius r assuming a perfect alignment. (1 parsec= 3.2616 lyr $\approx 3.0857 \times 10^{16}$ m) (Fig. 10.10).

Literature

1. M. Chaichian, I. Merches, D. Radu, A. Tureanu, *Electrodynamics. An Intensive Course* (Springer, Berlin, 2016). This is a very complete book, containing theory and applications, and the theoretical background and detection of gravitational waves are reported in a clear way
2. L.C. Epstein, *Relativity Visualized* (Insight Press, San Francisco, 1985). A highly appreciated intuitive approach to special and general relativity, replacing equations with diagrams and paper models
3. S. Hawking, *A Brief History of Time* (Bantam Books, New York, 1988). A unique book, recommended to all readers
4. L.D. Landau, E.M. Lifshitz, *The Classical Theory of Fields* (Pergamon, Oxford, 1975). The principles of general relativity are presented very clearly in this book
5. M.S. Longair, *Theoretical Concepts in Physics* (Cambridge University Press, London, 1984). There is a partial overlap in the scope of several sections of our book with that of this reference. The interested reader will find it useful to supplement several sections of the present chapter
6. T. Regge, *An Elementary Course on General Relativity* (CERN, Geneva, 1985). An excellent introduction to the topic, based on lectures given by the author
7. E.G. Thomas, D.J. Raine, *Physics to a Degree* (Gordon and Breach Science Publishers, 1998). Some of the proposed problems are adapted from those in this excellent book
8. S. Weinberg, *Gravitation and Cosmology* (Wiley, New York, 1972). This is a valuable book containing a large amount of information and adopting a very original approach, where the point of view of a particle physicist plays a more important role than in other books dealing with general relativity

Chapter 11
Unification of the Forces of Nature

In 1979, the Nobel Prize in Physics was awarded to Sheldon Glashow (b. 1932), Abdus Salam (1926–1996), and Steven Weinberg (b. 1933), for formulating a theory which unified the electromagnetic and weak interactions. By then, there was already enough experimental evidence concerning the predictions of the theoretical model they had built. As a coincidence, in 1979 was the centennial of the death of James Clerk Maxwell, who formulated a theory that clearly demonstrated the unified character of electric and magnetic phenomena. Also in 1979, the scientific world celebrated the centennial of the birth of Albert Einstein. Einstein devoted his last years to the search for a unified theory of electromagnetic and gravitational interactions.

The work by Glashow, Salam, and Weinberg partially achieved this goal when unifying electromagnetic and weak interactions into a common theory. But the way they did this was very different from Einstein's attempts. Einstein followed a classical (not quantum) approach, starting from general relativity and searching for a unification of gravitation and electrodynamics. In contrast, the electroweak theory was constructed within the framework of the modern renormalizable quantum theory of non-Abelian gauge fields with spontaneous symmetry breaking.

11.1 Theory of Weak Interactions

The discovery of radioactivity in 1896 by Henri Becquerel was the starting point of the study of what is now known as the decay of particles and transmutation of chemical elements. In 1899, Ernest Rutherford classified the radioactive emissions into two types: alpha and beta, followed by the addition in 1900 of the gamma rays, by Paul Villard (1860–1934). The power of penetration into matter increases from alpha to beta and finally to gamma radiation. Subsequent research showed that alpha radiation is composed of helium nuclei, beta radiation is composed of electrons and gamma rays are made up of very energetic photons. While gamma emission, being electromagnetic radiation, was obviously produced by the well-known electromagnetic interactions, the other two emissions turned out to be ways of probing two new

© The Author(s), under exclusive license to Springer-Verlag GmbH, DE, part of Springer Nature 2021
M. Chaichian et al., *Basic Concepts in Physics*, Undergraduate Lecture Notes in Physics, https://doi.org/10.1007/978-3-662-62313-8_11

fundamental interactions: the strong and the weak force. Alpha decay was explained in 1928 by George Gamow via quantum mechanical tunneling, the process being governed by the interplay of the strong and electromagnetic interactions.

Beta decay, which we briefly discussed in Sect. 9.4, is a manifestation of the weak interaction. The energy spectrum of the electrons in the beta emission is continuous, not discrete, as it would be expected. This seemed to show that the phenomenon was violating the energy-momentum conservation. Moreover, angular momentum seemed also not to be conserved in this process. The solution proposed by Wolfgang Pauli in 1930 came in the shape of a so-far undetected particle, very light, with spin 1/2 and no electric charge. Pauli called it a "neutron". In 1932, James Chadwick (1891–1974) discovered a chargeless nuclear particle, with a mass close to the mass of the proton, and named it also neutron. Since Pauli's "neutron" had to be extremely light, Enrico Fermi (1901–1954) cleared the confusion by naming Pauli's particle *neutrino*, i.e. "little neutron", and included the new particle in his theory of the beta decay.

Up until 1967, the weak interactions were described satisfactorily by means of a phenomenological model proposed in 1934 by Enrico Fermi. This used the notion of weak current $J_\mu(x)$, which contains a hadronic part (as mentioned earlier, hadrons are any particles that interact strongly, such as baryons and mesons) and a leptonic part (leptons are particles that do not interact strongly, including the electron and its associated neutrino, the muon and its neutrino, and the tau and its neutrino). Mathematically, the weak interaction was described by means of a Lagrangian density

$$\mathcal{L}_F(x) = \frac{G_F}{\sqrt{2}} J_h^{\dagger\mu}(x) J_{l\mu}(x) + \text{Hermitian conjugate}, \tag{11.1}$$

where J_h^μ is the hadronic current four-vector, J_l^μ is the leptonic current, and G_F is the weak (or Fermi) coupling constant. We do not enter into a more detailed analysis of (11.1). It is enough to point out that it is the product of two currents. Recall that the Lagrangian for the electromagnetic interactions has the form:

$$\mathcal{L}_{\text{em}}(x) = j^\mu(x) A_\mu(x), \tag{11.2}$$

where j_μ is the current four-vector and A_μ is the electromagnetic field four-vector. Fermi used this analogy in proposing the model (11.1) for weak interactions.

The Lagrangian (11.1) describes the beta decay process depicted in Fig. 11.1: a neutron decays into a proton, an electron, and an antineutrino. The hadronic current is composed of the proton and the neutron, $J_h^\mu = \bar\Psi_p \gamma^\mu \Psi_n$, and the leptonic current is composed of the electron and its antineutrino, $J_l^\mu = \bar\Psi_e \gamma^\mu \Psi_\nu$. The concurrence of the two pairs of particles corresponds to product of the two currents in the expression (11.1).

On the other hand, (11.2) implies several processes in quantum electrodynamics, one of them being indicated in Fig. 11.2: two electrons interact by exchanging a (virtual) photon and scatter on each other. But if we compare Figs. 11.1 and 11.2, it

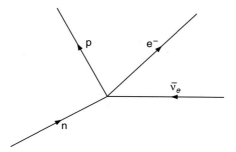

Fig. 11.1 Beta decay: decay of a neutron into a proton, an electron, and an antineutrino. The process is described as the interaction of two hadrons (proton and neutron) with two leptons (electron and antineutrino). At the vertex, we have four fermions and no boson. The direction of motion of an antiparticle is, by convention, opposite to the direction of the arrow on the diagram.

Fig. 11.2 Scattering of two electrons. In the diagram there are two pairs of fermions which exchange a boson, the photon γ.

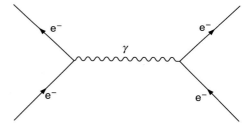

is striking that the first contains four fermions and no boson. The second contains four fermions, but the process is mediated by a vector boson, the photon.

Weak interaction processes were to some extent described satisfactorily by the model given by the expression (11.1). There were no phenomenological or experimental reasons for modifying this expression. However, the corresponding physical theory was not aesthetically pleasing. It did not satisfy the universally accepted principle that interactions between fermions should be mediated by bosons. In addition, scattering amplitudes calculated from (11.1) violate unitarity at very high energies and the theory is not renormalizable. By analogy with the expression (11.2), an interaction Lagrangian was proposed with the form

$$\mathcal{L}_w = g J_h^\mu W_\mu + g J_l^\mu W_\mu + \text{Hermitian conjugate},\tag{11.3}$$

where $W_\mu(x)$ is a new vector boson field, and g is a weak coupling constant. Instead of Fig. 11.1, we should have the diagram in Fig. 11.3, and between the neutron–proton and electron–antineutrino vertices, a virtual charged particle would appear, mediating the weak interaction.

There is an analogy between the model resulting from the expression (11.3) and quantum electrodynamics, but there are also important differences:

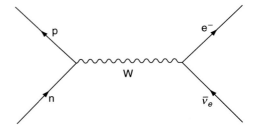

Fig. 11.3 Diagram that would replace the one in Fig. 11.1 if there existed a bosonic particle mediating the weak interactions. The particle W would differ from the photon in having electric charge and nonzero mass.

1. Electromagnetism is a long-range interaction, or equivalently, the photon has zero mass, while the weak interactions are short-range, whence the intermediate W bosons must be massive.
2. The photon is neutral, whereas the W bosons must be charged, in order to have charge conservation at each vertex.

These differences imply serious technical difficulties in constructing a satisfactory theory. A mathematical difficulty of prime importance was the non-renormalizability of the theory of weak interactions, with the consequence that the calculation of higher-order quantum corrections was inconsistent. This was in contrast to quantum electrodynamics, where the renormalization programme works.

The construction of a renormalizable theory of quantum electrodynamics was achieved due to the fact that the photon is a massless particle. This helped to relate divergences in different scattering probability amplitudes.

Now, concerning the massive W bosons, the situation was much more complicated. The theory was non-renormalizable, because it was not possible to conceive a formalism able to eliminate consistently the divergences. For this reason, it was not possible to use it for perturbative calculations, that is, to make an expansion in a series of diagrams, since the divergences in higher order terms were unrelated, creating as a whole an infinite number of divergences. So it was not possible to predict processes beyond the elementary ones.

Besides this renormalization problem, a number of other features of the weak interactions started to accumulated after Fermi's partially successful model. The most striking was the parity symmetry violation, which we discussed in Sect. 9.5. The conviction already existed that the electromagnetic and weak interactions should have been unified, and it was nontrivial to achieve parity violation only in the weak sector, while keeping the electromagnetic part parity-invariant, as all the experiments had shown.

Would there be some way of solving these difficulties? Would it be possible to modify the Lagrangian (11.3) to construct a unified renormalizable theory of electro- weak interactions? The answer was affirmative, and the credit for the creation of such a theory goes to Sheldon Glashow, Abdus Salam, and Steven Weinberg. Several other theoretical physicists, such as Yoichiro Nambu, Jeffrey Goldstone, Robert Brout, François Englert, Peter Higgs, Martinus Veltman, Gerard 't Hooft and others, contributed in a remarkable way to providing the theoretical support for its basic assumptions. Now, the idea was that the weak and electromagnetic forces are due to the existence of some vector fields which arise as the electromagnetic field, due to some gauge symmetry principle. In order to include such fields into the theory it was necessary to conceive new models of gauge fields and a mechanism for explaining the mass of the W bosons.

On conserved quantities and symmetries. We recall that there are several conservation laws in physics, and each conservation law is related to a symmetry by Noether's theorem. These are the conservation of energy–mass (due to invariance under time translation), conservation of linear momentum (due to invariance under space translation), conservation of angular momentum (due to invariance under spatial rotations), conservation of electric charge, conservation of color charge. In what follows, conservation of the baryonic number B is also assumed, which means conservation of the number of baryons minus the number of antibaryons ($N_B = n_B - \bar{n}_B$). Similarly, conservation of the lepton number L means conservation of the number of leptons minus the number of antileptons ($N_L = n_L - \bar{n}_L$). The conservation of these numbers is accidental. Some discrete symmetries, like invariance under parity (P), invariance under charge conjugation (C), invariance under time reversal (T), and CP symmetry are preserved in electromagnetic and strong interactions, but violated in weak interaction. CPT symmetry is universal in local relativistic quantum field theories.

11.2 Yang–Mills Fields

In 1954, Chen Ning Yang (b. 1922) and Robert L. Mills (1927–1999) proposed a generalization of the well-known gauge invariance of electrodynamics. In this model, each spacetime component of the gauge fields would also have components in an abstract space, the *isotopic space*. In 1932, Heisenberg had introduced a new quantum number, the isospin (or isotopic spin), in order to describe the similarity of properties of the proton and neutron under nuclear interaction. Heisenberg proposed to regard the proton and the neutron as two states of one particle, the *nucleon*, since their masses are almost equal and the strength of the strong interaction between any pair of them ($p - p$, $n - n$, or $p - n$) is the same. These two particles were included in a doublet, which was the fundamental representation of the isospin group of symmetry, $SU(2)$ (i.e. the *special unitary group of degree* 2). The group $SU(2)$ is locally isomorphic to $SO(3)$, the group of rotations in three-dimensional space. As a result, in isotopic space there are three independent "directions". One can consider the group $SU(2)$ as a higher-dimensional generalization of the group $U(1)$. The elements of the gauge group $SU(2)$, in the fundamental representation, are 2×2 matrices of the type

$$U(x) = e^{i \sum_{k=1}^{3} \alpha^k(x) T_k},$$

where $\alpha^k(x)$ are the real, coordinate-dependent parameters of the isospin transformations, and T_k are the 2×2 matrix generators of the transformations, which satisfy the relation

$$[T_i, T_j] = i f_{ijk} T_k. \tag{11.4}$$

This relation is general for any Lie group. The coefficients f_{ijk} are called structure constants and can be chosen to be antisymmetric in all indices. There is no difference if we place the internal symmetry indices up or down.

The complex matrices U are *unitary*, i.e. $U^\dagger = U^{-1}$, and *special*, meaning $\det U = 1$. One can show that the maximum number of independent 2×2 matrices satisfying these conditions is 3. The generators of the $SU(2)$ transformations are customarily chosen as $T_k = \sigma_k/2$, where σ_k are the Pauli matrices introduced in Chap. 7. In this case, $f_{ijk} = \epsilon_{ijk}$, the Levi-Civita symbol, or the completely antisymmetric unit tensor of rank 3.

One can define higher degree groups, denoted by $SU(n)$, whose elements are special unitary $n \times n$ matrices. For the general case, the number of generators is given by the formula $n^2 - 1$. This can be seen as follows: the total number of independent elements in a complex $n \times n$ matrix is $2n^2$. The unitarity conditions, written in components, are actually n^2 equations, which reduce the number of *independent* variables correspondingly. The unit determinant represents still one more condition, such that at the end we are left with the number $n^2 - 1$ mentioned above. The groups are non-Abelian because their generators do not commute, as shown by (11.4).

The special unitary groups, both global and local, are essential in particle physics, underlying the Standard Model and also various extensions of it. We shall frequently encounter them in this chapter. But let us return for the moment to the theory of Yang and Mills. Independently, the same idea of constructing gauge field theories in the general case, with any Lie group symmetry, was introduced by Ryoyu Utiyama around the same year of 1954. It should be emphasized that the most remarkable feature of a theory built on the principle of gauge invariance is that all the interactions and their corresponding interaction terms in the Lagrangian will appear automatically and in a unique way, as was the case of the electromagnetic interaction, which is an Abelian gauge theory.

In electrodynamics, the electromagnetic field is described using a four-vector potential A_μ, that is, a vector that has one component along each spacetime coordinate, and its mathematical transformation properties are dictated by the Lorentz group. The field proposed by Yang and Mills is such that in each direction of isotopic space there is a potential four-vector component. As the isotopic space has three dimensions, we have three four-vectors $A_\mu^1, A_\mu^2, A_\mu^3$, each one of them, in turn, with four spacetime components. The $SU(2)$ gauge transformations vary from point to point, and are as a consequence *local*, as we saw in Chap. 7. But due to the isotopic components, for such non-Abelian fields, the gauge invariance of electrodynamics does not work, and it is necessary to establish a new law relating the three iso-

Fig. 11.4 Trilinear and quadrilinear interactions of the non-Abelian gauge fields. Remark that such couplings cannot exist among photons in the Abelian gauge field theory of quantum electrodynamics.

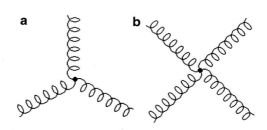

topic components A_μ^1, A_μ^2, A_μ^3 (which is beyond the scope of this book). This gives rise to new physical properties for these fields, as a consequence of their symmetry transformations.

In contrast to electrodynamics, the gauge transformations of Yang–Mills fields are non-commutative or non-Abelian (commutative transformations are also frequently called Abelian in honour of the Norwegian mathematician Niels Abel). For example, if we consider infinitesimal $SU(2)$ transformations, i.e. $\alpha_k(x) \ll 1$, the gauge fields transform as follows:

$$A_\mu'^k(x) = A_\mu^k(x) + \partial_\mu \alpha^k(x) + i\epsilon^{kij} A_\mu^i(x)\alpha^j(x). \tag{11.5}$$

Remark that the first two terms are identical as in the case of the Abelian $U(1)$ gauge field, while the last term is the mark of the non-Abelian transformation. In addition, a fundamental entity, the field-strength tensor, is nonlinear in the gauge fields having the expression

$$\mathcal{F}_{\mu\nu}^i = \partial_\mu A_\nu^i - \partial_\nu A_\mu^i + g\epsilon^{ijk} A_\mu^j A_\nu^k, \qquad i = 1, 2, 3.$$

In this formula, the summation convention for the indices j and k is used.

The tensor $\mathcal{F}_{\mu\nu}^i$ bears a close analogy to the Ricci tensor (10.24). The Yang–Mills fields $A_\mu^i(x)$ play the role of Christoffel symbols in general relativity. Actually, there are close similarities between the two theories.

In quantum electrodynamics, a photon has no direct coupling to other photons. The Lagrangian density for the $SU(2)$ gauge fields is

$$\mathcal{L}(x) = -\frac{1}{4}\mathcal{F}^{i\mu\nu}(x)\mathcal{F}_{\mu\nu}^i(x).$$

The above Lagrangian density physically implies that triple and quadruple interactions of non-Abelian gauge bosons are also possible, like those depicted in Fig. 11.4. We shall return to this property when we discuss the Standard Model of particle interactions.

Later on it became clear that the strong isospin symmetry $SU(2)$ remains all the way a global symmetry and there are in reality no gauge fields corresponding to it.

However, the theoretical framework proposed by Yang and Mills is the prototype for the description of any fundamental interaction, as we shall see later.

If a Yang–Mills field (or non-Abelian gauge field) interacts with some other field, such as a scalar or a fermionic field, the latter must also have several isotopic components. For instance, in the unified Glashow–Salam–Weinberg theory of electromagnetic and weak interactions, the electron and the neutrino behave like two (weak) isotopic components of one fermionic field.

One crucial aspect of the theory of Yang and Mills is that the gauge fields have to be massless, for the theory to be invariant with respect to the gauge transformations (11.5). But if the weak interactions are to be described by the Yang–Mills theory, the vector bosons have to be necessarily massive. We shall now discuss the mechanism by which the mass of the intermediate vector bosons is built into the electroweak theory.

11.3 Nambu–Goldstone Theorem

One of the simplest field-theoretical models is the scalar field, which is associated to particles of zero spin. But independently of its specific physical meaning, the scalar field can be studied as a simple and interesting model in which the Lagrangian has a potential term of the form

$$V(\Phi) = m^2 \Phi^\dagger \Phi + \lambda (\Phi^\dagger \Phi)^2, \qquad (11.6)$$

where Φ is a complex scalar field, $\Phi^\dagger$ its Hermitian conjugate, m^2 its mass squared, and $\lambda(\Phi^\dagger\Phi)^2$ describes its self-interaction, with λ a positive coupling constant. Such a Lagrangian is symmetric under the transformations of the group of phase transformations $U(1)$, i.e.,

$$\Phi(x) \rightarrow \Phi'(x) = e^{i\alpha} \Phi(x). \qquad (11.7)$$

One may ask what would happen if the potential term were

$$V(\Phi) = -m^2 \Phi^\dagger \Phi + \lambda (\Phi^\dagger \Phi)^2. \qquad (11.8)$$

One is tempted to answer that this is equivalent to having scalar tachyons, because m^2 corresponds to the square of the mass term, and if the term is negative, it leads to tachyons. But the presence of the term $\lambda(\Phi^\dagger\Phi)^2$ suggests a more careful interpretation. The potential (11.6) is represented in Fig. 11.5. The minimum of this potential corresponds to $\Phi = 0$. The expression (11.6) then describes particles of mass m. The ground state (or vacuum state) is non-degenerate.

The potential (11.8) takes the form depicted in Fig. 11.6. The minimum of the potential is reached at a value of $|\Phi|$ different from zero, which we may call ξ. In this case we say that spontaneous symmetry breaking (SSB) occurs, as discussed in Chap. 3. There are infinitely many vacuum states generated by transforming the field

Fig. 11.5 The potential of the scalar field $V(\Phi)$ for the *usual* case of positive m^2-term, as in (11.6). The minimum of $V(\Phi)$ corresponds to the value $\Phi = 0$.

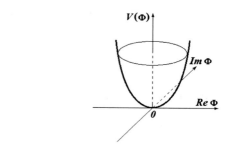

Fig. 11.6 The potential of the scalar field $V(\Phi)$ in the case of negative m^2-term, but with $\lambda > 0$, as in (11.8). The symmetry can be broken since the ground state is degenerate.

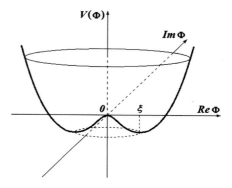

Φ according to (11.7). They are all different and a quantum mechanical superposition of them cannot exist—to different vacua correspond different worlds. More precisely, the Hilbert spaces constructed out of different vacua are all orthogonal to each other, so all interference effects in linear superpositions between the corresponding states vanish. Only one of these points can be used as the actual vacuum. The intercept with the real Φ axis is usually chosen. It can be demonstrated that, in this case, the Lagrangian describes a massive scalar field and another massless field.

The Nambu–Goldstone theorem states that when a global symmetry is spontaneously broken, massless particles appear. These scalar massless particles are called Goldstone bosons, or Nambu–Goldstone bosons.[1] If the symmetry is global, the Nambu–Goldstone boson should exist, but if the symmetry is local (i.e., if the parameter α is a function of x), this is not the case, according to the Brout–Englert–Higgs mechanism.

---

[1] The fact that massless particles are associated with a broken global symmetry was found in 1960 by Yoichiro Nambu (1921–2015) in the context of the Bardeen–Cooper–Schiffer (BCS) superconductivity mechanism. The idea was developed and elucidated by Jeffrey Goldstone (b. 1933). Thus, it is more proper to call them Nambu–Goldstone bosons.

11.4   Brout–Englert–Higgs Mechanism

What would happen if the scalar field interacted with some gauge fields, for instance, with non-Abelian Yang–Mills fields? In this case, if there is symmetry breaking, the would-be Nambu–Goldstone bosons will be eliminated from the theory, while some gauge fields become massive. This is equivalent to increasing their degrees of freedom by the same amount that the scalar fields have lost. For each would-be Nambu–Goldstone boson, a massless vector particle (gauge field) becomes massive.

This mechanism is also found in the theory of superconductivity. There, the scalar field is the Cooper pair of electrons and the vector field that becomes massive is the electromagnetic field. This 'mass' implies that the magnetic field has a short penetration depth.

The mechanism of spontaneous symmetry breaking in the context of gauge field theories was first proposed by Robert Brout (1928–2011) and François Englert (b. 1932) in 1964, and very soon after them, independently, by Peter Higgs (b. 1929). Gerald Guralnik, Carl Hagen, and Thomas Kibble also contributed to the elucidation of the mechanism. In this book we shall refer to it as the *Brout–Englert–Higgs mechanism*.[2] The 2013 Nobel Prize in Physics was awarded to François Englert and Peter Higgs for the theoretical discovery of this mechanism.

Actually, if we think deeper, the gauge symmetry is not really broken, but rather *hidden*. The vacuum structure is similar to the one in Fig. 11.6, only that now different vacuum states are related by gauge transformations. The gauge symmetry of the Lagrangian tells us that not all the degrees of freedom of the scalar field and of the vector fields are independent. To quantize the theory, one has to fix the gauge, i.e. to impose conditions among the fields, such that the number of degrees of freedom is decreased to only the physical ones. In usual gauge field theories, this is done by imposing relations between the components of the gauge fields, like, for example, $\partial_\mu A^\mu(x) = 0$, which is the so-called Lorenz gauge in quantum electrodynamics. In our case, we can fix the gauge also by imposing conditions on the scalar degrees of freedom. There is a gauge condition on the scalar degrees of freedom, called the *unitarity gauge*, which was originally used to show that the Yang–Mills fields acquire mass, and that no massless Goldstone boson appears. Since the gauge invariance is maintained, one may well use the Landau gauge (or more generally covariant gauges) as done by Brout and Englert, in which renormalization is strongly suggested but unitarity is not explicit. Alternatively, one may take the unitarity gauge used by Higgs (when translated in the language of field theory), in which unitarity is obvious but renormalization is hidden. Actually, what we regard as a symmetry breaking, is a gauge fixing procedure, and the gauge symmetry is in no way broken, but just hidden. The terminology "spontaneous gauge symmetry breaking", though not quite adequate, refers to such a situation. A remarkable consequence of the Brout–Englert–Higgs mechanism applied to the interaction of a Yang–Mills field with a scalar field

---

[2] Very often this mechanism has simply been referred to as the *Higgs mechanism*.

with spontaneously broken gauge symmetry is that the model is renormalizable. Consequently, the mechanism became a basic element in the theory describing the unification of electromagnetic and weak interactions.

11.5 Glashow–Salam–Weinberg Model

The Standard Model of particle interactions expresses our present knowledge of the electromagnetic, weak and strong interactions. A summary of the elementary particles which are included in the Standard Model is given in Fig. 11.7. On the left-hand side of the figure are the so-called matter particles, the quarks and the leptons, which have all spin 1/2. On the right-hand side are the carriers of the electromagnetic, weak and strong forces, which have all spin 1 (vector bosons). In the middle is the scalar (spin 0) Brout–Englert–Higgs particle, which arises from the spontaneous breaking of the gauge symmetry $SU(2)_L \times U(1)$. The antiparticles of all the elementary particles should be also included, although they are not explicitly mentioned.

The Glashow–Salam–Weinberg model is the unified theory of the electromagnetic and weak interactions. The story of its conception is a wonderful example of scientific theoretical creativity based on scarce experimental evidence and leading to spectacular predictions. Glashow, Salam, and Weinberg were awarded the Nobel Prize in 1979 for this theory, after many experimental confirmations had been

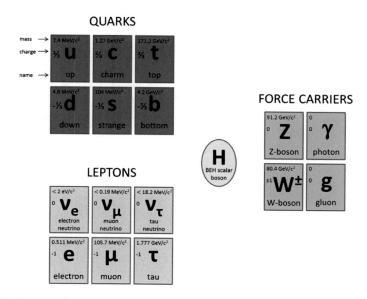

Fig. 11.7 Elementary particles in the Standard Model of particle interactions.

achieved. The story started in 1961, when Glashow proposed as electroweak unification structure the gauge group $SU(2) \times U(1)$. At that time, it was not known
how to make the gauge bosons massive. The Brout–Englert–Higgs mechanism was
proposed in 1964 and Weinberg incorporated the mechanism into Glashow's model
in 1967. By 1968, Glashow, Salam, and Weinberg had formulated in detail their theoretical model of electroweak unification. In 1971–1972, Gerard 't Hooft (b. 1946)
and Martinus Veltman (b. 1931) proved that this theoretical model with spontaneous
symmetry breaking was renormalizable; they were awarded the Nobel Prize in 1999.

The model is governed by the gauge group $SU(2)_L \times U(1)_Y$ (the sign $\times$ is read
"direct product of groups" and it means that the transformations of the two groups act
independently). The subscript L means "left" and it is an indication of the fact that
parity is broken in the weak interactions and the observed neutrinos are left-handed
spinors. The group $U(1)_Y$ with which we start does not represent the electromagnetic
interaction, but another symmetry, whose associated quantum number is called *weak
hypercharge* and is denoted by Y. Only after the symmetry breaking the usual electromagnetic interaction will arise in the form of a residual gauge symmetry, $U(1)_{em}$.
It should be emphasized that this $SU(2)_L$ gauge group has nothing to do with the
strong isospin global symmetry group $SU(2)$, which we have mentioned earlier.

In general, a usual spinor, like the electron field, can be decomposed in a sum of
the left-handed and right-handed helicities,

$$e(x) = e_L(x) + e_R(x).$$

In this formula, by $e(x)$ we denoted the field of the electron, which previously in
Chap. 7 was denoted by $\Psi(x)$. In the Standard Model there are so many matter
particles, that it is most convenient to denote the corresponding fields simply by the
symbols for the particles.

In the Glashow–Salam–Weinberg model, the leptons and quarks are grouped in
fundamental representations of the gauge groups $SU(2)$ and $U(1)$. In order to introduce the parity violation, only the left-handed parts of the spinors will be allowed to
interact with the three $SU(2)$ gauge bosons. This is achieved by distributing the left-
handed components in $SU(2)$ doublets and the right-handed components in $SU(2)$
singlets. However, both L and R components interact identically with the $U(1)_Y$
gauge boson. For the lepton sector we have the following assignment of representations

$$\begin{pmatrix} v_e \\ e \end{pmatrix}_L, \begin{pmatrix} v_\mu \\ \mu \end{pmatrix}_L, \begin{pmatrix} v_\tau \\ \tau \end{pmatrix}_L, \ e_R, \ \mu_R, \ \text{and } \tau_R, \tag{11.9}$$

while for the quarks we have

$$\begin{pmatrix} u \\ d' \end{pmatrix}_L, \begin{pmatrix} c \\ s' \end{pmatrix}_L, \begin{pmatrix} t \\ b' \end{pmatrix}_L, \ u_R, \ d'_R, \ c_R, \ s'_R, \ t_R, \ \text{and } b'_R. \tag{11.10}$$

The quark states d', s', and b', which take part in weak interactions, are linear combinations (mixtures) of the quark states d, s, and b, which interact strongly. The

necessity for mixing the quarks is dictated by experiment; this will be discussed in some detail later. Remark also that all the quarks have L and R components, while in the lepton sector the neutrinos do not have the R component, in accord with the experimental observations.

It should be also mentioned that this assignment of representations makes impossible the presence of usual mass terms for the matter particles, since such terms cannot be made gauge-invariant. The leptons (except neutrinos) and quarks in the Glashow–Salam–Weinberg model acquire their mass through the same Brout–Englert–Higgs mechanism of spontaneous symmetry breaking that gives masses to the carriers of the weak force.

Let us now turn to the gauge bosons. There are three gauge bosons corresponding to the $SU(2)_L$ symmetry, which we shall denote by W_μ^k, with $k = 1, 2, 3$; there is also one vector boson, B_μ, corresponding to the hypercharge symmetry $U(1)_Y$. Thus, in total, we have four massless gauge bosons. Each of them has only two possible polarizations, transverse to the direction of propagation; in other words, two physical degrees of freedom.

Upon the spontaneous symmetry breaking, the three carriers of the weak force have to become massive, thus justifying the short-range character of the weak interactions. The photon which mediates the electromagnetic interaction, however, has to remain massless. This implies that the breaking of the symmetry is not complete, but a residual $U(1)_{em}$ invariance remains:

$$SU(2)_L \times U(1)_Y \to U(1)_{em}.$$

To achieve this, we use the Brout–Englert–Higgs mechanism with a doublet of complex scalar fields,

$$\Phi(x) = \begin{pmatrix} \phi_1 + i\phi_2 \\ \phi_3 + i\phi_4 \end{pmatrix}. \tag{11.11}$$

The scalar doublet interacts with all the four gauge fields and its Lagrangian contains the potential part with a self-interaction term of the form

$$-m^2 \Phi^\dagger \Phi + \lambda (\Phi^\dagger \Phi)^2,$$

which produces the needed nonzero vacuum expectation value. The four distinct components of the scalar doublet correspond to four degrees of freedom.

The Lagrangian density of the Glashow–Salam–Weinberg model, containing all these matter and gauge fields with their respective interactions, is invariant under the gauge group $SU(2)_L \times U(1)_Y$. The vacuum of the theory, however, is not invariant: by gauge transformations one goes from one vacuum configuration to a distinct vacuum configuration. One has to make a choice of the vacuum state, which is equivalent to fixing the gauge. Any choice would be just as good physically, but one of them, called the *unitarity gauge*, is more convenient for the physical interpretation of the resulting theory. In the unitarity gauge, three of the scalar degrees of freedom disappear from the Lagrangian, and we are left with one massive chargeless scalar,

$H(x)$, which is the Brout–Englert–Higgs particle. Three combinations of the original gauge bosons become massive, while one combination remains massless. The vector bosons W_μ^k and B_μ mix as follows:

$$W_\mu^+ = \frac{1}{\sqrt{2}}(W_\mu^1 + i W_\mu^2),$$

$$W_\mu^- = \frac{1}{\sqrt{2}}(W_\mu^1 - i W_\mu^2) \tag{11.12}$$

and

$$A_\mu = B_\mu \cos\theta_W + W_\mu^3 \sin\theta_W,$$

$$Z_\mu = -B_\mu \sin\theta_W + W_\mu^3 \cos\theta_W. \tag{11.13}$$

The free parameter θ_W is called the Weinberg angle, or weak mixing angle, and it is related to the coupling constants of the $SU(2)_L$ and $U(1)_Y$ gauge symmetries. The fields $W_\mu^\pm$ have electrical charge and they form a particle–antiparticle pair. The field Z_μ is neutral and it is its own antiparticle. All three are now *massive* vector bosons, meaning that they have each three degrees of freedom (two transverse and one longitudinal polarizations). It is sometimes said that the gauge fields "ate" the would-be Goldstone bosons. Thus, the degrees of freedom lost by the scalar fields are gained by the vector boson fields. The mass of the Z_μ boson is 91.2 GeV/c^2 and the mass of the $W_\mu^\pm$ is 80.4 GeV/c^2 (about 100 times more than the mass of the proton). The Weinberg angle relates also these two masses: $m_W = m_Z \cos\theta_W$. The combination A_μ remains massless and it represents the photon of the electromagnetic interaction.

The electroweak unification means that at very high energies (above the unification scale), or at very tiny distances, the electromagnetic and the weak force are of comparable strength. The apparent "weakness" of the weak interaction in experiments performed at low energies reflects only its short range of action. As a result, the essential difference between the strengths of the two forces can be due only to the mass of the vector bosons which mediate the weak processes.

The processes in which the $W_\mu^\pm$ and Z_μ particles participate are weak interactions. The $W_\mu^\pm$ bosons interact only with the L components of the matter fields, while Z_μ bosons interact with both L and R components, but differently. The observed parity violation in weak interaction is thus perfectly well incorporated in the model. The photon A_μ interacts identically with the L and R components of the matter fields.

Since we have started with a non-Abelian gauge field theory, nonlinear couplings of gauge bosons appear. Naturally, the photon interacts with $W^\pm$, since they are charged particles. But also triple and quadruple couplings of $W^\pm$ occur, as well as other combinations of $W^\pm$, Z and γ. All possible vector boson vertices are represented in Fig. 11.8.

While the weak interactions involving the charged $W_\mu^\pm$ bosons were expected, the neutral boson Z_μ was a prediction of the theory. The Z_μ boson interacts with

Fig. 11.8 All possible vector boson vertices in the Glashow–Salam–Weinberg model of electroweak unification.

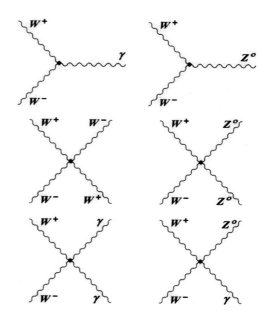

all the matter particles, including the neutrinos. They mediate in the neutral current processes, whose observation was one of the first confirmations of the Glashow–Weinberg–Salam model. The weak neutral currents were discovered in 1974 in a neutrino experiment with the Gargamelle bubble chamber at CERN. The existence of the neutral current means that a weak force also acts between electrons, in addition to the electromagnetic force mediated by photons. This force also acts between protons and electrons in an atom, and between the protons and neutrons inside the nuclei.

Early in 1983, in experiments carried out at CERN by the UA1 and UA2 collaborations, when beams of protons and antiprotons were made to collide at high enough energy, some very energetic electrons appeared whose presence was interpreted as being due to the decay of W_μ^- produced in the proton–antiproton collisions. Soon after this evidence of the $W_\mu^\pm$ bosons, very energetic electron–positron pairs were observed, whose origin could be attributed to the decay of Z_μ particles, also produced in the proton–antiproton collisions. In 1984, Carlo Rubbia (b. 1934) and Simon van der Meer (1925–2011) were awarded the Nobel Prize for this experimental discovery.

One particle predicted by the Glashow–Weinberg–Salam model eluded discovery for over 40 years: the Brout–Englert–Higgs scalar. It has been lately the most awaited discovery in high energy physics. On the 4th of July 2012, the ATLAS and CMS collaborations at CERN simultaneously reported the discovery of a previously unknown boson with a mass of about 125 GeV/c^2. The data analysis performed by December 2012 shows that the particle has spin 0 and its behaviour is consistent with the Standard Model scalar.

11.6   Electroweak Phase Transition

As already mentioned, there is a close analogy with the phenomenon of superconductivity, where the magnetic field acquires a mass inversely proportional to the
London penetration length, this implying the Meissner effect whereby the magnetic
field penetrating the superconductor decreases exponentially. Superconductivity is
destroyed by increasing the temperature above some critical temperature T_c, and in
1972 it was suggested by David Kirzhnitz and Andrei Linde that a similar mechanism
may occur in the electroweak theory. The point is that the symmetry breaking parameter ξ and, as a consequence, the effective masses of the $W_\mu^\pm$ and Z_μ bosons, should
decrease with temperature and vanish at some high enough critical temperature. At
this point, the weak interactions would become effectively long-range forces just
like the electromagnetic force. The critical temperature for this electroweak phase
transition is believed to be of order 10^{15} K. Such high temperatures are believed to
have occurred in the early stages of the Univers, after the Big Bang. Under these
extreme conditions, the symmetry, broken at lower temperatures, is restored, and all
the components of the electroweak field acquire equal status.

The basic idea is that the effective potential now takes a more complex form than in
(11.8), since quantum and temperature corrections must be taken into account. Actually, at very high temperature, the latter become dominant. The effective potential
takes the form

$$V(\xi) = \frac{\lambda \xi^4}{4} - \frac{a^2 \xi^2}{2} + V(\xi, T), \tag{11.14}$$

where $V(\xi, T)$ depends on ξ through the masses of the particles ($W^\pm$ and Z vector
bosons, Brout–Englert–Higgs scalar particle, electrons, and quarks), since all these
masses are assumed to be generated by spontaneous symmetry breaking and are
proportional to ξ. At high temperatures, the terms contributing to $V(\xi, T)$ can be
approximated by a series expansion containing terms proportional to $T^4, m^2 T^2, m^3 T$,
$\ln T^2/m^2$, etc., where m denotes the masses of the particle species. Fermions as well
as bosons contribute to these terms, except the last term linear in T, to which only
bosons contribute.

In oversimplified conditions, two possible models result, giving rise respectively
to first or second order phase transitions, in very close analogy to the cases discussed
in Chap. 3. In the case of a second order phase transition, the temperature of the
Universe decreased to the critical value for the electroweak phase transition T_c, at
which spontaneous symmetry breaking took place with the parameter ξ taking a small
value. But similarly to a ferromagnetic material (see Fig. 3.27 in Chap. 3, replacing
$G_1(M)$ by $V(\xi)$ and M by ξ), this parameter gradually increased as the temperature
decreased. This corresponds to the case where the effective potential has the form

$$V(\xi) = \frac{\lambda \xi^4}{4} - \frac{\gamma}{2}(T_c^2 - T^2)\xi^2, \tag{11.15}$$

with γ a constant and T_c the critical temperature of symmetry restoration.

But if the phase transition is first order, the situation looks like what happens in some ferroelectric materials (see Fig. 3.28 in Chap. 3, this time replacing $G_1(P)$ by $V(\xi)$ and P by ξ). When the free energy (or effective potential) of the symmetrical and non-symmetrical phases had the same value, spontaneous breaking of symmetry took place in some regions of the Universe. *Bubbles* were formed with a broken symmetry, and our visible Universe is one of them. An effective potential describing such a case is

$$V(\xi) = \frac{\lambda \xi^4}{4} + \frac{\gamma}{2}(T^2 - T_0^2)\xi^2 - \frac{1}{3}\alpha T \xi^3, \qquad (11.16)$$

where T_0 is the temperature above which the symmetrical phase $\xi = 0$ is metastable. This potential has two minima, one for $\xi = 0$ and another for $\xi_M \neq 0$. The critical temperature at which $V(0) = V(\xi_M)$ characterizes the discontinuous phase transition. The treatment of the electroweak phase transition given in this section is mainly of qualitative nature. For instance, the conditions may change drastically if the Brout–Englert–Higgs boson mass is more than 60 GeV/c^2, and this seems to be the case. The Large Hadron Collider (LHC) experiments suggest that it is around 125 GeV/c^2. In this case, numerical methods are necessary to get more realistic models of the electroweak phase transition.

11.7 Hadrons and Quarks

There is a fundamental difference between leptons and hadrons. The leptons do not interact strongly. On the other hand, they behave as particles without internal structure, that is, as pointlike particles, without perceptible dimensions. The hadrons differ from the leptons in many respects. In the first place, they have dimensions of order 10^{-13} cm, and in the collisions of these particles at very high energies, they exhibit an internal structure. The electric, magnetic, and strong fields of protons and neutrons seem to emanate from pointlike sources inside them—the quarks.

The quarks are fermions, that is, they obey Pauli's principle, and the strong interaction between them is mediated by the colour or gluon field.

There are two properties that are believed to be fundamental for the quarks: one is called confinement and the other is asymptotic freedom. Quark confinement means that these particles are not found in the free state: quarks live inside hadrons and there are no quark singlets. They only appear in doublets, triplets, or higher n-lets. Baryons are made from three quarks and mesons from quark–antiquark pairs.

Asymptotic freedom is a property of quarks wherein they behave as though free, or completely non-interacting, when they come close enough (or equivalently, when their energy is very high, e.g., when we observe them by colliding particles at very high energy). There are several types of quarks, known somewhat facetiously by physicists as *flavours*. At present it is believed that there are six *flavours*. There is a set of quantum numbers for any flavour and colour. These are baryonic number

B(=1/3), spin $J(=1/2)$, isotopic spin I and its projection I_3, strangeness S, charm C, bottomness (or beauty) B, topness (or truth) T, and electric charge Q (in units of e).

Quark flavor	I	I_3	S	C	B	T	Q
u (up)	$\frac{1}{2}$	$\frac{1}{2}$	0	0	0	0	2/3
d (down)	$\frac{1}{2}$	$-\frac{1}{2}$	0	0	0	0	$-1/3$
s (strange)	0	0	-1	0	0	0	$-1/3$
c (charm)	0	0	0	$+1$	0	0	2/3
b (bottom)	0	0	0	0	-1	0	$-1/3$
t (top)	0	0	0	0	0	$+1$	2/3

The following generalized Gell-Mann–Nishijima formula connects these quantum numbers characterizing each particle:

$$Q = (B + S + C + B + T)/2 + I_3. \tag{11.17}$$

Ordinary matter is made only of the elementary particles of the first generation, i.e. the quarks u and d and the leptons e and v_e, together with their antiparticles. All the members of the other two generations were discovered in high energy physics experiments (cosmic rays or particle accelerators). For instance, the proton has the structure (uud), with total electric charge $+1$. The neutron has the structure (udd), with zero net charge. Similarly, the π^+ meson has the structure $(u\bar{d})$, i.e., it is composed of a quark u of charge 2/3 and an antiquark $\bar{d}$, of charge 1/3, whence its charge is $+1$.

The mechanism of beta decay can now be understood as due to the process in which a d quark in the neutron emits a virtual W^- boson of charge -1, and then becomes a u quark, so that the original neutron passes from the neutral structure udd to uud, with charge $+1$, corresponding to a proton. The virtual W^- decays into an electron and an antineutrino (see Fig. 11.9).

The quarks assemble into baryons and mesons. The baryons, having baryonic number 1, are made of three quarks, as a result their spin can be either 1/2 or 3/2. The mesons, which are quark–antiquark pairs, can have either spin 0 (pseudoscalar mesons) or spin 1 (vector mesons). These two cases of spin states correspond to particles with lowest masses, when the angular momenta among the quarks are zero.

After the Second World War, with the development of the accelerator technology, a great number of particles—baryons and mesons—were produced in the laboratories. Some of them, the Λ^0 particle and the K mesons, had an unexpected, *strange* behaviour in their weak decays. What was strange about them? There is a direct connection between the strength of an interaction and the speed of interaction.[3] The Λ^0 particles, for example, have a relatively long lifetime on a nuclear time scale, of about 10^{-8} to 10^{-10} s, which is typical for a weak interaction decay. On the other hand, they are copiously produced, which is typical for the strong interaction. If they

---

[3]It is illuminating to think that, if a strong interaction process takes place on a given time scale in 1 s, then a weak interaction process takes place in one million years!

Fig. 11.9 The mechanism of beta decay as W boson exchange, with the nucleons represented using the quark model.

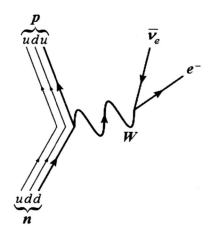

were to decay by strong interaction, their lifetime would have been about 10^{-21} to 10^{-22} s. Later, a *cascade particle* Ξ was produced, which decayed in 10^{-10} s into a Λ^0 and a pion, and subsequently the Λ^0 decayed further into a pion and a nucleon. To account for the nature of such particles, which were produced only in pairs in strong interactions, a new quantum number, *strangeness*, was proposed by Kazuhiko Nishijima (1926–2009) in 1955 and by Murray Gell-Mann (b. 1929) in 1956. This was the first quantum number to be introduced after Heisenberg's strong isospin, which characterizes the nucleons. The strangeness is conserved in strong and electromagnetic interactions, but not in weak interactions. A classification scheme of the hadrons, based on the flavour symmetry group $SU(3)$, was then introduced in 1961, independently, by Yuval Ne'eman (1925–2006) and Murray Gell-Mann. Gell-Mann called it "The Eightfold Way". This scheme is reminiscent of Mendeleev's periodic table of elements, which was put together based on the properties of atoms, without their underlying electronic structure being known at that time. Similarly, the $SU(3)$ flavour classification was an empiric scheme, which led to the prediction of particles that were subsequently found.

The $SU(3)$ classification of baryons and mesons is presented in Figs. 11.10 and 11.11. In 1961 were known only particles with single strangeness (like Λ^0 and the K mesons) and with double strangeness (the cascade particles, Ξ^0 and Ξ^-), but no particle with triple strangeness was known. However, the representations of the group $SU(3)$ required such a particle to complete the decuplet of baryons (Fig. 11.10b). The particle Ω^-, with strangeness -3, was discovered in 1964 at the Brookhaven National Laboratories, with the decay properties predicted by Gell-Mann and Ne'eman. In 1969, Murray Gell-Mann received the Nobel Prize for his contributions to the classification of elementary particles and their interactions.

Interesting enough, while the known particles fell into the higher representations of the group $SU(3)$, there were no particles to be ascribed to the smallest, or fundamental, representation, which should have had only three elements. In 1964, Murray Gell-Mann and George Zweig, independently, proposed a set of three elementary

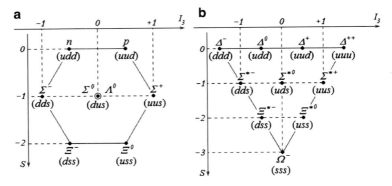

Fig. 11.10 **a** The spin $1/2$ baryons and **b** the spin $3/2$ baryons in the quark model based on $SU(3)$ flavour symmetry.

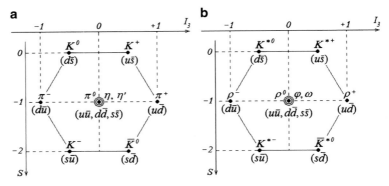

Fig. 11.11 **a** The pseudoscalar mesons and **b** the vector mesons in the quark model based on $SU(3)$ flavour symmetry.

particles as members of the fundamental representation, from which all the known baryons and mesons were supposed to be made up. Zweig called them "aces", while Gell-Mann called them "quarks" and denoted them by u, d, and s. The latter name became popular. In Figs. 11.10 and 11.11 the quark content of the particles is also indicated. The quarks were introduced as a mere mathematical device, to explain the hadron classification scheme. Even their inventors did not think of them as real particles. The physical reality of quarks was proven in the end of the 1960s, by experiments of deep inelastic scattering. The idea is inspired by Rutherford's experiments with atoms and alpha particles. In the deep inelastic scattering, the protons and neutrons of atomic nuclei were probed with very energetic leptonic beams. This experiments showed that the hadrons had internal structure, with three pointlike scattering centres in baryons and two in mesons. The experimental data also confirmed that the centres had the fractional electric charges assigned to quarks in the Standard Model.

Nevertheless, free quarks have never been observed. In high energy collisions, in which quarks and gluons are knocked free from a nucleon, for example, a process called hadronization takes place: the free quark or gluon combines with virtual quarks and antiquarks created spontaneously from the vacuum and tight jets of hadrons are formed. This is the case with all but one of the quarks: the most massive quark, the top, once produced, it decays by weak interactions before it has time to hadronize by strong interactions. The top is the only quark which has been studied "bare".

But let us return to the story of the quarks as it was in the 1960s.

Recall that strangeness is determined by the quarks s, $\bar{s}$, and is conserved in strong and electromagnetic interactions, but not in weak interactions. Among the set of strange particles are the baryon Λ^0 and the mesons $K^\pm$, K^0, $\bar{K}^0$. The quark structure of kaons is $K^0 = d\bar{s}$, $\bar{K}^0 = s\bar{d}$, $K^+ = u\bar{s}$, $K^- = s\bar{u}$.

It was found that the rates of weak decays in which the change in strangeness was $\Delta S = 1$ were much smaller than the rates for those decays in which $\Delta S = 0$. In 1963, Nicola Cabibbo (1935–2010) introduced a new mixing parameter, the Cabibbo angle θ_C, in order to salvage the universality of weak interactions. Using the quark model, this means that the d and s quarks participate in weak interactions in a mixed quantum state of the form $d' = d \cos\theta_C + s \sin\theta_C$. If the angle θ_C is small enough, this can explain the observed difference of results. The problem became more serious in the case of some neutral current decays in which the change $\Delta S = 1$ was absent, although theoretically they seemed to be allowed.

Then, in 1970, Sheldon Glashow, John Iliopoulos (b. 1940), and Luciano Maiani (b. 1941) suggested the existence of a fourth quark, *charm*, and of the mixed state $s' = s \cos\theta_C - d \sin\theta_C$. This eliminated the term $\Delta S = 1$ by what is called nowadays the GIM mechanism. Theoretically, now there were two quark doublets, (u, d') and (c, s'), where

$$\begin{pmatrix} d' \\ s' \end{pmatrix} = \begin{pmatrix} \cos\theta_C & \sin\theta_C \\ -\sin\theta_C & \cos\theta_C \end{pmatrix} \begin{pmatrix} d \\ s \end{pmatrix}. \tag{11.18}$$

The first particle containing the charm quark was observed in 1974 and simultaneously reported by the Stanford Linear Accelerator centre (SLAC) team led by Burton Richter (1931–2018) and the Brookhaven National Laboratory team led by Samuel Ting (b. 1936). Richter and Ting were awarded the Nobel Prize in 1976. The famous particle is the J/ψ vector meson, as it was given different names by the two experimental groups that discovered it. This meson is a bound state of charm and anticharm, and such bound states are also called *charmonium*. After the discovery of the c-quark, the flavour group $SU(3)$ was not enough anymore, and the composite particles had to be classified according to the representations of $SU(4)$.

By 1964, the CP violation in weak interactions had been discovered experimentally (see Chap. 9). However, there were no CP-violating terms in the Lagrangian of the Standard Model. After the proposal of the GIM mechanism, in 1972, Makoto Kobayashi (b. 1944) and Toshihide Maskawa (b. 1940) suggested to enlarge the number of quark families by one, formed of the quarks top and bottom, or t and b, such that the quarks d, s, and b are all mixed. Thus, they were enlarging the 2×2 Cabibbo matrix to a 3×3 unitary matrix. The crucial difference was that their matrix had four independent parameters: three mixing angles and one phase, $e^{i\delta}$. A simple argument shows that the phase in the Kobayashi–Maskawa matrix is responsible for the CP violations: the Lagrangian of the model remains CPT invariant, and indeed no departure from the CPT symmetry has yet been observed. Upon the time reversal transformation, $t \to -t$, all the complex numbers in the Lagrangian have to be replaced by their conjugates, therefore $e^{i\delta} \to e^{-i\delta}$. Due to the change in phase, the Lagrangian will violate T parity, but this is equivalent to the violation of the CP-parity, since the Lagrangian is CPT-symmetric. With the 2×2 Cabbibo matrix, the CP violation could not be introduced, since in this case there is no complex coupling. Thus, a new family of quarks was predicted. In 2008, Kobayashi and Maskawa shared the Nobel Prize with Yoichiro Nambu, "for the discovery of the origin of the broken symmetry which predicts the existence of at least three families of quarks in Nature".

Of the third family, the b-quark was the first to be discovered, in 1977, at Fermilab, by the E288 experiment team led by Leon Lederman. The t-quark, more than 40 times heavier than the bottom, was discovered in 1995 also at Fermilab, by the CDF and D0 teams, using the Tevatron collider.

In 1967, Andrei Sakharov (1921–1989)   realized that CP violation was an essential ingredient for explaining the matter–antimatter asymmetry observed in our Universe.

In 2001, important experiments were reported by two multinational groups, one in the Belle collaboration, at the KEK laboratory in Tsukuba, Japan, and another in SLAC, using the BaBar detector. In these experiments, a difference or asymmetry was shown between the decays of mesons $B^0 = d\bar{b}$ and those of their antiparticles $\bar{B}^0 = b\bar{d}$. This was an important discovery after the experiments with K^0—$\bar{K}^0$ carried out by Cronin and Fitch in 1964 concerning the violation of CP invariance. It was of great importance for the question of matter–antimatter asymmetry and baryogenesis (see Sect. 11.11).

11.8   Neutrino Oscillations and Masses

Although the electron neutrino was theoresized by Pauli in 1930 to explain the puzzling spectrum of the beta decay, the direct observation of neutrinos was reported only as late as 1956. Neutrinos are elusive particles, with no electric charge and such a tiny mass, that they propagate in practice at the speed of light. They interact only weakly, and it took a lot of ingenuity to devise an experiment to observe them. The method chosen by Frederick Reines (1918–1998) and Clyde Cowan (1919–1974) to detect free antineutrinos was the so-called *inverse beta decay*. In this process, an antineutrino interacts with a proton, giving in the final state a neutron and a positron:

$$\bar{\nu}_e + p \to n + e^+.$$

In the Reines–Cowan experiment, a significant flux of antineutrinos was produced by a nuclear reactor. They interacted with the protons from a water tank. The resulting positrons immediately combined with electrons, leading to two photons ($e^+ + e^- \rightarrow \gamma + \gamma$) detectable by scintillation. The occurrence of the inverse beta decay was made certain by the detection of the final state neutrons, using cadmium bars. Cadmium absorbs neutrons with high probability, passing to an excited state. When it returns to the lowest energy state, the atom of cadmium releases a photon. The coincidence of the electron–positron annihilation with the neutron capture was a double signature of the detection of free antineutrinos $\bar{\nu}_e$.

The muons were detected in 1936 by Carl Anderson in experiments with cosmic rays, which were in those times natural accelerators for particles. Their properties were very similar to those of the electrons, for which reasons they were called also "heavy electrons" (having the mass some 200 times larger than the electron). It was expected that the pairing with neutrinos would happen in this case as well. The muon neutrino was discovered in 1962 at the Brookhaven National Laboratory by a team led by Leon Lederman (1922–2018), Melvin Schwartz (1932–2006), and Jack Steinberger (1921–2020). They were awarded the Nobel Prize in 1988.

The last lepton to be discovered was the tau or the "superheavy electron", discovered in 1975 at SLAC, by a team led by Martin Perl (1927–2014). In 1995, by the Nobel Prize awarded to Frederick Reines and Martin Perl were honoured the discoveries of the first neutrino and of the last lepton. The tau neutrino discovery followed in the year 2000 at Fermilab, by the DONUT collaboration. The tau neutrino events are so rare, that observation of even one of them makes the headlines of newspapers.

In the Standard Model, neutrinos are massless particles. However, the experiments show that the situation is quite different—they have masses and they oscillate from one flavour into another. The idea of neutrino oscillations was proposed in 1957 by Bruno Pontecorvo (1913–1993), in analogy with the oscillations of the K mesons presented in Chap. 9. Basically, it is assumed that the *weak eigenstates* ν_e, ν_μ, and ν_τ are quantum mechanical superpositions of the *mass eigenstates* ν_1, ν_2, and ν_3:

$$\begin{pmatrix} \nu_e \\ \nu_\mu \\ \nu_\tau \end{pmatrix} = V \begin{pmatrix} \nu_1 \\ \nu_2 \\ \nu_3 \end{pmatrix}, \tag{11.19}$$

where V is a 3×3 unitary matrix called the Pontecorvo–Maki–Nakagawa–Sakata matrix. It is misleading to speak about the masses of the weak eigenstates. The limits given in Fig. 11.7 refer to mass expectation values of the weak eigenstates. However, direct mass measurements cannot be made, but the experiments show unequivocally that the neutrinos oscillate and mass eigenstates must exist.

The number of electron neutrinos detected on Earth as coming from the Sun is only about one third of what is expected according to the theory of thermonuclear reactions occurring in the Sun. This is known as the solar neutrino problem, and it was discovered in the end of 1969, in the Homestake experiment led by Raymond Davis (1914–2006) and John Bahcall (1934–2005). The idea that neutrinos may oscillate, i.e., the electron neutrino may change into μ or τ neutrinos, is a convincing explanation for this peculiar phenomenon. Actually, before the Homestake experiment, Pontecorvo had predicted that the observed flux of solar neutrinos might be two times smaller than what was predicted (in those times, only two species of neutrinos were known). Observing neutrino oscillation was a great task for the experimentalists over many years. The conclusive evidence for neutrino oscillation was provided starting with 1998 by the Super-Kamiokande team in Japan led by Takaaki Kajita (b. 1959). Before that, the Kamiokande and Super-Kamiokande experiments had been used for the discovery of cosmic neutrino, by a team led by Masatoshi Koshiba (1926–2020). In 2002, Davis and Koshiba were awarded the Nobel Prize "for pioneering contributions to astrophysics, in particular for the detection of cosmic neutrinos". In 1985, Stanislav Mikheyev and Alexei Smirnov, using previous work by Lincoln Wolfenstein, suggested that flavour oscillations could be modified when neutrinos propagate through matter. This is the so-called MSW (Mikheyev–Smirnov–Wolfenstein) effect. This oscillation, which is similar to refraction, requires neutrinos to have a small mass, of the order of a few eV/c^2. In 2002, measurements of neutral currents produced by neutrinos of different families were observed in the Sudbury Neutrino Observatory (SNO) in Ontario, Canada, and the results confirmed that there were twice as many non-electron neutrinos as electron neutrinos. This observation indicated that there are no missing solar neutrinos, and that neutrinos are in fact massive. In 2015, the Nobel Prize for Physics was awarded to Takaaki Kajita (b. 1959) and Arthur B. McDonald (b. 1943) "for the discovery of neutrino oscillations, which shows that neutrinos have mass."

Dirac, Weyl, and Majorana Fermions. A Dirac fermion is a fermion which is not its own antiparticle. Most fermions appearing in Nature fall under this category, as they are not their own antiparticles. With the possible exception of neutrinos, they are Dirac fermions and are modeled by the Dirac equation.

Hermann Weyl (1885–1955) showed that the massless Dirac equation could be reduced to a two-component equation. The solutions of that equation are called Weyl spinors, or Weyl fermions, and have well defined chirality. A Dirac fermion is equivalent to two Weyl fermions. A Majorana fermion is such that it is its own antiparticle. It has been conjectured that neutrinos are candidates for Majorana fermions. If this were so, it would violate lepton number conservation, and this would change several basic ideas of present day physics. The idea of double beta decay was first proposed by Maria Goeppert-Mayer (1906–1972) in 1935. In 1937, Ettore Majorana (1906–1938) demonstrated that all the results of beta decay theory (see Chap. 9) would remain unchanged if the neutrino were its own antiparticle. In 1939, Wendell H. Furry (1907–1984) showed that if neutrinos were Majorana particles, then double beta decay could proceed without the emission of any neutrinos, via the process now called neutrinoless double beta decay. The nature of neutrino as Dirac or Majorana particle is one of the fundamental standing problems in elementary particle physics. The experimental effort for discovering the signature of the Majorana neutrinos, namely the neutrinoless double beta decay, has not yet yielded any conclusive results.

11.9 Quantum Chromodynamics

The strong interaction between quarks is characterized by new quantum numbers, called *colours*. Each quark of a given *flavour*, for instance u, appears in three colours, say, *red*, *green*, and *blue*.

The colour field is mediated, according to the Standard Model of particle interactions, by massless particles called gluons. As mediators of the strong interactions, they are the analogs of photons as mediators of the electromagnetic interactions. The existence of three colours for quarks provides the foundation for the current theory of strong interactions, known as quantum chromodynamics (QCD).

The necessity for having a colour quantum number for quarks was noticed first in 1965, in connection with the baryon Ω^-. This particle contains three strange quarks, all with spins aligned, since the total spin is $3/2$ (see Fig. 11.10). The particle was produced in the lowest energy state of the three quarks, which is a symmetric state under the interchange of any two quarks. However, by Pauli's exclusion principle the state *had to* be antisymmetrical. With the known quantum numbers, it was impossible to achieve such an antisymmetric state, therefore it was necessary to introduce another quantum number, such that each s-quark composing the particle Ω^- has a different colour and then the state is obtained by antisymmetrization. The solution for saving Pauli's principle in the cases of the particles Ω^- and Δ^{++} was given by various physicists in 1965. Among them, Moo-Young Han and Yoichiro Nambu proposed that the colour group $SU(3)$ were a gauge group, thus including in the picture the dynamics of strong interactions.

For a quark of given flavor, the three colours, red, green, and blue, form a triplet: $u = (u_r, u_g, u_b)$, $d = (d_r, d_g, d_b)$, etc. Antiquarks have *anticolours*, such that colour plus anticolour gives white (no colour). All observable hadrons are "white" or colourless—the baryons are composed of three quarks of three different colours, while the mesons are composed of quark–antiquark pairs carrying colour–anticolour. The *colour gauge group SU(3)*, has eight generators. There is therefore a set of eight independent gauge fields A_μ^a, which are called *gluon* fields. The gluon fields are Yang–Mills fields with eight components. But in contrast to the case of the electroweak field, the symmetry of these Yang–Mills fields is not broken: gluons, like photons, are massless particles. Being non-Abelian gauge fields, they also carry colour charge and interact among themselves.

One might therefore think that the strong interaction mediated by gluonic field would have long range. However, quantum chromodynamics has the property of confinement, mentioned previously. The gluons have a remarkable feature, different from the electromagnetic field, concerning screening. For example, we have seen that an electric charge attracts opposite charges, so that the original charge is screened, and the net effect is a smaller charge. With the colour field the opposite happens: a quark of given colour attracts colour charges of the same polarity. As a result of this, the colour charge decreases at short distances from the quark, and it increases with increasing distances.

This anti-screening effect has an important consequence on the running of the coupling constant of quantum chromodynamics. Denoting by k^2 the square of the exchanged momentum, the running of the strong coupling constant $\alpha_s(k^2)$ happens according to the formula:

$$\alpha_s(k^2) = \frac{\alpha_s(\mu^2)}{1 + \left(11 - \frac{2n_f}{3}\right)\ln\frac{k^2}{\mu^2}}. \tag{11.20}$$

In the case of quantum chromodynamics, due to confinement, we cannot observe free quarks and gluons, therefore one cannot experimentally measure the probability of a scattering of Compton type, in order to find the value of α_s at the zero-momentum limit, i.e. $\alpha_s(m_{quark})$ (see Sect. 7.4.6). It is therefore necessary to introduce an arbitrary renormalization scale represented by the parameter μ, which provides the initial condition $\alpha_s(\mu^2) = \alpha_s$. The number of quark flavors is n_f. For $n_f = 6$, the coefficient of the logarithm is 7, then the denominator is positive and it grows with k^2. When $k^2 \to \infty$ (the distance tends to zero), $\alpha_s(k^2) \to 0$. This property is called asymptotic freedom, and its discovery in quantum chromodynamics was honoured with the Nobel Prize in 2004 awarded to David Gross (b. 1941), David Politzer (b. 1949), and Frank Wilczek (b. 1951).

One can re-write the running coupling constant of quantum chromodynamics in the form:

$$\alpha_s(k^2) = \frac{4\pi}{\left(11 - \frac{2n_f}{3}\right)\ln(k^2/\Lambda_{QCD}^2)}. \tag{11.21}$$

In this formula, instead of an arbitrary renormalization point μ, we have a quantity with the dimension of energy, Λ_{QCD}. This gives the scale at which α_s becomes strong as k^2 decreases. Experimentally, Λ_{QCD} was found to be of the order of 150–200 MeV. The perturbation theory in quantum chromodynamics is valid only above this scale, for example above energies of 1 GeV, where $\alpha_s(k^2) \approx 0.4$, and the strong interaction can be treated indeed as a perturbation. When the distances are more than $1/\Lambda_{QCD}$ (in natural units), the interactions become too strong for the perturbation theory to give reliable results. It is not accidental that distances of the order of $1/\Lambda_{QCD}$ are roughly the size of the light hadrons, r_h. This procedure is called *dimensional transmutation*: instead of describing the strength of the interaction by the *dimensionless* coupling constant, we describe it by the *dimensionfull* energy scale.

For asymptotic freedom, the gluons play a key role. Gluons have colour charge, in contrast to photons, which do not carry electric charge. Their effect is to increase the effective colour charge of the quarks with the distance, instead of shielding it (anti-screening effect). They contribute the factor $+11$ in the denominator of (11.20), while the term $2n_f/3$, due to the quark–gluon interaction, comes with a minus sign and is the analog of the electron–photon interaction term in quantum electrodynamics. In Fig. 11.12 is presented comparatively the running of the coupling constant in quantum electrodynamics and quantum chromodynamics.

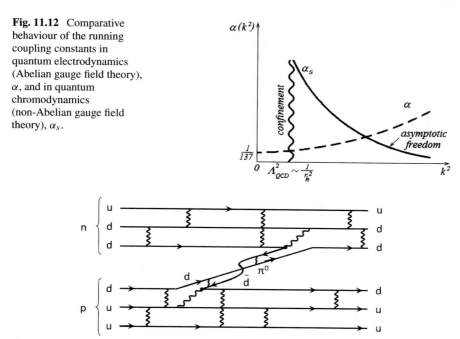

Fig. 11.12 Comparative behaviour of the running coupling constants in quantum electrodynamics (Abelian gauge field theory), α, and in quantum chromodynamics (non-Abelian gauge field theory), α_s.

Fig. 11.13 The nuclear force as pion exchange between nucleons, represented by virtual processes of quantum chromodynamics.

In this way, if the condition $(11 - 2n_f)/3 > 0$ holds true, an anti-screening effect is produced. As an example, if one takes a pion and tries to separate it into a quark and an antiquark, as the exerted force increases with distance, the increase in potential energy would cause the formation of a new quark–antiquark pair, and the final result would be two pions. Indeed, only colourless bound states have ever been observed.

The interaction force between nucleons about which we talked in Chap. 9 is understood in quantum chromodynamics as a residual strong force, and can be explained in terms of exchanges of quarks and gluons, as it is shown in Fig. 11.13.

Is it possible to produce quarks and gluons by colliding particles which do not interact strongly? The answer is affirmative. The first quark–antiquark production events were reported in 1975, at the e^+e^- SPEAR collider at SLAC. The electron and positron annihilate into a virtual photon (or Z boson), which forms in the final state the quark–antiquark pair. As we mentioned before, bare quarks cannot be observed— what is really observed are two jets of hadrons, whose structure and angular distribution shows that they are obtained from the hadronization of two spin 1/2 particles, with fractional charge.

In 1979, the first three-jet event was reported from the PETRA e^+e^- accelerator of Deutsches Elektronen-Synchrotron (DESY) in Hamburg. This contained the expected signature of a gluon: when the energy is higher, one of the quarks in the

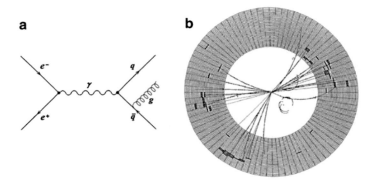

Fig. 11.14 A three-jet event: **a** the Feynman diagram; **b** an actual event observed in the JADE experiment at DESY.

final state emits one gluon, which also hadronizes and produces its own jet. The kinematical configuration changes and a sort of planar "star" is observed (see Fig. 11.14).

The presence of a spin 1 gluon was unequivocally determined in the experiments. Naturally, by increasing the energy of the e^+e^- collision, multi-jet events appear, involving pairs of quarks and various numbers of gluons. They were important also as a proof of the trilinear coupling of gluons, g (*virtual*) $\rightarrow gg$. The precision tests of quantum chromodynamics have been essentially based on electron–positron scattering experiments.

11.10 Grand Unification

The quarks interact among themselves by exchange of gluons, and with leptons through the electroweak field. It thus seems natural to look for models that unify strong and electroweak interactions. One of the first models believed to be promising was the grand unified theory (GUT) based on the simple gauge group $SU(5)$, proposed by Howard Georgi and Sheldon Glashow in 1974. The unification of all three forces had been initiated the same year by Abdus Salam and Jogesh Pati.

This kind of unified theory of electroweak and strong forces predicts mechanisms transforming hadrons into leptons. The model contains some supermassive bosons X and Y, with masses of order 10^{14}–10^{15} GeV/c^2. These would act as intermediaries in the decay of a proton into a positron and a pion. Proton decay, however, has not been observed. The model also predicts a tiny neutrino mass and the existence of magnetic monopoles. In addition to $SU(5)$, there are by now an impressive number of grand unified models, based on gauge groups, such as $SO(10)$, $SU(6)$, superstring-inspired $E(6) \times E(6)$, and others. None of them is currently considered to be satisfactory.

The idea of grand unification stems from the fact that the coupling constants depend on the energy (or momentum) of the interacting particles, as either increasing or decreasing functions. Thus, as seen in Chap. 7, the fine structure constant characterizing the electromagnetic interactions is a constant at low energies, but its value increases with momentum at high energies. We have also seen in the present chapter that in strong interactions, the coupling constant decreases with decreasing distances (increasing energy and momentum). The weak interaction coupling also decreases with increasing energy, but more slowly.

The grand unification theories predict the unification of the electroweak interactions with the strong ones at extremely high energies, of order 10^{15} GeV, which correspond to a wavelength of order 10^{-29} cm. Above these energies, the values of the three coupling constants coincide (Fig. 11.15).

This predicted energy of unification of strong and electroweak interactions is 10^{-4} times smaller than the Planck energy, characteristic of quantum gravitational effects. For energies lower than 10^{15} GeV, symmetry breaking occurs, resulting in two separate fields: the strong or colour field and the electroweak field. At energies of the order of 100 GeV, the latter also separates into two components. The electromagnetic field remains until low energies, but the weak force is manifest only in virtual processes of creation of $W^{\pm}$ and Z particles.

The fact that the values of the energies determining these changes of symmetry are 100 and 10^{15} GeV means that, in this model, there is a very large interval of energies between these two significant physical changes, called "desert". Their ratio is 10^{13}, and this "desert" is considered a shortcoming of the theory. It is believed that there could be many remarkable new phenomena in the range between 10^{2} and 10^{15} GeV.

Grand unification theories have immediate interest in cosmology. In Chap. 10 we discussed that, at the beginning of the expansion of the Universe, the average temperature was of the order of 10^{32} K. At this temperature, all the fundamental interactions had the same status. After going below 10^{28} K, the separation between the strong and electroweak interactions took place. The phase transition leading to the breaking of electroweak symmetry would have occurred between that temperature and 10^{15} K.

At the present average temperature of the Universe, the forces of Nature have the characteristics described in Chap. 9. From the initial equality between all the

Fig. 11.15 Qualitative behaviour of the strong, weak, and electromagnetic coupling constants as functions of energy. The unification, if it exists, would take place at around 10^{15} GeV.

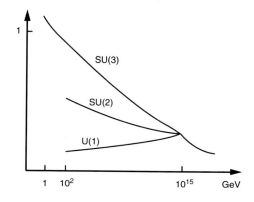

fundamental interactions, reached at the very high temperature of 10^{32} K, the cooling of the Universe has brought us to the hierarchy in the forces of Nature observed today.

11.11 Inflation

Looking in different directions in space, we observe similar physical properties. In particular, we find a background radiation with nearly the same temperature. It is thus natural to wonder how these phenomena, separated by enormous spacelike intervals, can look so similar? This has led to the postulation of a theory called inflation, which assumes that the very early Universe expanded at an accelerated rate, or inflated. Typically, during the inflationary stage the cosmic scale factor $R(t)$ grows exponentially with time.

If there existed a scalar field ϕ displaced from its equilibrium position, with a scalar field potential $V(\phi)$ flat enough for the field to slowly roll down toward the minimum of the potential, the Universe may have been dominated by the energy density of the scalar field, which would have acted as vacuum energy, and this would have caused an exponential expansion of the Universe.

Let $R(t)$ be the Friedmann scale factor of the Universe. The energy density of the field ϕ is

$$\frac{1}{2}\dot{\phi}^2 + V(\phi) = \rho_\phi. \tag{11.22}$$

Let us assume that the total density ρ is dominated by the energy density of the scalar field ϕ, i.e. $\rho \approx \rho_\phi$. We can write the Friedmann equation in the form

$$\frac{\dot{R}^2}{R^2} = \frac{8\pi G\rho}{3} - \frac{K}{R^2}. \tag{11.23}$$

If the scalar field potential $V(\phi)$ is sufficiently flat, we find that ϕ evolves slowly, $\dot{\phi}/M_P \ll \dot{R}/R$. That is, ϕ (in units of Planck mass) evolves more slowly than the expansion of the Universe, and $\rho_\phi \simeq V(\phi) \sim$ const. Then the Friedmann equation (11.23) gives an exponential solution:

$$R(t) \sim e^{t/t'}, \tag{11.24}$$

where $t' \simeq (8\pi G V(\phi)/3)^{1/2}$. This process continues until ϕ reaches the vicinity of the minimum of the potential. After that, ϕ evolves more quickly, and the production of particles heats the Universe.

After this phase transition, the Universe continues its evolution in the era dominated by radiation, where $R(t) \sim t^{1/2}$. But the consequences of the brief inflationary period are many. Let us mention two:

• It explains the flatness of the observable Universe;

- It solves the so-called horizon problem of explaining how causally disconnected regions of the Universe can show similar properties.

The flatness of the Universe can be understood if the matter density is critical. The duration of order $100t'$ for the inflationary period is enough to make the curvature term K/R^2 in (11.23) in negligibly small, and to make the Universe grow to its present dimensions. The second problem of the apparent regularity of the Universe in all directions can also be understood: the distances between the initially causally connected regions grew exponentially during the inflationary period.

Baryogenesis and Nucleosynthesis. As we saw in Chap. 7, the Dirac equation predicts the existence of antiparticles of the corresponding particles. Also, the CPT symmetry states that a particle and its antiparticle have exactly the same mass and lifetime, and exactly opposite charge. Then why does our Universe have a matter–antimatter asymmetry? Did this exist at the beginning of its evolution, or did it appear later? These questions have not yet found convincing answers.

Andrei Sakharov pointed out in 1967 that three conditions are required for the baryon asymmetry: (i) baryon number non-conservation; (ii) C and CP violation, and (iii) interactions out of thermal equilibrium. We have seen that CP violation indeed occurs in Nature. However, there is no evidence of baryon number violation, although it would be a necessary condition to produce more baryons than antibaryons. The C symmetry violation would act to avoid that the interactions producing more baryons than antibaryons be compensated by some C-symmetric interactions producing more antibaryons than baryons. To be out of equilibrium is also necessary, since otherwise the CPT symmetry would erase any previous baryon asymmetry.

Nucleosynthesis is the process of formation of the chemical elements. Some light elements like hydrogen were formed in the earlier stages of our Universe. This is called Big-Bang nucleosynthesis. Heavier elements, from helium to carbon to iron, were formed in stellar nuclear fusion processes. Even heavier elements are formed in supernova explosions. This has been observed in gamma-ray spectral lines coming from supernova.

Elements like gold and platinum are now known to be formed in collisions of neutron stars, as already mentioned in Chap. 10. This has been witnessed in the merger of two neutron stars to form a black hole, observed in August 2017.

11.12 Supersymmetry and Superstrings

Supersymmetry has arisen over the last few decades as a symmetry between fermions and bosons. The idea started from the consideration of models in which the variables representing bosons (which commute) may have transformation properties involving products of anticommuting quantities (Grassmann variables). Following this idea, the notion of *superspace* was proposed as an extension of spacetime. This involves the addition of four anticommuting coordinates θ_α to the four spacetime coordinates x_μ.

In this way transformations mixing bosonic and fermionic fields can be implemented, and the supersymmetric models are invariant under these transformations.

The idea of supersymmetry in the context of four-dimensional quantum field theory was put forward in the beginning of the 1970s independently by several groups: Yuri Gol′fand and Evgeny Likhtman in 1971, Dmitri Volkov and Vladimir Akulov in 1972, and Julius Wess and Bruno Zumino in 1974. Supersymmetric particle physics models have been continuously developed ever since, and nowadays the search for the supersymmetric partners of the known particles is one of the priorities of the high energy physics experiments at the Large Hadron Collider at CERN.

One consequence of supersymmetry is the possibility of building theories without divergences (these theories remain finite at any order of perturbative expansion). A supersymmetric theory of special interest is the one whose local supersymmetric transformation properties (supersymmetries depend on the spacetime coordinates) lead to *supergravity*, i.e., a supersymmetric theory of gravity. Supergravity, coupled with supersymmetric Yang–Mills theories, gives rise to interesting unified phenomenological models. However, they do contain divergences. Since in supersymmetry, a supersymmetric bosonic partner would correspond to each fermion, and vice versa, the photon would have a partner called the photino, the electron would have a scalar partner, the selectron, the graviton with spin 2 would have as partner the gravitino with spin 3/2, and so on. No supersymmetric partner of any of the known elementary particles has been detected experimentally, but the idea of supersymmetry is interesting and promising.

The supersymmetric models have particular interest in string theory, the theory of extended elementary objects, in which what is quantized is not located at spacetime points, but on curves that can be open or closed. Superstring theory (or the supersymmetric theory of strings) is considered a promising model for the unification of the four fundamental forces of Nature.

The theory of strings evolved from an old theory of dual models for strong interaction. In 1968, Gabriele Veneziano constructed a dual model amplitude for strong interaction scattering of mesons which was interpreted as a theory of oscillating strings, independently, by Yoichiro Nambu (1968), Holger Bech Nielsen (1969), and Leonard Susskind (1969). In 1970, fermionic excitations were added to the string theory by Pierre Ramond, and later by André Neveu and John Schwarz. Supersymmetry in the context of string theory was discovered in 1971 by Jean-Loup Gervais and Bunji Sakita. The research in superstring theory has been one of the most vigurously developing branches of theoretical physics ever since.

It is interesting, as an example, to write the action corresponding to a string. Above all, when the string moves in spacetime, it sweeps out a two-dimensional surface, the so-called *worldsheet*, having as coordinates two variables: σ, defined in the interval $(0, \pi)$, and τ, which can be any real number. The spacetime coordinates are defined in a D-dimensional space as $X_\mu(\tau, \sigma)$, where $\mu = 0, 1, \ldots, D - 1$. If $h^{\alpha\beta}$ is the metric on the worldsheet, and $g_{\mu\nu}(X)$ is the spacetime metric, the action of the bosonic string is written as

$$S = \frac{T}{2} \int d\tau d\sigma \sqrt{h} h^{\alpha\beta} g_{\mu\nu}(X) \partial_\alpha X^\mu \partial_\beta X^\nu, \qquad (11.25)$$

where T is the tension of the string, proportional to the reciprocal of the gravitational constant G, $h = -\det(h_{\alpha\beta})$, and the partial derivatives are taken with respect to the worldsheet variables τ and σ. As a consequence, in the quantized theory, the typical length of the string is of the order of the Planck length, $l \sim l_P = \sqrt{G\hbar/c^3}$. This formula is called the Polyakov action, in honour of Alexander Polyakov, who quantized it by the path integral method in 1981. The action was introduced in 1976 by Stanley Deser and Bruno Zumino, and independently by Lars Brink, Paolo di Vecchia, and Paul Howe.

In the case of the bosonic string theory, the number of spacetime dimensions is $D = 26$. In the supersymmetric theory, which would be a modification of (11.25) including fermionic fields, $D = 10$. The six additional coordinates (compared to our 4-dimensional spacetime) would have been coiled up, or compactified, at the time of the Big Bang. The strings themselves can be either open or closed. The theory of superstrings leads to the only known finite theory of quantum gravity, and it seems to reproduce all the interactions found in Nature. However, the theory is not directly verifiable experimentally.

Five principal string theories were developed, up to the mid-1990s, each one having different mathematical properties, in particular, the number of dimensions, and each best describing different physical circumstances. All these theories looked equally correct. Then Edward Witten proposed that these five theories might be describing the same phenomenon viewed from different points of view. The essential new feature was that, by invoking certain symmetry operations sometimes called dualities, these different string theories turned out to bear such deep relations to one another that they could actually be taken to be equivalent string theories.

Each of the string theories is a special case of the M-theory.

M-theory also incorporates a number of other string-related and supersymmetry-related ideas. Strings are actually a special case of a more general notion which includes higher dimensional structures called p-branes, or simply branes, which have p spatial dimensions plus one temporal dimension, the worldsheet being $(1 + p)$-dimensional. These structures are embedded in an 11-dimensional space. Of particular interest are the D-branes (or D_q-branes) which are timelike structures of $1 + q$ spacetime dimensions (q space dimensions and time). The two ends of an open string are supposed to reside on a D-brane.

Due to its logical consistency, and the fact that it includes the Standard Model, many physicists believe that string theory is the first candidate for a Theory of Everything (TOE), a manner of describing the known fundamental forces (gravitational, electromagnetic, weak, and strong interactions) and matter (quarks and leptons) in a mathematically complete system. However, some other prominent physicists do not share this view, because it does not provide quantitative experimental predictions.

A closed string looks like a small loop, so its worldsheet will look like a pipe or, more generally, a Riemann surface (a two-dimensional oriented manifold) with no boundaries (i.e., no edge). An open string looks like a short line, so its worldsheet will

Fig. 11.16 Worldsheet representation of some open and closed superstring interactions.

look like a strip or, more generally, a Riemann surface with a boundary. Interactions in the subatomic world are described by worldlines of pointlike particles in the Standard Model and by a worldsheet swept up by closed strings in string theory.

Strings can split and connect (see Fig. 11.16). This is reflected by the form of their worldsheet. If a closed string splits, its worldsheet will look like a single pipe splitting into (or connected to) two pipes. If a closed string splits and its two parts later reconnect, its worldsheet will look like a single pipe splitting into two and then reconnecting, which also looks like a torus connected to two pipes. An open string doing the same thing will have a worldsheet that looks like a ring connected to two strips.

Brane Cosmology. It is assumed that the visible, four-dimensional Universe is restricted to a D-brane inside a higher-dimensional space, called the bulk. The additional dimensions are compactified, so the observed Universe contains extra dimensions. Other branes may be moving through this bulk. Interactions with the bulk, or with other branes, can introduce effects not seen in more standard cosmological models.

This model proposes an explanation for the weakness of gravity as compared to the other fundamental forces, by assuming that the other three forces are localized on the brane. Not imposing this constraint on gravity, a large part of its attractive power 'leaks' into the bulk. As a consequence, the force of gravity should appear significantly stronger at small scales, where less gravitational force has 'leaked'. Various experiments have been suggested to test this hypothesis.

Problems

Problem 11.1 Discuss the decay of μ^+. By considering the fact that $\nu_\mu + n \to e^- + p$ is forbidden, find the possible lepton number assignments that satisfy additive quantum number conservation laws.

Problem 11.2 Express the muon decay $\mu^+ \to e^+ + \nu_e + \bar{\nu}_\mu$ using a Feynman diagram.

Problem 11.3 Decide whether the following particles can exist or not according to the quark model:

(i) A baryon with spin 1.
(ii) An antibaryon with electric charge $+2$.
(iii) A meson with charge $+1$ and strangeness 1.

Literature

1. A.D. Linde, *Inflation and Quantum Cosmology* (Academic, San Diego, 1990). An exposition of the inflationary model by one of its most active researchers
2. M. Chaichian, A. Demichev, *Path Integrals in Physics*. Quantum Field Theory, Statistical Physics and Other Modern Applications, vol. 2 (IOP, Bristol, 2001). Readers interested in path integral methods as applied to the formulation of quantum field theory, statistical physics, and other areas will find a careful introduction to these topics in this book
3. M. Chaichian, N.F. Nelipa, *Introduction to Gauge Field Theories* (Springer, Berlin, 1984). This book is recommended to those readers interested in the basic concepts of gauge field theories in particle physics
4. S. Coleman, *Aspects of Symmetry* (Cambridge University Press, Cambridge, 1985). The concept of spontaneous symmetry breaking is masterfully treated in this book
5. L.D. Faddeev, A.A. Slavnov, *Gauge Fields. Introduction to Quantum Theory*, 2nd edn. (Addison-Wesley, New York, 1991). The analogy of the basic concepts of non-Abelian gauge theories with general relativity is pointed out in this advanced book
6. M. Kaku, *Quantum Field Theory* (Oxford University Press, Oxford, 1993). A modern text on quantum field theory that embraces the fundamental topics in a coherent and complete form
7. I.D. Lawrie, *A Unified Grand Tour of Theoretical Physics* (Adam Hilger, Bristol, 1990). The book contains a good summary of the topics of quantum chromodynamics, models of grand unification, and the early Universe
8. V. Mukhanov, *Physical Foundations of Cosmology* (Cambridge University Press, 2005). An excellent advanced monograph expounding the modern theories of inflationary cosmology, by one of the pioneers of the theory
9. Y. Ne'eman, Y. Kirsh, *The Particle Hunters*, 2nd edn. (Cambridge University Press, Cambridge, 1996). A fascinating account of the search for the fundamental building blocks of matter, by a world-famous elementary particle physicist and a physicist specializing in popular science writing
10. D.H. Perkins, *Introduction to High Energy Physics* (Addison-Wesley, California, 1987). An excellent book on high energy physics
11. S. Weinberg, *The First Three Minutes*, 2nd edn. (Basic Books, New York, 1993). A book giving a modern vision of the origin of the Universe, starting from the models of gauge theories
12. S. Weinberg, *Cosmology* (Oxford University Press, Oxford, 2008). An updated account of the observational and theoretical advances in modern cosmology, by a leading authority in particle physics and cosmology

Chapter 12
Physics and Life

Life sciences are among the most interesting fields of contemporary scientific research. The biological world has such a wide range of complexities that any attempt to present it in any depth would take us outside the scope of the present book. In the present chapter, we shall thus only refer to a few specific physical problems in the biological world.

12.1 Order and Life

Biological systems have the property of being highly complex and highly organized, and from the simple bacterium to the human being, living organisms perdure and reproduce thanks to a continuous exchange of energy and matter with the environment. The maintenance of this high level of biological order in living organisms in comparison with the non-living environment can be understood in terms of their property of being thermodynamically open systems. (Viruses are apparent exceptions, but they cannot multiply unless they infect a cell, that is, unless they become parts of an open system.) In contrast with the non-living phenomena in which the entropy tends to grow, there is a tendency for the entropy to decrease in living systems. Put another way, they have a tendency to increase their information or negative entropy content (or information density, see Chap. 2). But while living systems increase both structural and functional biological order, they continuously produce biochemical reactions with the opposite consequence of increasing the entropy. The capacity of self-replication makes life a unique process in Nature. It is also characterized by the essential properties of feed-back and other control and communication mechanisms. These mechanisms have been extended to technology and evolved to complex adaptive systems such as computers and robots.

 An essential component of living systems is DNA—the deoxyribonucleic acid. DNA is the depository of the genetic information, and the cells of each organism

M. Chaichian et al., *Basic Concepts in Physics*, Undergraduate Lecture Notes in Physics, https://doi.org/10.1007/978-3-662-62313-8_12

Fig. 12.1 The two DNA strands showing the adenine–thymine and guanine–cytosine links.

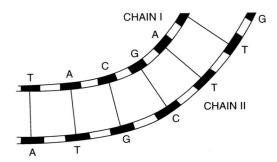

(except certain cells related to reproduction) contain all the information for its development as a whole individual of a given species. DNA is composed of two long polymer strands made up of four basic nucleotides, adenine, guanine, cytosine, and thymine (see Fig. 12.1). These provide a code of four symbols in terms of which the blueprint for any living organism can be written. There are some $n = 10^9$ molecules in a part of a cell volume of order $10^{-15} \mathrm{cm}^3$, and this implies the immense number of 4^n possible states. The information density is estimated as 1 bit per cubic nanometer. This means approximately 10^{12} Gb cm^{-3}.

The two DNA strands are anti-parallel, that is, they run in opposite directions to each other. The DNA information is read by means of the genetic code by copying stretches of DNA into RNA (ribonucleic acid) in a process called transcription.

James Watson (b. 1928) and Francis Crick (1916–2004) discovered the helical structure of DNA in 1953. For this and subsequent work they were jointly awarded the Nobel Prize in Physiology or Medicine in 1962, with Maurice Wilkins. Both Watson and Crick were influenced on their early interest in genetics by Erwin Schrödinger's classical book "What is life?".

At this point we should mention Rosalind Franklin (1920–1958). For her precursory work, she undoubtedly deserves an important place among the names behind the discovery of the structure of DNA. She is best known for her work on the X-ray diffraction images of DNA, particularly the so-called "Photo 51", at King's College, London, which led to the discovery of the DNA double helix. Unfortunately, Nobel Prizes cannot be awarded posthumously.

Evolution is an essential characteristic of life. According to Charles Darwin, the struggle for life and the survival of the fittest is the key mechanism which produces evolution. This is strongly dependent on environmental changes. Changes in a species are due to genetic mutations. Those making the fittest individuals will survive in the species. Those which do not will disappear. The mechanisms triggering mutations are not yet fully understood.

Biological order is maintained thanks to the permanent exchange of substance and radiation with the environment, so that there is a net inflow of information or negative entropy toward the living matter. For open thermodynamical systems, the change in entropy over an infinitesimal time interval is given by

$$dS = d_e S + d_i S, \tag{12.1}$$

where $d_e S$ is the entropy flow from the exterior and $d_i S$ is the change in entropy produced by the irreversible processes which occur inside the system. The entropy variation $d_i S$ is always positive, but $d_e S$ can have arbitrary sign, and for this reason, during its evolution, a system can reach a state in which its entropy is lower than in the initial state. Such a state can be maintained indefinitely if a condition is satisfied in which the flow of negative entropy compensates the entropy produced inside the system:

$$d_e S \le -d_i S, \tag{12.2}$$

that is, $dS \le 0$. This expresses in a quantitative way what was said before about the maintenance of biological order.

The expression (12.2) is valid under conditions of non-equilibrium. Near equilibrium, the tendency to destroy order dominates over the tendency to create it. But under certain favourable non-equilibrium conditions, order can be created. A simple physical example is the convective (ordered) motion of a layer of liquid heated from below.

Another interesting example is the emergence of order in a laser. The coherent light coming from the laser is highly ordered. However, the light from the lamp exciting the atoms is incoherent and highly disordered. This can be interpreted as a phase transition in a non-equilibrium system. Below a transition threshold, one has the incoherent regime. Above the threshold, one has the coherent state.

Similarly, the complex chemical reactions and regulatory processes that allow the maintenance of biological order take place under conditions of non-equilibrium. These ideas have been widely discussed by Hermann Haken (b. 1927) and Ilya Prigogine (1917–2003).

If we return to the content of (12.2), there is a question concerning the ultimate source of the negative entropy of living systems. This problem is not yet completely solved. In animals, the source of negative entropy is the food coming from other animals or from plants. In the case of the vegetal world, George Gamow (1904–1968) and Wesley Brittin (1917–2006) suggested in 1961 that solar radiation could be the source of negative entropy, and the process of photosynthesis is compatible with the laws of thermodynamics.

Basically, the argument goes as follows: the Sun emits high-energy photons at the temperature $T_{Sun} = 6,000$ K. While traveling toward the Earth, the solar radiation becomes very "diluted" due to the expansion in space and it reached the Earth at the temperature $T_{Earth} = 300$ K, with much lower energy density. The process that takes place at the impact of the diluted high-energy radiation with the Earth is essential: the contact with a material surface allows the exchange of energy between different frequencies, and the high-frequency photons are transformed irreversibly in a much larger number of low-energy photons. This irreversible process is accompanied by a significant increase in the entropy of the radiation. If the material surface which engenders this process is organic, specifically—the cloroplasts in the leaves of the plants, the increase in entropy is used to compensate for the decrease of entropy which takes place in creating organic molecules out of water (H_2O) and carbon dioxide (CO_2). According to the second law of thermodynamics, the increase in radiation entropy in the transition from a high-energy non-equilibrium state to a low-energy equilibrium state has to be larger than the decrease of entropy accompanying the organic matter formation.

The physical process that takes place when energy is exchanged between different frequencies is equivalent to a flow of heat from a hot source at the temperature T_{Sun} to a cooler reservoir at the temperature T_{Earth}. The total energy E is conserved, therefore the entropy variation is basically

$$\Delta S \cong E \left(\frac{1}{T_{Earth}} - \frac{1}{T_{Sun}} \right).$$

If we neglect the term containing the temperature of the Sun, $1/T_{Sun}$, and we put $E = nh\nu$, where n is the number of photons participating in the synthesis of one organic molecule, then

$$\Delta S \cong E \frac{1}{T_{Earth}}.$$

Numerically, considering red light with $\nu = 4 \times 10^{14}$ Hz and $T_{Earth} = 300$ K, we obtain

$$\Delta S \cong 10^{-14} n \text{ erg K}^{-1}/\text{molecule} = 150 \, n \text{ cal K}^{-1} \text{ mol}^{-1}.$$

This increase in entropy has to be compared with the decrease taking place in the basic photochemical reaction

$$6\,CO_2 + 6\,H_2O \xrightarrow[energy]{light} C_6H_{12}O_6 + 6\,O_2.$$

This reaction requires at least three photons and corresponds to an entropy decrease of 40 cal K^{-1} mol^{-1}. Thus, Brittin and Gamow concluded that the growing of plants in sunlight is consistent with the second law of thermodynamics, if the process of photosynthesis has at least 10% efficiency in entropy conversion. This means that negative entropy can be extracted from solar radiation and used for the construction of biological order by photosynthesis.

In this way one can justify the primary origin of biological order. However, some researchers consider that this is not enough to completely justify the increase in order (e.g., in the division of a cell, where information is doubled).

A pair of hot and cold sources around which biological order is created has been found in the cold and dark conditions on the deep ocean floor. During the last few decades, very active colonies of shrimps and other organisms have been discovered around hydrothermal vents. Before the discovery of these underwater vents, all life was believed to be driven by sunlight. However, these organisms seem to get their nutrients directly from the Earth's mineral deposits, in extreme conditions of pressure (of order 1000 kg/cm^2), salinity, and temperature (in the range 150–400 °C). It is believed that they derive their basic sustainability from hydrogen sulfide (which is otherwise toxic for terrestrial life).

It is an interesting fact that, around the hydrothermal vents containing hydrogen sulfide, there is the cold water of the ocean (around 3–4 °C). We thus have two sources at different temperatures, which create conditions for the cyclic production of work, or creation of order from disorder, as in a heat engine, or in the photosynthesis process. But in this case it is a new mechanism of *chemosynthesis* that mediates the creation of biological order.

12.2   Life and Fundamental Interactions

From the point of view of the fundamental interactions, when intracellular biological processes take place by means of chemical reactions, electromagnetic forces and quantum laws play the main role. The energies involved in these reactions is of the order of a few electron-volts per particle. Electromagnetic interactions are manifested in most cell processes. The gravitational interaction is important from the point of view of their mass and weight, and for their motion and external location as a whole. However, it does not to play an important role in intracellular processes. It also seems that strong interactions are unimportant due to the scale of energies associated with them, which is much higher than the average energies involved in the biochemical reactions of living organisms.

On the other hand it would appear that the electroweak interactions do play a significant role, related to the microscopic asymmetry observed in living organisms with regard to chirality.

12.3   Homochirality: Biological Symmetry Breaking

A mirror exchanges left and right: if you are right-handed your mirror image is left-handed, and vice versa (Fig. 12.2). Right to left exchange is a mathematical transformation called inversion. Thus, a left glove is an inversion of a right glove, i.e., it is its mirror image. Recall that the mirror inverts the space in the direction perpendicular to it: if we come close to it, the image also comes closer, that is, it moves in the opposite direction to us. However, directions parallel to the mirror do not change and if we move up or down, or left and right, the image moves in the same directions. But the image given by the mirror, if *alive*, could not exist in our world. And this would be so, not because of any important external difference (in general bodies have some approximate bilateral symmetry), or because its organs were disposed in the opposite way. The reason is that this living mirror image would differ microscopically. At the cellular level, its substance would be inverted with respect to ours, with dramatic consequences.

Fig. 12.2 A mirror inverts an image. If you are right-handed, your mirror image is left-handed.

The components of the basic macromolecules of life (DNA, RNA, proteins and polysaccharides) are amino acids and sugars. These molecules can have two independent spatial structures, called enantiomeric forms, which are the mirror images of one another—the D and L forms (Fig. 12.3). This type of chirality at molecular level manifests itself as birefringence (or optical activity) at macroscopic level: these chiral forms have the propriety of rotating the plane of polarized light. The substances which rotate clockwise the plane of polarized light when viewed toward the source are caller *dextrorotatory* (dextrogyre) or right-handed and symbolized by "+" or "d". Those which have the opposite effect are called *levorotatory* (levogyre) or left-handed and denoted by "−" or "l". By D-(L-)amino acid and D-(L-)sugar is designated that form of the molecule which can be synthesized from D-(L-)glyceraldehyde, the simplest chiral sugar molecule. The D/L nomenclature is not directly linked to the optical activity of the enantiomer: while D-glyceraldehyde itself is dextrorotatory, the same is not valid for all D-amino acids or all D-sugars. For example, the amino acid D-alanine and the sugar D-gulose are levorotatory. Of the 20 standard proteinogenic amino acids, only glycine is symmetric or achiral. Most of the more important macromolecules of life are built from chiral pieces linked together in such a way that all the pieces have the same chirality. This is a biological breaking of symmetry, and we call this property *homochirality*.

The proteins of living substance are formed almost without exception from L-amino acids (some D-amino acids appear in the cell walls of some bacteria). For this reason, our twin in the mirror could not survive in our world: her amino acids and proteins would have opposite chirality to ours, and if she ate our food, she could not digest it. To feed her, we would have to synthesize artificial nutrients in a laboratory.

Molecular chirality reminds us of the chirality of neutrinos, and it suggests that, if worlds of antimatter existed with living organisms in them, their anti-amino acids would be of type D. The fact that most living substance is composed essentially from L-amino acids and other substances of definite chirality has always been an enigma,

Fig. 12.3 Enantiomers D and L of the glyceraldehyde. Each molecule is the mirror image of the other. The chirality is due to the fact that the middle carbon atom, in the sp^3 hybridization state, is asymmetric—it is attached to four different groups of atoms.

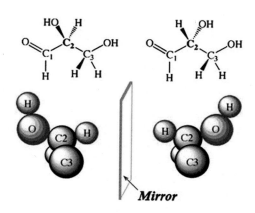

and several hypotheses have been formulated to explain this. Among them, it has been assumed that circularly polarized light coming from synchrotron radiation produced in pulsars may have had an influence on the selection mechanism. Another hypothesis assumes that homochirality is due to the chiral asymmetry of weak interactions, leading to the non-conservation of parity discussed in Chap. 9.

12.4   Neutrinos and Beta Decay

As pointed out earlier, neutrinos are extremely light particles, and millions of them may pass through us without interacting with a single atom of our bodies (they interact weakly with other particles). We have also explained that neutrinos are usually created in the beta decay of neutrons (or by quarks inside the neutron) inside atomic nuclei: the neutron emits an electron and an antineutrino, and in the process the initial nucleus increases its atomic charge Z by one, and becomes another element. For instance, if a carbon 14 ($Z = 6$) nucleus suffers beta decay, it emits an electron and an antineutrino and becomes a nitrogen ($Z = 7$) nucleus.

Up until 1954 it was believed that all phenomena in elementary particle physics had the property of P-symmetry (parity), that is, for any process, its mirror image would occur equally often in Nature. But it was discovered that beta decay does not have such a property. When it decays, the nucleus, with a spin that is a half integer multiple of $\hbar$, as indicated by the arrow in Fig. 9.1, emits an electron and an antineutrino in opposite directions. The electron and the antineutrino also have spin $\hbar/2$. But the helicity, which combines the *direction of rotation* with the direction of the velocity of the particle, is as in a right-hand screw for the antineutrino, while for the electron it is as in a left-hand screw. However, the *symmetrical* process or mirror image of beta decay, that is, the case of a nucleus which emits a right-handed electron and a left-handed antineutrino, simply does not occur in Nature. This property, as pointed out earlier, is called non-conservation of parity. On the other hand, an antimatter nucleus, composed of antineutrons and antiprotons (the latter with negative charge), in the process of beta decay, would emit a right-handed positron and a left-handed neutrino.

A natural question arises: Is parity non-conservation related to biological homochirality? There are reasons for believing that it is. The field maintaining the electrons around the nucleus in an atom is the electromagnetic field. But as we have already seen, it has been established that the electromagnetic and weak fields (associated with neutrino interactions) form a unified electroweak field. This means that, besides the electric attraction between protons and electrons in the atom, there is an additional force, extraordinarily smaller, which is exerted not only between protons and electrons, but also between neutrons and electrons.

This force, though it is much weaker than the electromagnetic force, may give an extremely small energy advantage to the L-enantiomer over the D-enantiomer, equivalent to one part in 10^{-16}, as suggested in 1978 by Vladilen Letokhov. This is because this force has left-hand selectivity. In spite of the smallness of this number, if one bears in mind that the evolution of life on Earth has been produced over several billion years, such a period of time may guarantee the complete dominance of organisms based on L-amino acids over those composed of D-amino acids, as a consequence of a cumulative effect. That is to say, the energy difference would be imperceptible in a reaction happening over a short time, but would lead to an absolute supremacy in biochemical evolution extended over billions of years.

The problem of homochirality has attracted the attention of many scientists, like Dilip K. Kondepudi and G. W. Nelson, Abdus Salam, Cyril Ponnamperuma, Julian Chela-Flores. In particular, Salam suggested considering the problem in terms of symmetry breaking, as a consequence of a phase transition at a certain critical temperature T_c. This temperature would be higher than 300 K, unless the so-called prebiotic conditions occurred at lower temperatures. There are various candidates for the mechanism producing the phase transition, and this remains an open question.

On the other hand, experiments support both parity non-conservation and the influence of circularly polarized light as a cause for homochirality.

An experiment at Stanford University provides evidence in favour of the first hypothesis. A mixture of equal amounts of L-leucine and D-leucine was prepared, then bombarded with left-handed electrons (like those produced in the beta decay of nuclei). It was observed that the left-handed electrons decomposed a larger amount of D-leucine than L-leucine. The opposite experiment, bombarding the mixture with right-handed electrons, destroyed a larger amount of L-leucine than D-leucine. But a similar effect was achieved in 1976 by irradiating leucine with circularly polarized ultraviolet light.

In connection with the first mechanism, such a process could have occurred naturally in living matter throughout its evolution, due to the influence of carbon 14 formed when atmospheric nitrogen is bombarded with neutrons produced by cosmic rays. This carbon 14 is subsequently assimilated by plant leaves via photosynthesis, and thereby gets incorporated into living matter. Later, the carbon 14 nuclei suffer beta decay, transforming back into nitrogen and emitting right-handed antineutrinos and left-handed electrons. Throughout the evolution of life, organic molecules having been bombarded continuously by left-handed electrons from the decay of carbon 14 contained in their own bodies, type L-amino acids would have had better chances of survival.

If on Earth there was initially a biological sector based on L-amino acids and another based on D-amino acids, they would have been independent, because the exchange of amino acids and proteins between them would have been impossible. It is not difficult to imagine that, with any small imbalance between the populations of the two sectors, after a sufficiently long time, the smaller one would have been reduced to extinction. In other words, the existence of a mechanism that favoured the formation of L-amino acids over D-amino acids, even in extraordinarily small amounts, would have been enough to ensure its absolute prevalence over the other, producing life as we know it today. D-protein-based life would have been wiped out in the early stages of its evolution, since it would have been less apt than L-protein-based life.

Concerning the existence of extraterrestrial homochirality, the Murchison meteorite, studied since 1970, contains several amino acids. At first it seemed that both enantiomers occurred in the same amounts, but detailed studies by Cronin et al. (1997) discovered that this meteorite contained a significant excess of two L-amino acids: isovaline and α-methylnorvaline. The extent to which this extraterrestrial matter coming from meteorites has influenced the evolution of terrestrial homochirality is a current subject of research, but the evidence seems to indicate that homochirality is universal, and that its presence is an indication of the existence of life in other regions of the Universe.

12.5 Anthropic Principle

Modern science has been characterized by an evolutionary process in which the role of Man as an observer has been gradually getting a more and more central position. As we saw in Chap. 1, Man was displaced from his position (central, but subordinated to religious dogma) in the medieval Aristotelian–Christian philosophy by the Copernican model of the Universe, and even more, by Newtonian mechanics, in which the basic laws of motion are independent of the observer and of divine action. Classical electromagnetic theory, optics, and thermodynamics further contributed to this line of thought.

However, the special theory of relativity started to locate the observer in a singular position due to new concepts such as the relativity of the simultaneity, proper time, and others. Quantum mechanics then attributed an outstanding role to the observer, starting with the concept of wave function and the observer's role in the concept of measurement and the collapse of the wave function. And even this goes further, enlarging the concept of reality by conceiving of virtual or non-observable processes as being part of reality, and capable of leading to observable physical effects.

Man created science with the purpose of having a description and a logical and systematic knowledge of objective reality, but since he is part of this objective world, he cannot avoid applying this description to himself. On the other hand, this description is conditioned by the fact that the physical, chemical, and cybernetic processes determining the functioning of his own brain do not escape the laws and principles that he tries to identify for the external world. He can observe the Universe because he exists, and this constrained existence as an open thermodynamical system, based on carbon chemistry and other essential substances for life, is possible because the Universe is old enough to have allowed a long process of stellar evolution, of the order of 10^{10} years.

As a consequence of its age, the Universe is also sufficiently large due to expansion to have given rise to the formation of thousands of millions of galaxies, each containing thousands of millions of stars inside which nuclear combustion has occurred, and with it the synthesis of elements heavier than hydrogen. In this way, evolution could give rise to the formation of planets and to the appearance of exceptional conditions for the advent of life on some of them, as on our own planet, but probably on millions of other planets too.

Many thinkers have wondered whether one can conceive of a Universe eternally without life. Does it make sense to conceive of a Universe without somebody able to speak about its existence? Several physicists—the pioneers being Gerald J. Whitrow, Brandon Carter, and John A. Wheeler suggested an answer: the anthropic principle. In its weak form, it can be stated as suggested by John D. Barrow and Frank J. Tipler:

> The observed values of all the physical and cosmological quantities are not equally probable, but they take values restricted by a first requirement, that places exist where carbon-based life can evolve, and by a second one, that the Universe be sufficiently old so that it has already happened.

In its strong form, the statement becomes:

> The Universe should have properties that allow life to develop in some stage of its history.

An implication of the strong anthropic principle is that physical constants and laws of Nature should be such that life can exist. This gives rise to different interpretations.

Evidently, the anthropic principle is a speculative hypothesis in the framework of physics, and in a certain sense, it claims to answer in an affirmative way questions like: Is life a universal phenomenon? In other words, is the appearance of life, in particular, of intelligent beings, the manifestation of universal and unavoidable laws, like those governing other phenomena, e.g., physical laws? These ideas can encourage different attitudes, one of which is the search for other manifestations of life in our Universe, in particular, the search for intelligent extraterrestrial life.

12.6   Search for Extraterrestrial Life

There is increasing interest in finding extraterrestrial life. Up to 2018, several organic molecules, the building blocks of living systems, have been identified in some places, such as on Mars by NASA's rover Curiosity. Thiophenes, benzene, toluene, and small carbon chains, such as propane or butane, have been found. Methane (which may also have an inorganic origin) had been found previously, but now observations over three Martian years indicate a seasonal variation. Furthermore, on Saturn's moon Enceladus, analysis of data from NASA's Cassini spacecraft indicates the presence of large organic molecules ten times heavier than methane. And in a mass of gas and dust called the Taurus Molecular Cloud 1, at 430 light-years from Earth, researchers have used the Green Bank Telescope in West Virginia to identify signatures of the molecule benzonitrile, a building block for polycyclic aromatic hydrocarbons (PAH). Although PAH molecules are usually carcinogenic, they also contain the ingredients for the seeds of life. Obviously, these findings do not yet mean that we have found extraterrestrial life itself, but they are perhaps a first step towards it. Concerning the search for intelligent life, several projects have been carried out over many decades, such as the SETI project (Search for Extraterrestrial Intelligence). Its purpose is to try to detect intelligent extraterrestrial life from Earth, as opposed to looking for it from outer space. Several attempts have been made and the names Paul Horowitz (b. 1942), Carl Sagan (1934–1996), and Frank Drake (b. 1930) should be mentioned, among others, as pioneers of signal detection research. However, after many years, this research has not succeeded in producing any positive result.

In 1950, Enrico Fermi suggested that, if technologically advanced civilizations are common in the Universe, then they should be detectable one way or another. However, it may be that our assumptions are flawed, since we assume technological development comparable with our own. Can we be sure that more advanced communication technologies could not be found, e.g., not based on electromagnetic signals? Can we be sure that means of transportation not imagined by ourselves could not be found in future?

A typical mistake made when thinking about extraterrestrial beings is to assume that they are similar to us, but just a few hundred years more advanced. In that case, we may simply observe that, even the most imaginative thinkers of the Middle Ages were unable to guess the progress to be made by our present communications and travel technologies. It is very important also that we are distant from them in scientific language and concepts. We must extrapolate these ideas to the future: we are surely unable to guess the technological possibilities and scientific advances of a civilization hundreds of thousands, or million years ahead of us.

Michio Kaku has suggested travel through wormholes. If we fold a sheet of paper and punch a hole through it, we realize that a wormhole is the shortest distance between two points, rather than the straight line joining them on the unfolded surface. According to Kaku, a civilization able to harness the power of stars might perhaps use such shortcuts through spacetime, and bridge the vast distances of space to reach Earth. Kaku believes that only civilizations millions of years more advanced than us, and capable of using wormholes as shortcuts, could reach Earth and perhaps visit us, coming from unbelievably remote regions of space and time.

Here arises the main question: to what extent is communication possible, according to our present standards, between societies separated from us by such enormous lapses of time, and scientific and technological gaps, if they are ahead us as we are, for instance, from Australopithecus or Homo erectus in the past?

Literature

1. E. Schrödinger, *What is Life?* (Cambridge University Press, Cambridge, 1944). An excellent study of the thermodynamics and statistical physics of biological systems. Several topics dealt with here remain valid today
2. N. Wiener, *Cybernetics or Control and Communication in the Animal and the Machine* (MIT Press, 1948). This was a pioneering book on cybernetics, by one of the creators of the field
3. G. Nicholis, I. Prigogine, *Self Organization in Non-Equilibrium Systems* (Wiley, New York, 1977). The thermodynamics of open systems, like biological systems, is very clearly discussed in this book
4. G. Gamow, W. Brittin, Negative entropy and photosynthesis. Proc. Natl. Acad. Sci. **47**, 724 (1961). This paper supplies a quantitative argument regarding the role of low entropy sunlight as the mechanism driving the development of biological order
5. R.P. Feynman, *The Feynman Lectures on Physics*, The Definitive edn., vol. I. (Pearson-Addison Wesley, Reading, 2006). One of the first references in a physics book to the homochirality problem was made in the first edition of this book, in 1965
6. J. Chela-Flores, Terrestrial microbes as candidates for survival on Mars and Europa, in *Journey to Diverse Microbial Worlds: Adaptation to Exotic Environments* (Kluwer Academic Publishers, Dordrecht, 2000). Contains a very interesting discussion on the topic of exobiology
7. J.D. Barrow, F.J. Tipler, *The Anthropic Cosmological Principle* (Clarendon, Oxford, 1986). In this book a deep and exhaustive analysis is made of the physical and biological basis for the anthropic principle
8. M. Gell-Mann, *The Quark and the Jaguar: Adventures in the Simple and the Complex* (W. Freeman and Co., New York, 1995). A stimulating and uncommon book, dealing with diverse topics, from elementary particle physics to complex adaptive systems
9. D. Abbot, P.A.W. Davies, A.K. Pati (eds.), *Quantum Aspects of Life* (Imperial College Press, London, 2008). This book, with a foreword by Roger Penrose, presents the hotly debated question of whether quantum mechanics plays a non-trivial role in biology

Appendix
Solutions of the Problems

Solutions for Chap. 1

Solution 1.1 From Sect. 1.11, $x(t) = A \cos(\omega t + \phi)$, so we can write

$$K = \frac{1}{2}m\dot{x}^2 = \frac{1}{2}mA^2\omega^2 \sin^2(\omega t + \phi), \tag{A.1}$$

and since $\omega = \sqrt{k/m}$, (1.87) follows by calculating $K + V$. We have that the potential energy $V(x)$, as well as the absolute value of the displacement, get their maximum values for $\omega t + \phi = 0, \pi$. Moreover, the force $F(x) = -\partial V/\partial x = -kx$ also reaches its maximum absolute value at these points. The kinetic energy, however, is zero at both points, and its maximum value is reached when $\omega t + \phi = \pi/2, 3\pi/2$, where x as well as $V(x)$ and $F(x)$ also vanish. Note that, for a given oscillator, the total energy depends explicitly on the amplitude A and the constant k, and we can write $E = \frac{1}{2}m\omega^2 A^2$. In the quantum case (Chap. 6), we will see that the harmonic oscillator energy depends linearly on the frequency according to $E = \hbar\omega(n + \frac{1}{2})$, where $\hbar$ is the reduced Planck constant and n is an integer describing the quantum state of the oscillator.

Solution 1.2 Assume the position of the planet, moving on a circular orbit around the centre of forces, is given by the radius vector **r** forming an angle θ with the diameter of the orbit. The projection, taken along the x axis, is a point $x = r \cos\theta$. The velocity component along x is $\dot{x} = -r \sin\theta\dot{\theta}$, leading to the kinetic energy $T = \frac{1}{2}m\dot{x}^2 = \frac{1}{2}mr^2\omega^2 \sin^2\theta$, where $\omega = \sqrt{GM_\odot/r^3}$. The oscillator potential energy is $V = \frac{1}{2}kx^2$, where $k = m\omega^2$. Thus, simple harmonic motion is equivalent to the projection of uniform circular motion on a diameter. This property might be useful, for instance, in the study of the motion of satellites around their planets. b) Note that r can be expressed in terms of $| E |$, and the solution is obtained by substituting v and r into $L = mrv$. The expression indicates that, for a given planetary mass, if we

© Springer-Verlag GmbH Germany, part of Springer Nature 2021
M. Chaichian et al., *Basic Concepts in Physics*, Undergraduate Lecture Notes in Physics,
https://doi.org/10.1007/978-3-662-62313-8

increase the orbital radius, which is equivalent to decreasing the absolute value of the total energy, the angular momentum increases.

Solution 1.3 Write the equation

$$\frac{T_V^2}{T_E^2} = \left(\frac{1.080}{1.496}\right)^3 = 0.376. \tag{A.2}$$

Then, $T_V \approx 0.613 T_E$, and for $T_E = 365.256$ days, this leads approximately to $T_V \approx 224.04$ days. The observed period is 224.7 days.

Solution 1.4 In most cases in what follows, we shall assume circular orbits as a good approximation for low eccentricity. (a) We get $\alpha = m_2/M_\odot = 3 \times 10^{-6}$, and the reduced mass is $m = m_2/(1 + \alpha) \approx (1 - 3 \times 10^{-6})m_2$. The attractive force is exerted along the vector $\mathbf{r}$. Thus, by calling r_{1E} the distance from the Earth–Sun centre of mass to the centre of the Sun,

$$r_{1E} = \alpha r \sim 3 \times 10^{-6} r \approx 450 \text{ km}. \tag{A.3}$$

Thus $r_{1E} \approx 6.46 \times 10^{-4} R_\odot$, where $R_\odot = 6.96 \times 10^5$ km is the solar radius.
(b) For the Sun–Jupiter system, the distance from its centre of mass to the centre of the Sun is $r_{1J} = m_J r/(M_\odot + m_j) \approx m_J r/M_\odot$. Inserting the various quantities, this gives

$$r_{1J} = 7.02 \times 10^5 \text{ km} = 1.01 R_\odot,$$

slightly larger than the solar radius and 1560 times larger than r_{1E}.

Solution 1.5 (i) First calculate $r_1 = m_J r/M_\odot(1 + \alpha_J) \approx \alpha_J r$, where $\alpha_J = m_J/M_\odot = 0.9 \times 10^{-3}$. Use also $r_2 = r/(1 + \alpha_J)$. By equating the Sun's centripetal (with respect to its centre of mass) and gravitational forces in its interaction with Jupiter, i.e., $M_\odot v^2/\alpha_J r = GM_\odot^2 \alpha_J/r^2$, it follows that $v_\odot \approx \alpha_J \sqrt{GM_\odot/r}$, where $\sqrt{GM_\odot/r} \approx 1.3 \times 10^4$ ms^{-1}, leading to $v_\odot \sim 0.954 \times 10^{-3} \times 1.30 \times 10^4 = 12.4$ ms^{-1}.

Thus, adistant observer, by using Doppler spectroscopy, may deduce from the Sun's wobble that it has (at least) one companion planet. (ii) Jupiter's centripetal force is $\alpha_J M_\odot v_J^2/r_2 = GM_\odot^2 \alpha_J/r^2$, from which, neglecting terms in α_J^2, we get $v_J \approx \sqrt{GM_\odot/r} \approx 13.02$ km s^{-1}, more than 1000 times greater than that of the Sun.

Solution 1.6 (a) By imposing the equality of the centripetal and gravitational forces for the Earth in its orbit around the Sun, we have $mv^2/r = GM_\odot m/r^2$, from which $v = \sqrt{GM_\odot/r}$. If t is the period of rotation of the Earth around the Sun, it is given by $t = 2\pi r/v$. It follows that

$$r^3 = \frac{GM_\odot}{4\pi^2} t^2. \tag{A.4}$$

Similarly, assuming that the Sun describes a circular orbit around the centre of the Milky Way, and estimating the galactic mass inside a sphere of radius R to be given by $M_G = N M_\odot$, we get

$$R^3 = \frac{GNM_\odot}{4\pi^2}T^2. \tag{A.5}$$

Dividing (A.5) by (A.4) and substituting in the data, we find $N = (t/T)^2 \times (R/r)^3 = 1.83 \times 10^{11}$, from which $M_G = NM_\odot \sim 3.6 \times 10^{41}$ kg is the part of Milky Way mass enclosed in a sphere of radius R measured from its centre O.

(b) If we assume the Sun is located at a distance of $R = 0.55R_G$ from O, where R_G is the average galactic radius, and if we assume that the Milky Way is spherical with uniform density (of course, neither of these assumptions is accurate, but it helps to have an approximate answer), the total mass of the Milky Way is proportional to R_G^3, and in fact approximately $(1/0.55)^3 \sim 6$ times greater than M_G, that is, $1.1 \times 10^{12} M_\odot$.

Solution 1.7 The linear force acting on the train can be written at each point of its trajectory in terms of the mass $M(r) = 4\pi\rho r^3/3$ of a sphere of radius $r \leq R$ concentric with the Earth, and the angle θ between the radius and the chord. It is $F = -GM(r)m_T \cos\theta/r^2 = -GM(r)m_T x/r^3$, where $x = R\cos\theta$, $R = \sqrt{x^2 + y^2}$, and y is the distance from the centre of the Earth to the chord. Thus, the maximum force is exerted at the extremes, since it is the projection of the gravitational force on the chord, and it is zero at its centre, $x = 0$, where $\theta = \pi/2$. Here we use $GM/R^2 = 9.8$ m s^{-2}. The potential energy is $U = \frac{1}{2}GM(R)m_T x^2/R^3$ and we have $T + U = E$, where $T = \frac{1}{2}m_T \dot{x}^2$ and E is the total energy. The train will move as a linear oscillator. The maximum value of U is reached at the extremes $x_m = \pm 160$ km, where the kinetic energy is zero, and $E = U(x_m)$. By equating this to the maximum kinetic energy of the oscillating train $T = \frac{1}{2}m_T v^2$, reached at $x = 0$, we get

$$v = \sqrt{\frac{GMx_m^2}{R^3}} = 199 \text{ m/s} = 715 \text{ km/h.} \tag{A.6}$$

Solution 1.8 We use spherical coordinates. The Earth rotation frequency ω is assumed to be approximately constant. Around a point r, ϕ, θ, take a volume element $dV = r^2 dr \sin\theta d\theta d\varphi$. Its mass is $dM = \rho dV$. The angular momentum is $dS_E = dM_E r^2 \sin^2\theta$. Integrating r over $(0, R)$, we obtain

$$S_E = \frac{8\pi\rho R^5 \omega}{15} = I_E \omega,$$

where the quantity $I_E = 2M_E R^2/5$ is the moment of inertia of the Earth around its axis. (b) Assuming circular motion of the Moon around the Earth and equating the gravitational and centripetal forces, we easily find $L_M = M_M\sqrt{GM_E r}$. As

$$S_E + L_M = const.,$$

a decrease in S_E is compensated by an increase in L_M, i.e., $\delta S_E = -\delta L_M$. We consider r as a function of the Earth's period of rotation T to find δr in terms of

δT. We assume $dT = \delta T$, from which $\delta \omega = -2\pi \delta T / T^2$ and assume that the day lengthens by $\delta T = 1.75 \times 10^{-5}$ s every year. We also take the following values: the Moon's mass $M_M = 7.34 \times 10^{25}$ g, the Earth's mass $M_E = 5.90 \times 10^{27}$ g, $G = 6.67 \times 10^{-8}$ cm^{3}g^{-1}s^{-2}, the Earth–Moon distance $r = 3.84 \times 10^{10}$ cm, the Earth's radius $R = 6.35 \times 10^8$ cm, and the length of the day $T = 8.62 \times 10^4$s. We get

$$\delta r = \frac{8\pi R^2}{5 M_M T^2} \sqrt{\frac{M_E r}{G}} \delta T.$$

Finally, we obtain approximately $\delta r = 3.78$ cm per year for the increase in the radius of the lunar orbit. The observed value is around 3.80 cm per year.

Solutions for Chap. 2

Solution 2.1 (a) After removing the partition, the gas flows continuously to the right-hand side and finally reaches equilibrium. The second half of the container being empty, the expanding gas does not do work. Since also no heat is exchanged, from the first law of thermodynamics, it follows that the internal energy of the system U is unchanged. Since U depends only on the temperature T for an ideal gas, the equilibrium temperature is still T. (b) The probability of the system returning spontaneously to be confined in the volume V_1 is negligibly small, and this indicates that the process is irreversible. But from the thermodynamic point of view, this is confirmed if we show that the entropy is increased. Let us calculate the change in entropy, assuming the equation of state $pV = NkT$ and no change in the internal energy, i.e., $dU = TdS - pdV = 0$, which implies $T \int dS = \int pdV$. (The work done by the gas is equal to the heat absorbed.) We have

$$\Delta S = Nk \int_{V_1}^{V_1+V_2} \frac{dV}{V} = Nk \ln \frac{V_1 + V_2}{V_1} > 0. \qquad (A.7)$$

Thus, the process is irreversible.

Solution 2.2 The elementary change in the free energy is $dF = -SdT - pdV$, so if $dT = 0$, we have $dF = -pdV$, and for a finite change of volume, we have $\Delta F = -p\Delta V$. In other words, in such a restricted case, we can write $dF = -dW$, and $F = -W$ behaves as a function of the thermodynamic state.

Solution 2.3 The elementary change in enthalpy is $dH = TdS + Vdp$. If $dp = 0$, we have $dH = TdS$. In other words, the heat exchanged at constant p is a function of the thermodynamic state, since it is equal to the variation of the enthalpy, which is a function of the thermodynamic state.

Solution 2.4 (a) At constant temperature T_0, the work done by the system is

$$\Delta W = \int_{V_0}^{2V_0} p \, dV = RT_0 \int_{V_0}^{2V_0} \frac{dV}{V} = RT_0 \ln 2. \qquad (A.8)$$

As the change in the internal energy $dU = 0$, the work done is equivalent to the amount of heat absorbed by the gas to increase its volume at $T = const.$, which is given by

$$\delta Q = \Delta W = RT_0 \ln 2. \qquad (A.9)$$

However, the work done *on the system* is $\delta W = -\Delta W$. Note that increasing the volume at $T = const.$ involves keeping the system in contact with a source at constant temperature, $dT = 0$, and the system absorbs an amount of energy equal to $dU = \delta Q + \delta W = 0$.

(b) At constant pressure p, the work done is

$$\Delta W = \int_{V_0}^{2V_0} p \, dV = 2RT_0, \qquad (A.10)$$

which implies that, if the system has the same pressure but twice its volume, its temperature will have been doubled. In consequence, if the internal energy was initially $U_1 = \frac{3}{2}RT_0$, it will now be $U_2 = 3RT_0$, i.e., it will also have been doubled, $U_2 = 2U_1$.

Solution 2.5 At the surface of the sea,

$$\frac{\Delta H}{RT_i} = \frac{40700}{8.31 \times 373} \approx 13.13, \qquad (A.11)$$

and deep inside the sea, where the boiling temperature of water is $400\,^\circ$ C,

$$\frac{\Delta H}{RT_f} = \frac{40700}{8.31 \times 673} \approx 7.28. \qquad (A.12)$$

The pressure increases by a factor $e^{-(7.28-13.13)} = e^{5.85} \approx 347$. If we assume that at sea level the atmospheric pressure is around 1 kg cm^{-2}, the pressure increases by approximately 1 kg cm^{-2} for every 10 m we go down in the sea. Thus, the factor 347 corresponds to a depth of around 3.5 km. In deep oceanic hydrothermal vents, the water boiling point is raised high enough to allow living organisms to exist in this very high pressure, high temperature environment (we shall refer to this again in Chap. 12).

Earlier in this chapter we mentioned that the Clausius–Clapeyron equation provides a theoretical basis for the fact that the inner core of the Earth is solid, as discovered in 1936 by the Danish seismologist Inge Lehmann (1888–1993). She observed that seismic waves were reflected on the boundary of the inner core.

Solution 2.6 The Helmholtz free energy is $F = -kT \ln Z$. Equation (2.29) thus implies $F = -kTN \ln \frac{eV}{N\lambda^3}$. This in turn implies that the entropy is

$$S = -\frac{\partial F}{\partial T} = kN \ln \frac{eV}{N\lambda^3} + \frac{3}{2}kN.$$

Hence, $U = \frac{3}{2}NkT$. By calculating $p = -\partial F/\partial V$, the reader will find that the equation of state $pV = NkT$ is obtained for the ideal gas.

Solutions for Chap. 3

Solution 3.1 (a) The charge density is the total charge divided by the volume of the ball, i.e.,

$$\rho = \frac{3Q}{4\pi R^3}.$$

(b) The electric field inside ($r < R$) is obtained from Gauss' law by integrating over a sphere of radius r, whence

$$\oint \mathbf{E} \cdot d\mathbf{S} = 4\pi \int_0^r \rho dV,$$

implying $4\pi r^2 E = \frac{3Q}{4\pi R^3}\frac{4}{3}\pi r^3$, from which

$$\mathbf{E} = \frac{Q\mathbf{r}}{4\pi R^3}.$$

For the electric field outside, we assume a sphere of radius $r > R$. We have $4\pi r^2 E = Q$, from which

$$\mathbf{E} = \frac{Q\mathbf{r}}{4\pi r^3}.$$

Solution 3.2 In a polarizable medium the free charge density is given by $\nabla \cdot \mathbf{D} = \nabla \cdot \epsilon \mathbf{E} = 4\pi\rho$, where $\epsilon = 1 + 4\pi\chi$, and χ is the electric susceptibility.

Using the relation $\nabla \cdot a\mathbf{b} = \nabla a \cdot \mathbf{b} + a\nabla \cdot \mathbf{b}$, we have

$$\nabla \cdot \mathbf{E} = q\left[\nabla(e^{-\lambda r}) \cdot \frac{\mathbf{r}_0}{r^2} + e^{-\lambda r}\nabla \cdot (\frac{\mathbf{r}_0}{r^2})\right], \tag{A.13}$$

from which we get

$$\frac{4\pi\rho}{\epsilon} = q\left[\frac{-\lambda e^{-\lambda r}}{r^2}\mathbf{r}_0 \cdot \mathbf{r}_0 + e^{-\lambda r}\nabla \cdot \left(\frac{\mathbf{r}_0}{r^2}\right)\right] \tag{A.14}$$

$$= q\left[\frac{-\lambda e^{-\lambda r}}{r^2} + e^{-\lambda r}\nabla \cdot \left(\frac{\mathbf{r}_0}{r^2}\right)\right]$$

$$= q\left[\frac{-\lambda e^{-\lambda r}}{r^2} + 4\pi\delta(\mathbf{r})\right].$$

This in turn implies

$$\rho = \frac{\epsilon q}{4\pi}\left[\frac{-\lambda e^{-\lambda r}}{r^2} + 4\pi\delta(\mathbf{r})\right].$$

Since the density comprises two terms of opposite charge, if we assume q positive, we have a positive charge at the origin, surrounded by a spherically symmetric negative charge. The total screened charge is obtained by integrating the first term over spherical shells, and the second term by using Cartesian coordinates, both over the whole space:

$$Q = \int \rho dV = \int \frac{\epsilon q}{4\pi}\left[\frac{-\lambda e^{-\lambda r}}{r^2}4\pi r^2 dr\right] + \epsilon q\int \delta(\mathbf{r})d^3\mathbf{r} \tag{A.15}$$

$$= \epsilon q e^{-\lambda r}]_0^\infty + \epsilon q = -\epsilon q + \epsilon q = 0.$$

The same result is obtained by calculating the flux of $\mathbf{E}$ through a sphere of radius R, then taking $R \rightarrow \infty$. This is an apparent effect on a probe charge, which does not "see" the fixed charge q at all if it is placed at infinity. Naturally, the total charge of the neutral polarizable medium with the fixed charge remains q.

Solution 3.3 Taking the electron mass as 0.91×10^{-27} g and substituting other data into (3.24) and (3.23), we find $r = 427$ cm and $f = 28$ kHz.

Solution 3.4 For $r > R$, assuming the system is in vacuum, what we observe at the distance r is the magnetic field $\mathbf{B}$. Then Ampère's law gives $2\pi r B = \frac{4\pi}{c}I$, implying

$$\mathbf{B} = \frac{2}{cr}I\mathbf{e}_\theta$$

outside the wire, where $\mathbf{e}_\theta$ is in the direction tangent to circles orthogonal to the direction of the current I. For $r < R$, we need the current density. Using cylindrical coordinates with the z-axis parallel to the current I, we can write the current density as

$$\mathbf{j} = \frac{I}{\pi R^2}\mathbf{e}_z.$$

Then Ampère's law is expressed in terms of the effective field $\mathbf{H}$ and gives $2\pi r H = \frac{4\pi}{c}I$, for $r < R$. This leads to

$$\mathbf{H} = \frac{2r}{cR^2}I\mathbf{e}_\theta$$

and $\mathbf{B} = \mu\mathbf{H}$ inside the wire.

Solution 3.5 The electric field generated by a fixed charge is such that lines of force begin or end at that charge, that is, charges are sources and/or sinks of electric fields. An electric field produced by the time variation of a magnetic flux leads to closed electric lines of force. In most media, the converse cannot occur, due to the absence of free magnetic monopole particles and magnetic currents.

Solution 3.6 The coil rotates with angular velocity $\omega = 2\pi f$. Thus the flux through the surface bounded by the coil is $\phi_B(t) = \mathbf{B} \cdot \mathbf{S}(t)$ and the projection of the field B on the plane containing the N loops (we assume them identical and tightly wound) at any time t is given by $\phi_B(t) = NabB \cos 2\pi ft$. Faraday's law states that the electromotive force is given by the rate of change of the magnetic flux:

$$\mathcal{E} = -\frac{d\phi_B}{dt} = 2\pi NabB \sin 2\pi ft, \tag{A.16}$$

where $\mathcal{E} = \oint \mathbf{E} \cdot d\mathbf{l}$ is the electromotive force obtained finally and $\Phi_B = \int \mathbf{B} \cdot d\mathbf{S}$ is the magnetic flux across $\mathbf{S}$. The direction of the electromotive force is such that it induces a current creating a magnetic field opposite to $\mathbf{B}$, as demanded by Lenz's law. This exercise illustrates the basis of the alternating current generator.

Solutions for Chap. 4

Solution 4.1 (i) The first minimum is determined from $d \sin \theta = \pm(2n + 1)\lambda/2$, by taking $n = 0$. Thus

$$\sin \theta = \frac{\frac{1}{2}5.46 \times 10^{-5}}{10^{-2}} = 0.0027.$$

As $D \gg d$, we take $\sin \theta \sim \theta$. This is approximately $0.153°$. (ii) The fifth maximum, not counting the first one at the origin, is determined from $d \sin \theta = \pm 2n\lambda$ as

$$\sin \theta = \frac{54.6 \times 10^{-5}}{10^{-2}} = 0.027,$$

which is approximately $3.06°$.

Solution 4.2 By defining the integration variable $x = h\nu/kT$, we get $U = aT^4V$ where

$$a = \frac{8\pi k^4}{h^3 c^3} \int_0^\infty \frac{x^3 dx}{e^x - 1} = \frac{8\pi^5 k^4}{15c^3 h^3} = 7.56 \times 10^{-15} \text{ erg} \cdot \text{cm}^{-3}\text{K}^{-4}. \tag{A.17}$$

One can write $a = 4\sigma/c$, where $\sigma = 2\pi^5 k^4/15h^3c^2$ is the Stefan–Boltzmann constant, related to the black body emitting power by the expression $E_b = \sigma T^4$. The latter is the black body energy emitted per unit surface per unit time.

Solution 4.3 (i) $N \sim 5.12 \times 10^{12}$ cm^{-3}.

(ii) As $U = F - T(\partial F/\partial T)$, we conclude that both F and U are homogeneous functions of T^4. Thus, we have $U = F - 4F = -3F$, and finally, $F = -aT^4V/3$. Thus, $S = 4aT^3V/3$ and $C_V = \partial U/\partial T = 4aT^3V$. For S and C_V, a similar dependence on temperature is found in solids when the temperature T is small compared with the so-called Debye temperature Θ (see Chap. 8).

Solutions for Chap. 5

Solution 5.1 The expression (5.6) relates the set of coordinates x, y, z, t in the rest frame K to the coordinates x', y', z', t' in the moving frame K'. As K' is moving with velocity V with respect to K, this implies that K moves with velocity $-V$ with respect to K'. Thus, the transformation is obtained by swapping primed and unprimed coordinates, and setting $V \to -V$. Thus, we get

$$
\begin{aligned}
x &= \frac{x' + Vt'}{\sqrt{1 - V^2/c^2}}, \\
y &= y', \\
z &= z', \\
t &= \frac{t' + (V/c^2)x'}{\sqrt{1 - V^2/c^2}}.
\end{aligned}
\tag{A.18}
$$

Solution 5.2 If the source is moving at a speed V parallel to the x axis of the K_0 system, the K system is moving with speed $-V$ with respect to K_0. The frequency ω in the observer frame K is obtained from the inverse Lorentz transformation (A.18) applied to the four-vector k_i, with speed $-V$. Define $\beta = V/c$. Then,

$$
k_4^{(0)} = \frac{k_4 - i\beta k_1}{\sqrt{1 - \beta^2}} = \frac{i\omega(1 - \beta\cos\alpha)}{\sqrt{1 - \beta^2}}.
\tag{A.19}
$$

This implies

$$
\omega = \omega_0 \frac{\sqrt{1 - \beta^2}}{1 - \beta\cos\alpha}.
\tag{A.20}
$$

For a source approaching the observer, if we take $\alpha = 0$ (for instance, $\mathbf{k} \parallel \mathbf{V}$), the frequency is increased to

$$\omega = \omega_0 \frac{\sqrt{1-\beta^2}}{1-\beta} = \omega_0 \sqrt{\frac{1+\beta}{1-\beta}}.$$

For a receding source for which $\alpha = \pi$ (for instance, $\mathbf{k} \parallel -\mathbf{V}$), then $1 - \beta \cos \pi = 1 + \beta$ and the frequency is decreased to $\omega = \omega_0 \sqrt{\frac{1-\beta}{1+\beta}}$. In the case $\alpha = \pi/2$, the wave is moving parallel to the y axis. In this case, only time dilation affects the frequency.

The Doppler effect has several practical uses in technology and in science. In particular, Doppler spectroscopy is an indirect method for finding extrasolar planets and brown dwarfs from radial velocity measurements via observation of Doppler shifts in the spectrum of the planet's parent star.

Solution 5.3 Momentum conservation means that $\mathbf{p}_1 + \mathbf{p}_2 = 0$. As a consequence, $p_1^2 = p_2^2 \neq 0$. Consequently,

$$E_1^2 - m_1^2 c^4 = E_2^2 - m_2 c^4. \tag{A.21}$$

By squaring (5.60) and substituting, say, E_2 from (A.21), we easily obtain $E_1 = \frac{M^2 + m_1^2 - m_2^2}{2M} c^2$ and $E_2 = \frac{M^2 + m_2^2 - m_1^2}{2M} c^2$.

Solution 5.4 We use the expression obtained in Problem 2, Chap. 4, for the total energy density of a black body at temperature T. By using the factor $a = 7.56 \times 10^{-15}$ erg $\cdot$ cm^{-3}K^{-4} and multiplying by $T^4 = 3.421 \times 10^{28}$, we get the energy density $u = 2.586 \times 10^{14}$ erg $\cdot$ cm^{-3}. We must calculate

$$\frac{V_\odot}{125} \frac{u}{c^2} = \frac{1.422 \times 10^{33} \times 2.586 \times 10^{14}}{125 \times 9 \times 10^{20}} = 3.27 \times 10^{24} \text{ g.}$$

A lower bound on the mass of the Sun's core is $M_\odot' \approx 2 \times 10^{33}/125 = 1.6 \times 10^{31}$ g. Thus, the radiation mass is at least of order 2×10^{-7} the matter mass in the process of nuclear fusion in the Sun's core.

Solution 5.5 The deviation of the photon from a straight line is twice the angle formed by the asymptotes with the vertical axis (perpendicular to the polar axis) $\phi = \pi/2$. Let $\phi = \pi/2 + \varphi$ and take $\phi \sqrt{1 - K^2/c^2 L^2} \approx \phi$. Note that, by expanding the square root, the first correction to unity is $-K^2/2c^2 L^2 \sim -0.88 \times 10^{-11}$, which can be neglected. For very large r and using (5.59), we get

$$r_g/2R_\odot \approx \arccos(\phi\sqrt{1 - K^2/c^2 L^2}) \approx \arccos(\pi/2 + \varphi) = \arcsin(\varphi) \approx \varphi. \tag{A.22}$$

The total deviation is $2\varphi = \frac{r_g}{cL/E}$. Taking $L = R_\odot E/c$, where $R_\odot = 6.96 \times 10^{10}$ cm is the solar radius and $r_g = 2.93 \times 10^5$ cm is the gravitational radius for the Sun, we obtain (see Chap. 1)

$$2\varphi = \frac{r_g}{6.96 \times 10^{10}} = 0.42 \times 10^{-5} \text{ rad} = 0.87'', \tag{A.23}$$

which is the same as the result obtained using Newtonian theory. It is also one half of the value predicted by general relativity. Special relativity essentially unified space and time, and served as an appropriate basis for the formulation of electromagnetic theory. But it did not deal with the gravitational field in its basic postulates. This was achieved by the general theory of relativity, which is necessary to account for the full deviation of light by a gravitational field. It does so by including the curvature of space.

Solutions for Chap. 6

Solution 6.1 Let v be the average speed of the electron and m its mass. Denote the radius of the first Bohr orbit by a_0. Then $\frac{mv^2}{a_0} = \frac{Ze^2}{a_0^2}$, where $a_0 = \frac{\hbar^2}{Ze^2m}$, and we get

$$v = \frac{Zce^2}{\hbar c} = Zc\alpha \approx 0.0073Z = 2.2 \times 10^8 Z \text{ cm s}^{-1}.$$

Solution 6.2 (a) We start from the state described by $\psi(x, t) = C[\psi_1(x, t) + \psi_2(x, t)] = \sqrt{\frac{2}{a}}\left[e^{i\frac{E_1 t}{\hbar}} \sin \frac{\pi x}{a} + e^{i\frac{E_2 t}{\hbar}} \sin \frac{2\pi x}{a}\right]$. It is easy to see that normalization requires $C = 1/\sqrt{2}$. The probability density is

$$|\psi(x, t)|^2 = \frac{1}{a}\left[\sin^2 \frac{\pi x}{a} + \sin^2 \frac{2\pi x}{a} + \left(e^{-i\frac{E_1 - E_2}{\hbar}t} + e^{i\frac{E_1 - E_2}{\hbar}t}\right) \sin \frac{\pi x}{a} \sin \frac{2\pi x}{a}\right]$$

$$= \frac{1}{a}\left[\sin^2 \frac{\pi x}{a} + \sin^2 \frac{2\pi x}{a} + 2\cos\left(\frac{3\hbar\pi^2}{2ma^2}t\right) \sin \frac{\pi x}{a} \sin \frac{2\pi x}{a}\right].$$

The average energy is

$$\langle E \rangle = -i\hbar \int \psi^*(x, t) \frac{\partial}{\partial t}\psi(x, t) = \frac{1}{2}(E_1 + E_2) = \frac{5}{4}\frac{\hbar^2\pi^2}{ma^2},$$

whereas the average squared energy is

$$\langle E^2 \rangle = -\hbar^2 \int \psi^*(x, t) \frac{\partial^2}{\partial t^2}\psi(x, t) = \frac{1}{2}(E_1^2 + E_2^2).$$

The standard deviation of the energy is

$$\sigma(E) = \sqrt{\langle E^2 \rangle - \langle E \rangle^2} = \frac{1}{2}(E_2 - E_1) = \frac{3}{4}\frac{\hbar^2\pi^2}{ma^2}.$$

(b) The average position can be expressed as

$$\int_0^a x|\psi(x,t)|^2 dx = A + B\cos\omega t,$$

which is an oscillatory motion in which

$$A = \frac{1}{a}\int_0^a x\left[\sin^2\frac{\pi x}{a} + \sin^2\frac{2\pi x}{a}\right]dx = \frac{a}{2}$$

and

$$B = \frac{2}{a}\int_0^a x\sin\frac{\pi x}{a}\sin\frac{2\pi x}{a}dx = -\frac{16a}{9\pi^2},$$

and the frequency can be written in the form

$$\omega = (E_2 - E_1)/\hbar = \frac{3\hbar\pi^2}{2ma^2} = \frac{2\sigma(E)}{\hbar}.$$

The average particle position oscillates around the midpoint $x = a/2$ of the well with amplitude $B = \frac{16a}{9\pi^2} \approx 0.18a$.

(c) The period of the oscillation is

$$T = \frac{2\pi}{\omega} = \frac{h}{2\sigma(E)},$$

and we get $T\sigma(E) = h/2$ as an expression of the energy-time uncertainty relation.

Solution 6.3 We have

$$E_\gamma = \frac{3 \times (6.6261 \times 10^{-27})^2}{8 \times 1.6726 \times 10^{-24} \times 4 \times 10^{-24}} = 2.457 \times 10^{-6} \text{ erg}.$$

Since 1 MeV $= 1.6022 \times 10^{-6}$ erg, we have

$$E_\gamma = 1.5335 \text{ MeV},$$

which is about 3 times the electron rest energy (equal to 0.51099906 MeV). Note that 1 MeV $= 10^6$ eV, 1 eV $= 1.602 \times 10^{-12}$ erg, and 1 keV $= 10^3$ eV.

Solution 6.4 Neutron mass $= 939.56563$ MeV/c^2, proton mass $= 938.27231$ MeV/c^2.

Solutions for Chap. 7

Solution 7.1 We can write the time-independent Schrödinger equation in terms of the unit 2×2 matrix **1** and the Pauli matrix σ_3 as

$$\left[\frac{d^2}{dy^2} + \frac{2m}{\hbar^2} [E - \frac{p_z^2}{2m} - \frac{m}{2} (\frac{eB}{mc})^2 (y - y_0)^2] \right] \mathbf{1} \psi + \left[\frac{eB\hbar}{2mc} \sigma_3 \right] \psi = 0, \quad (A.24)$$

which has two sets of eigenvalues $E_{n,\mp} = p^2/2m + (n + \frac{1}{2})eB\hbar/mc \mp eB\hbar/2mc$, corresponding to the two-component wave functions (spinors) $\psi_1 = (\psi_{n,-1}, 0)$ and $\psi_2 = (0, \psi_{n,+1})$.

Solution 7.2 (i) From the set of Pauli matrices (7.2), it is straightforward to check this relation. (ii) We find that $\{\sigma_l, \sigma_j\} = 2\delta_{lj}$.

Solution 7.3 We start from

$$\frac{e}{c} \oint (\mathbf{r} \times \mathbf{B}) \cdot d\mathbf{r} + \frac{e}{c} \int \mathbf{B} \cdot d\mathbf{S} = nh.$$

By permuting $\mathbf{r}$ and $\mathbf{B}$ and also $\cdot$ and $\times$, we get

$$-\frac{e}{c} \oint \mathbf{B} \cdot (\mathbf{r} \times d\mathbf{r}) + \frac{e}{c} \int \mathbf{B} \cdot d\mathbf{S} = nh.$$

Note that $\mathbf{r} \times d\mathbf{r} = 2d\mathbf{S}$. From this and setting $\int \mathbf{B} \cdot d\mathbf{S} = \Psi$, we can write the quantization rule in the form

$$-2\Psi + \Psi = -\Psi = n\frac{hc}{e}.$$

Putting $|e| = -e$, $e < 0$, we finally arrive at $\Psi = 4\pi n\Phi$ for the flux quantization given in terms of the flux quantum $\Phi = \frac{hc}{2|e|}$. Note that, although this problem is solved using the methods of the old quantum theory, it does give a result in agreement with the one obtained from the Aharonov–Bohm effect (except by a factor 2π). (b) There is a basic difference between the two problems: in the Bohm–Aharonov case the field $\mathbf{B}$ is confined inside the solenoid, and $\mathbf{A}$ assumes a constant value outside it, leading to $\mathbf{B} = 0$. This implies a non-local effect of the magnetic field on the phase of the electron wave functions, leading to the quantized flux, whereas the present problem, although it also leads to the quantization of the magnetic flux, is based on the assumption of a constant field $\mathbf{B}$ everywhere, and would only be valid for electrons moving inside the solenoid in the Aharonov–Bohm problem.

Solutions for Chap. 8

Solution 8.1 $N/V \approx 6 \times 10^{49}$ cm^{-1}.

Solution 8.2 (i) Substituting in numerical values, we have $L = 5.67 \times 10^{-8} \times 6.09 \times 10^{18} \times 1.108 \times 10^{15} = 3.82 \times 10^{26}$ W.

(ii) We must first calculate $3.82 \times 10^{26} \times 4.6 \times 10^9 \times 3.15 \times 10^7 = 5.53 \times 10^{43}$ J as the amount of energy radiated by the Sun throughout its life. Dividing by $c^2 = 9 \times 10^{16}$ m^{2}s^{-2}, we get the equivalent mass $m = 6.15 \times 10^{26}$ kg, of order $3 \times 10^{-4} M_\odot$.

Solution 8.3 From the expression $d\Omega = -SdT - pdV - Nd\mu$, we get $\Omega = -\int Nd\mu$ at constant T and V, where $N = \Sigma n_i$. If we divide by $e^{\frac{\varepsilon_i - \mu}{kT}}$ the numerator and the denominator of the boson density (g_i is a degeneracy factor)

$$n_i = \frac{g_i}{e^{\frac{\varepsilon_i - \mu}{kT}} - 1}$$

we have

$$\Omega = -\int Nd\mu = \sum g_i \int \frac{e^{-\frac{\varepsilon_i - \mu}{kT}} d\mu}{1 - e^{-\frac{\varepsilon_i - \mu}{kT}}} = -kT \sum g_i \ln(1 - e^{-\frac{\varepsilon_i - \mu}{kT}}). \quad \text{(A.25)}$$

Assuming $\mu = 0$, which is true for photons, and expressing as an integral, over energies viz.,

$$\Omega = \frac{8\pi TV}{h^3 c^3} \int_0^\infty E^2 \ln(1 - e^{-\frac{E}{kT}}) dE = -\frac{8\pi k^4 T^4 V}{3h^3 c^3} \int_0^\infty \frac{x^3 dx}{e^x - 1}, \quad \text{(A.26)}$$

then comparing with (8.46), we get

$$\Omega = -U/3. \quad \text{(A.27)}$$

Since $\Omega = -pV$, this implies $p = U/3V = aT^4/3$ for the photon gas.

Solution 8.4 The adiabatic expansion means that we can write $dU = -pdV = -UdV/3V$. We have $dU/U = -dV/3V$, leading to $\ln UV^{1/3} = const$. Thus, $U = KV^{-1/3}$. As $u = U/V = aT^4$, and also $R = CV^{1/3}$, we have $aT^4 = K'V^{-4/3} = K''R^{-4}$ (here K, K', K'', and C are constants). We conclude that $TR = const$. Thus T decreases as R^{-1}. The Universe cools as R expands.

Solution 8.5 Denoting the energy density by $u = U/V$, we have $dU = Vdu + udV$. Then

$$dS = \frac{V}{T} du + \frac{4u}{3T} dV = (4VaT^2)dT + \frac{4aT^3}{3} dV = d\left(\frac{4}{3}aT^3 V\right).$$

Thus, we have

$$S = \frac{4a}{3} T^3 V.$$

Solutions for Chap. 9

Solution 9.1 (a) By charge conservation, (b) by lepton number conservation, (c) by baryon number and spin conservation.

Solution 9.2 Yes, charge is conserved and so are the baryon and lepton numbers. This is called the proton–proton process, and it is a basic step in the formation of deuterium in stars, itself an intermediate step in hydrogen–helium fusion, according to the nuclear reactions:

$$_1H^1 +_1 H^1 \rightarrow_1 H^2 + e^+ + \nu,$$

$$_1H^1 +_1 H^2 \rightarrow_2 He^3,$$

$$_2He^3 +_2 He^3 \rightarrow_2 He^4 + 2_1H^1,$$

where the first reaction corresponds to p and n binding in a deuterium nucleus.

Solution 9.3 (i) $N_\odot = M_\odot / m_B = (1.98/1.67) \times 10^{57} = 1.18 \times 10^{57}$.
(ii) Nuclear fusion in the Sun preserves the baryon number. However, baryon number will not actually be conserved, because there is a loss of massive particles, for instance, in the form of solar wind, a stream of charged particles released from the upper atmosphere of the Sun. This is a plasma consisting largely of electrons, protons, and alpha particles at temperatures between 10^5 and 10^6 K, and mass is ejected at a rate of order 10^9 kg·s^{-1}. The Sun also loses mass through the radiation of electromagnetic energy, but this does not decrease the baryon number. Assuming the Sun age as 5×10^9 years, we have the mass lost by solar wind as $\Delta M_\odot = 3.15 \times 10^7 \times 10^9 \times 5 \times 10^9 \approx 1.57 \times 10^{25}$ kg. This means that

$$\frac{\Delta M_\odot}{M_\odot} = 0.8 \times 10^{-4}.$$

Thus, a small fraction of the solar mass has been lost due to solar wind throughout its life. We may therefore conclude that the number of baryons in the Sun is in fact *approximately* conserved.

Solutions for Chap. 10

Solution 10.1 We need the velocities of both the satellite and the Earth. By equating centrifugal and gravitational forces, we obtain the velocity of the satellite (the orbital radius is 26 600 km). We find

$$\frac{mv^2}{r} = \frac{GMm}{r^2},$$ (A.28)

from which

$$v = \sqrt{\frac{GM}{r}} = \sqrt{\frac{6.67 \times 10^{-8} \times 5.974 \times 10^{27}}{2.6600 \times 10^9}} \text{ cm s}^{-1} = 3.870 \times 10^3 \text{ m s}^{-1}.$$ (A.29)

This is the velocity of the satellite relative to the static Earth. It also satisfies the relation $v^2 = c^2 r_g/2r$. We take the velocity of the Earth's surface to be the equatorial velocity, $V = 2\pi R/T$, where T is one day:

$$V = \frac{2\pi \times 6.378 \times 10^3}{24 \times 3600} = 464 \text{ m s}^{-1}.$$ (A.30)

We also find $r_g = 0.8866$ cm.

We use (10.19) and the equations after it to write

$$d\tau_S - d\tau_E = \left(-\frac{r_g}{2r} - \frac{v^2}{2c^2} + \frac{r_g}{2R} + \frac{V^2}{2c^2} \right) d\tau_E.$$ (A.31)

From this, expressing v^2 in terms of r_g, we have $\frac{r_g}{r} + \frac{v^2}{c^2} = \frac{3}{2}\frac{r_g}{r}$, and

$$d\tau_S - d\tau_E \approx \left(-\frac{3}{4}\frac{r_g}{r} + \frac{r_g}{2R} + \frac{V^2}{2c^2} \right) d\tau_E = \left(4.451 \times 10^{-10} \right) d\tau_E.$$ (A.32)

In one day, the satellite time advances with respect to the Earth time by the amount

$$(\Delta\tau_S - \Delta\tau_E)_{day} \approx \left(4.451 \times 10^{-10} \right) \times (24 \times 3600) = 38.4 \times 10^{-6} \text{s}.$$ (A.33)

We thus observe that the clock in the satellite goes faster than the one on the ground by more than 38 microseconds per day. Actually, this contains both the general and special relativistic contributions. The special relativistic effect contributes an amount

$$(d\tau_S - d\tau_E)_{sp} = \left(-\frac{v^2}{2c^2} + \frac{V^2}{2c^2} \right) d\tau_E = (-0.832 \times 10^{-10}) d\tau_E,$$ (A.34)

so due to this effect, the satellite clock loses

$$(-0.832 \times 86400 \times 10^{-10}) \, s = -7.18 \times 10^{-6} \, s$$

every day. Thus, the general relativistic effect alone makes the clock run faster by around $38 + 7 \sim 45$ microseconds/day.

Corrections like this are necessary if we are to synchronize Earth-based and satellite-borne clocks, and this in turn is essential for precise position measurements.

Solution 10.2 (i) Assume fixed φ, θ. We have

$$l_2 - l_1 = \int_{r_1}^{r_2} \frac{\partial l}{\partial r} dr = \int_{r_1}^{r_2} \frac{dr}{(1 - r_g/r)^{1/2}} = r_2 g(r_2) - r_1 g(r_1) < r_2 - r_1,$$

where $\qquad g(r) = (1 - r_g/r)^{1/2} + (r_g/r) \ln \left((r/r_g)^{1/2} \left[1 + (1 - r_g/r)^{1/2}\right]\right).$ For $r_g = 0$, we have $g(r) = 1$, and for $r \gg r_g$, we have $g(r) = 1 + (r_g/r) \ln \left(2(r/r_g)^{1/2}\right)$. Note that the last inequality, due to the curvature of space, is a consequence of $(1 - r_g/r)^{1/2} < 1$.

(ii) For $r_2 > r_1 \gg r_g$, we have

$$g(r) = 1 + (r_g/r) \ln 2(r/r_g)^{1/2}.$$

We then have

$$\frac{r_2 - r_1}{l_2 - l_1} = \frac{1}{1 + \frac{r_g}{2(r_2 - r_1)} \ln \frac{r_2}{r_1}} \approx 1 - \frac{r_g}{2(r_2 - r_1)} \ln \frac{r_2}{r_1}.$$

(iii) We get $1 - \frac{r_g}{2(r_2 - r_1)} \ln \frac{r_2}{r_1} = 1 - 4.33 \times 10^{-7}$, which is a very small correction.

Solution 10.3 We take the black hole diameter to be the uncertainty in the length $\Delta x = 2r_g = 4GM/c^2$, and the momentum uncertainty of a particle inside it is of order $\Delta p = \frac{\hbar}{2\Delta x} = \hbar/4r_g$. This leads to an average energy per particle Δpc. Equating this with kT, we find $T = \frac{\hbar c^3}{8GMk}$, which is the Hawking–Bekenstein temperature for the black hole, except for a factor of π in the denominator.

Solution 10.4 The estimated radius of Sagittarius A^* is $r_g = 2 \times 6.674 \times 10^{-11} 8 \times 10^{36}/9 \times 10^{16} \approx 1.2 \times 10^{10}$ m. This is approximately ten times the solar radius. From (10.42), we get the density $\rho = 3c^2/8\pi GR^2 = 4.57 \times 10^6$ kg/m^3 = 4.57 kg/cm^3. This density is around 10^{-3} times smaller than the density of a white dwarf, estimated to be of thes order 10^3 to 10^4 kg/cm^3.

Solution 10.5 (a) $r_g \sim 10^6$ km. (b) $r_g \sim 10^{10}$ light-years. This is the same order of magnitude as the size of the observable universe.

Solution 10.6 By expressing the Hawking–Bekenstein black hole entropy formula in terms of M, and since the black hole internal energy must satisfy the equation $U = Mc^2$, we can write the entropy as

$$S = \frac{4\pi k G U^2}{\hbar c^5}.$$

From (A.35), we have $1/T = \partial S/\partial U$, and from this we get the Hawking–Bekenstein temperature

$$T = \frac{\hbar c^3}{8\pi k G M}.$$

Then, expressing M in terms of T, we write finally $U = \hbar c^5/8\pi k G T$.

Solution 10.7 Defining $C_V = \partial U/\partial T$, we get immediately

$$C_V = -\frac{\hbar c^5}{8\pi k G T^2}.$$

This result is a consequence of U being proportional to T^{-1}.

Solution 10.8 We must calculate the length r from the data. Let θ and φ, respectively, be the angles formed by the line from the observer to the ring, and from the star to the ring, with the line joining the observer and the star. This line is divided by the lens into two segments, h_{Ol} and h_{Sl}. We neglect the effects of the curvature of space and take $h_{Ol} + h_{Sl} = H$. As θ and φ are small, the approximations $\sin x \approx \tan x \approx x$ can be made and we may write $\theta = r/h_{Ol}$, $\varphi = r/h_{Sl}$. They satisfy the equation

$$\theta + \varphi = \frac{2r_g}{r}, \qquad (A.35)$$

where r_g is the gravitational radius of the lens. As $\theta + \varphi = r\left(\frac{1}{h_{Ol}} + \frac{1}{h_{Sl}}\right)$, equating with (A.35), we can write

$$\theta = \frac{r}{h_{Ol}} = \sqrt{\frac{2r_g}{(1 + \frac{h_{Ol}}{h_{Sl}})h_{Ol}}} - 2.6 \times 10^{-9}\text{rad}. \qquad (A.36)$$

As $h_{Ol} \sim 6.2 \times 10^{20}$ m, we have $r \sim 1.6 \times 10^9$ km. The ring has a radius of the same order as the average radius of Saturn's orbit around the Sun. The total deviation of light is $\varphi + \theta = 2r_g/r \sim 3.75 \times 10^{-9}$ rad.

Solutions for Chap. 11

Solution 11.1 The positive muon μ^+ decays according to $\mu^+ \rightarrow e^+ + \nu_e + \bar{\nu}_\mu$. It follows that $\bar{\nu}_e + \mu^+ \rightarrow e^+ + \bar{\nu}_\mu$. From this, we see that, for allowed reactions involving leptons, if there is a lepton of one family in the initial state, there must be a lepton of the same family in the final state. Thus, we must define an electron lepton number

Fig. A.1 Feynman diagram
for the decay of the positive
muon

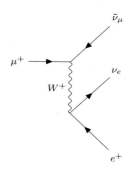

$L_e = 1$ for e^-, ν_e (and $L_e = -1$ for their antiparticles) and a muon lepton number
$L_\mu = 1$ for μ^-, ν_μ (and $L_\mu = -1$ for their antiparticles). Such numbers are separately
conserved in any reaction.

For instance in the pion decay

$$\pi^+ \to \mu^+ + \nu_\mu,$$

$L_\mu = 0$ on the left and $L_\mu = -1 + 1 = 0$ on the right. The decay is allowed, whereas

$$\mu^+ \to e^+ + \gamma$$

is such that $L_\mu = -1$ on the left and $L_e = -1$ on the right, so both these lepton
numbers change and the decay is forbidden. The latter reaction is the subject of
experimental research for the detection of lepton number violation, suggested by the
existence of neutrino oscillations.

Solution 11.2 See Fig. A.1.

Solution 11.3 (i) According to the quark model, a baryon consists of three quarks
and each quark has spin $1/2$, so they cannot combine to form a baryon of integer
spin. In consequence, it has to be a fermion and cannot have spin 1.

(ii) An antibaryon is composed of three antiquarks. To combine three antiquarks
to form an antibaryon of electric charge $+2$, we require antiquarks of electric charge
$+2/3$. However, there is no such antiquark in the quark model and such a particle
cannot exist.

(iii) A meson is composed of a quark and an antiquark. If we choose an s antiquark
($S = +1$ and $Q = 1/3$) and a quark with $Q = 2/3$ like u, we get the positively
charged kaon $u\bar{s} = K^+$. This particle is therefore allowed.

Subject Index

© Springer-Verlag GmbH Germany, part of Springer Nature 2021
M. Chaichian et al., *Basic Concepts in Physics*, Undergraduate Lecture Notes in Physics,
https://doi.org/10.1007/978-3-662-62313-8

Author Index

© Springer-Verlag GmbH Germany, part of Springer Nature 2021
M. Chaichian et al., *Basic Concepts in Physics*, Undergraduate Lecture Notes in Physics,
https://doi.org/10.1007/978-3-662-62313-8